Fisheries Ecology and Management

Fisheries Ecology and Management

CARL J. WALTERS AND STEVEN J.D. MARTELL

PRINCETON UNIVERSITY PRESS

PRINCETON AND OXFORD

Library of Congress Cataloging-in-Publication Data

Walters, Carl J., 1944–
    Fisheries ecology and management / Carl J. Walters and Steven J. D. Martell.
      p. cm.
    Includes bibliographical references and index.
    ISBN: 0-691-11544-3 (cl: alk. paper) — ISBN 0-691-11545-1 (pbk.: alk. paper)
      1. Fishery management. 2. Fish stock assessment. 3. Marine ecology. I. Martell, Steven
J. D., 1970– II. Title.

SH328.W36 2004
333.95'6—dc22   2003064804

British Library Cataloging-in-Publication Data is available

This book has been composed in PostScript Sabon

Princeton University Press books are printed on acid-free paper and meet the guidelines for
permanence and durability of the Committee on Production Guidelines for Book
Longevity of the Council on Library Resources

Printed on acid-free paper. ∞

www.pup.princeton.edu

Printed in the United States of America

10 9 8 7 6 5 4 3 2

ISBN-13: 978-0-691-11545-0 (pbk.)

ISBN-10: 0-691-11545-1 (pbk.)

THIS BOOK IS DEDICATED TO C. S. HOLLING,
FRIEND AND MENTOR

CONTENTS

x

LIST OF TABLES

THIS BOOK IS intended as a text in upper division and graduate classes on fisheries-stock assessment and management. It aims to provide a broader review of assessment methods and policy issues than is available in existing texts on fisheries harvest management, like Hilborn and Walters (1992) and Quinn and Deriso (1999). It is not just a book of recipes for the analysis of fisheries data, and it is deliberately critical of the "science" of fisheries stock assessment as we have taught it in the past. It warns students that we have often been confused about nearly every aspect of that science, from its basic aims and objectives to the trust that we should place in our results and recommendations. That confusion has led fisheries scientists to do much work that is either irrelevant or damaging to the world's obvious fisheries management crises. The book begins by asking the student to think about what we are trying to accomplish by presuming or pretending to "manage" fisheries, and it emphasizes that management is a process of making choices. There is no way to make choices without making at least some predictions about the comparative outcomes of the choices, and these predictions cannot be made without some sort of "model" for how the world works. Usually, management choice requires trade-offs among objectives, and these trade-offs need to be quantified in some way before we can make an intelligent choice or recommendation. This means that much of the book has to be about quantitative, mathematical modeling, and we make no apologies for demanding that people who would engage in fisheries assessment and management should at least be able to read and understand some basic mathematics.

There have been four really important developments in the decade since Hilborn and Walters wrote *Quantitative Fisheries Stock Assessment*. First, we have been able to look back more clearly on some spectacular fisheries collapses, like the cod of Newfoundland, and to understand a bit about the role that poor data and scientific assessment mistakes played in actively promoting those collapses. Second, there have been some substantial technical advances in the statistical and computational machinery of assessment, which allow scientists to build much more complex assessment models to account for more of the causes of variation in fisheries data and to measure uncertainty in assessments more accurately. But in our view, these advances have not dealt effectively with the causes of failure in the first place and, in fact, have diverted much working time and attention from the real problems. Third, thanks largely to the efforts of nongovernmental conservation groups, there is now strong public demand for sustainable fisheries and for protection of nontarget organisms and ecosystem functions, backed by the powerful threat of market sanctions. Fourth, we have finally begun to develop trophic interaction, food web, and ecosystem models that appear capable of making useful predictions about policy issues like marine mammal protection that were simply ignored in single-species assessment and policy recommendations. These models arise from the marriage of ideas from evolutionary biology, in the form of what we are now calling "foraging arena theory," with methods from ecosystem analysis about mass balance and historical reconstruction models, like Ecopath, that have helped organize rich sources of information on trophic interaction rates.

We think it is clear from these developments that students of fisheries assessment and management need to be conversant with a much broader set of issues and tools than have traditionally been provided in fisheries ecology courses. They need to learn to think in hierarchic terms about both broad strategic trade-offs and about the frustratingly detailed tactics involved in achieving these trade-offs. They need to understand the many things that can and will go wrong with analysis and modeling in a world where there will never be enough funding to get all the scientific answers that the public will demand. And perhaps most important, they need to understand how to embed analyses of fish dynamics in a broader analysis of the dynamics of the complex systems created by the linkage between fish, fishers, and the ecosystems that support them, and to construct such analyses by deliberately looking over a wide range of spatial and temporal scales. As Ludwig, et al. (2001) have eloquently argued, fisheries scientists need to be much more broadly educated in order to contribute more effectively to the development of sustainable fisheries policies; we hope that this book will contribute at least to the scientific side of that broadening.

> *I haven't asked for much in this life, and Lord knows I've got it.*
> (Al Bundy, in *Married with Children*)

ACKNOWLEDGMENTS

FINANCIAL SUPPORT to free the senior author's time for writing this book came from a Pew Fellowship in Marine Conservation and from the Mote Eminent Scholar Program, Florida State University and Mote Marine Laboratory. Felicia Coleman, FSU, and Ken Leber, Mote Marine Lab, provided much support in relation to the Mote Scholar work. We are also grateful for financial assistance and much intellectual assistance from James Kitchell, University of Wisconsin, through his National Science Foundation and National Center for Ecological Analysis and Synthesis "Apex Predators" projects. Additional funding was from the Natural Sciences and Engineering Research Council Operating Grant to Walters. Much of the model development would have been impossible without financial and personal support from Daniel Pauly and Villy Christensen, UBC, through the "Sea Around Us" project, funded by the Pew Charitable Trust. A uniquely brilliant crew of graduate students at UBC provided much critical, creative review and discussion, particularly Sean Cox, Bob Lessard, and Nathan Taylor. Rob Ahrens contributed much to this crew, in addition to taking over the senior author's teaching duties. We could not have even begun the work without much strong support and understanding from Sandra Buckingham and Dawn Cooper.

PART ONE

Changing Objectives and Emerging Assessment Methods

Introduction

MUCH OF THIS BOOK is about the derivation, use, and abuse of various mathematical models used to make decisions about how to manage harvested aquatic ecosystems. There is a long tradition of such modeling, and many biologists still look upon that tradition with much puzzlement and even contempt. Anyone who has taken even a bit of time to look at any aquatic ecosystem cannot help having seen that such systems are incredibly complex in their spatial, temporal, and trophic organization. Further, the complexity is not just a matter of structural diversity (lots of kinds of creatures). It also involves dynamic complexity in the form of a rich variety of feedback effects. Changes in the abundance of any creature due to natural or human factors are likely to result in a cascade of changes in the vital statistics (birth, death, growth rates) of other creatures in the food web, which in turn can feed back to impact further changes in the abundance of that creature. In the face of this complexity, it often seems both arrogant and foolish to pretend that we can make any useful predictions about what will happen when people selectively harvest some species that are fun to catch or good to eat, or change ecosystem fertility through deliberate or inadvertent changes in nutrient loading, or alter the physical habitat of an ecosystem.

After much experience in the field, we would be the first to agree that it is indeed impossible to capture fully the rich behaviors of ecosystems in mathematical models, particularly when we try to include unregulated human activities (humans as dynamic predators) in the calculations. But in this chapter, we offer three main arguments about why it is important to keep trying to build useful models. The first, which we will not discuss any further because it is so obvious, is that modeling is a great and perhaps necessary way for scientists to force themselves to think clearly and to put their claims to understanding on the table in the form of specific predictions. The second, which we discuss in the following three sections, is that prediction in some form is *required* for management choice, i.e., the issue in management-policy design is not whether to model but rather how to go about it. The third, which we discuss in a closing section, is that there are some predictable regularities in the way natural populations and ecosystems respond to human disturbance, so that at least some kinds of useful predictions are not as likely to fail as they may initially appear.

1.1 THE ROLE OF PREDICTIVE MODELS

If the people in a fisheries management agency watch some fishery change while asserting that they are powerless to implement regulations that might alter the path of change, then that agency is not really a management agency at all; at best, it is a monitoring agency. The very word "management" implies some capability for making choices among options that might make some difference. That is, management *is* making choices. But what is involved in making any choice among alternatives? If we can choose either option A or option B, then we must either toss a coin or consciously construct arguments in the form "we believe that the outcome of A will be X while the outcome of B will be Y, and we prefer X to Y." Such sentences contain two kinds of assertions: (1) about the outcome (or range of outcomes, or probabilities of various outcomes) for each choice, i.e., *predictions about what will happen in the future*, and (2) about management objectives, i.e., *which future outcomes would be preferred*.

So making choices necessarily involves some method for predicting the future. This means the issue in management decision-making is not whether to model the future somehow (that is inevitable) but, rather, what model to use in making the prediction(s). Here there are two basic choices: to predict using the sometimes wonderful intuitive (and largely subconscious) capabilities of the human mind, or to resort instead to some explicit model or "deductive engine" for piecing together known elements of the prediction in some conscious way.

It is worth noting in passing that scientific research also necessarily involves making predictions, whether or not these predictions are stated as explicit alternative hypotheses about the outcomes of alternative experimental treatments. Even purely "observational" or "natural history" research programs cannot be designed and implemented without making some very strong assumptions (predictions) about where, when, and what variables or factors are worth observing, i.e., are likely to carry useful information about causal relationships. The experimental scientist can escape some responsibility for making specific predictions by constructing treatments (choices) that give clear, qualitatively different predictions about directions of response under alternative hypotheses. And the scientist has another advantage in terms of being able to choose the questions (options) to be addressed without much regard for whether those questions are of general interest to anyone else. So it is perhaps not surprising that scientists are much more likely than managers to make misleading assertions, i.e., "prediction is impossible in complex systems" or "it is not necessary to construct quantitative models in order to make useful predictions." Scientists who make such claims are clearly not the people to provide guidance about policy choices, nor are they likely to have much experience with the agonies of having to make hard choices.

Given that natural ecosystems are very complex and will be "driven" to future change by unpredictable environmental changes as well as human activities, so that we cannot possibly produce good unconditional or "open loop" predictions of future change, how can we hope to manage ecosystems if management choices require prediction? Or how can we hope to compare policy choices until we "understand" all the interactions and external forces that drive change? The answer to these questions is actually quite simple, if we look carefully at the character of the policy predictions required for decision-making: to choose between policy A and policy B, we do not, in fact, require unconditional predictions about the future, or even about most of the causes and patterns of variability that the future will bring. Rather, we need only to be able to predict whether policy A *will do better than* policy B for a sufficiently wide range of possible futures to make it a "better bet" than policy B. That is, policy predictions need not be about the future in general but, rather, only about those aspects of future change that could be directly impacted by the specific actions/interventions involved in the policy, and even in relation to these changes we generally require only predictions of *relative* performance. This means, e.g., that when someone asks, "How can you manage the fish when you do not even know how many there are?", we can answer by pointing out that we can compare policy choices for a wide range of possible actual numbers of fish, to find choices that are at least somewhat robust despite the uncertainty about the numbers. Further, we can generally specify policy choices as *rules for response to change* rather than absolute degree of impact. Consider the following example: suppose policy choice A is to allow a particular, fixed quota of fish to be harvested in perpetuity (i.e., a quota property right), and policy choice B is to allow some fixed proportion of the fish to be harvested each year (this proportion is called the exploitation rate). It is easy to show with practically any population or ecosystem accounting model that policy A is prone to catastrophic failure: under natural variation, the stock is bound to get low enough so that the quota looms larger and larger as a factor of change, driving the stock down faster and faster as the number of remaining fish (and hence the basis for future population growth) declines. On the other hand, policy B has built-in "feedback" to adjust harvests downward during stock declines (and hence help reverse the declines) and to take advantage of higher harvest opportunities when the stock is large. In this example, only a fool would advocate policy A, whether or not we can predict specifically what variation the future will bring.

1.2 The Distinction between Fish Science and Fisheries Science

We can provide useful predictions and advice about some kinds of management choices without resorting to precise, quantitative models that are bound to be incorrect to at least some degree. For example, it is easy to explain in qualitative terms why fixed-quota harvest policies are dangerous compared to feedback policies in which harvests are varied in response to unforeseen change. But most management decisions involve quantitative choices: *How many* fish should be harvested this year? *What sizes* of fish should be caught? *How large* should a protected area be? *How many* licenses should be issued? *How much* unregulated fishing effort will occur this year if a given regulation is imposed on catch or size of fish or location of fishing? *How much* can we harvest without "impairing" the ability of the ecosystem to support other creatures that depend on the ones we harvest?

Somebody has to provide the answers to these difficult questions, i.e., somebody has to do some quantitative modeling and prediction, whether the work is done well and systematically or instead by some seat-of-the-pants calculation. In a way, it has been really unfortunate in the historical development of fisheries management that there has been a general assumption that the right people to answer such questions are fish biologists. There have been no real professional standards in fish or fisheries biology, and a high proportion of us got into the field in the first place because we could do so without a lot of distasteful quantitative training. We were taught to study biological process and pattern from a largely qualitative perspective, and we never expected to be "bean counters." Furthermore, most of us never imagined that many of the questions that we would be asked would not even be about fish at all but would, instead, be about the behavior of people (fishers). This state of affairs is changing rapidly, with recognition that there is a lot more to fisheries science than just studying biological processes and counting fish. But a new pathology is accompanying the change: the top levels of management agencies are dominated by people with the traditional training (and cunning as institutional players) who now have to turn to younger people for help when there is no way to sidestep the difficult quantitative questions. This means that as demands for improved, quantitative management prescriptions have grown in order to deal with more complex management options and trade-offs, key fisheries managers have had to rely more and more on people and methods (modeling) that they do not understand and certainly do not trust. Such specialization of capabilities and functions leads in turn to increased opportunities for misinterpretation and misunderstanding, among all stakeholders involved in management (fishers, managers, scientists, representatives of conservation interest groups, etc.).

1.3 APPROACHES TO PREDICTION OF POLICY IMPACT

Given that predictions are an inevitable part of making management choices, what options does a fishery manager have for making these predictions? Surely there are alternatives to the rather complicated mathematical modeling described in this book; indeed, there are at least five alternative approaches that can be (and have been) used. These approaches are not mutually exclusive; each uses or is derived from at least some components or results of the others.

Appeal to Conventional Wisdom and Dogma

In a surprising variety of decision situations, fisheries managers have ignored empirical data and past experience in favor of essentially dogmatic assumptions about the responses to particular policy options and system disturbances. For example, it is routine to presume that habitat alterations to natural ecosystems always cause reduced productivity (because the organisms are "adapted to" the natural circumstances). Another common assumption is that harvesting always causes a reduction in the abundance of target species, even if/when the harvesting selectively removes individuals that differentially drive away or kill other conspecifics (e.g., cannibalism). When field evidence is found that contradicts such assumptions—e.g., evidence that coho salmon may actually be enhanced by forest harvesting in some watersheds of the Pacific northwest (Holtby 1988; Thedinga et al. 1989; see discussion in Walters 1993)—this evidence is either ignored entirely or is rejected as "nonrepresentative" or "atypical." When this happens, managers are essentially indicating their willingness to behave essentially as though some principles or assumptions were equivalent to religious dogma, i.e., were impervious to scientific invalidation.

Trend Extrapolation

A time-honored way of making fisheries management predictions has been by simple trend extrapolation: plot the historical data, and "eyeball" alternative projections forward in time while making some intuitive guess about the likely impact of policy change on the trend. We can, of course, formalize the eyeball part of this approach by using formal time-series analysis models, but that is unlikely to produce a better result (except perhaps in multivariate systems) than the remarkable integrative and pattern-finding abilities of the human eye.

This approach has failed in modern fisheries, for a variety of reasons. (1) It is really only valid for systems that exhibit *incremental*, slow change; modern fisheries can change very rapidly. (2) It is easy to confuse wishful thinking with good intuition in making predictions about the effects of policy change on trends, and to keep applying small Band-Aids to gaping wounds. (3) It is all too common to use misleading trend indicators, especially catches. In any fishery, catch results from three factors: the area "swept" by fishing, the size of the stock, and the area over which the stock is distributed. So an apparently "healthy" increase in catch can mean either that the stock is healthy, that the fishing effort (the area swept by gear) has increased, or that the range occupied by the stock has decreased. It does not help matters to use catch per effort, since this commonly used trend index can even increase during stock declines due to contractions in the range area used by the fish.

Empirical Models Based on Past Experience and/or Experience with Similar Systems

For many policy issues there is a rich range of historical and spatial comparative data upon which to base predictions about the responses to any particular new circumstances. Some fish stocks (e.g., Pacific herring off British Columbia) have been severely overfished, then allowed to recover, so that we have good information about likely stock response as a function of stock size. There are large data sets on how lakes and coastal areas respond to eutrophication, and strong regression relationships have been found between nutrient loading and performance measures such as chlorophyll concentration, so that the likely response in almost any new case can be "interpolated" from the regressions. For fish populations that are maintained through artificial stocking (hatcheries), there are large data sets on the effects of factors such as time and size of release and stocking density on performance measures such as survival rate and growth.

Unfortunately, most of the important policy issues in fisheries today involve options and performance measures for which there are no historical precedents. We have not yet tried to manage aquatic ecosystems in any holistic way, and in particular, we have not systematically gathered information on the abundances and spatial distributions of the wide variety of organisms (beyond harvested fish) in an expanded view of what would constitute a "healthy" managed ecosystem. Existing reviews of comparative data, e.g., May (1984) and Hall (1999), show mainly a confusing variety of fragmentary patterns.

Experimental Components Analysis (Reductionist) Modeling

This is the basic approach taken in most fisheries modeling. The idea is to try to break prediction problems into more manageable components, using some basic "tautologies" (statements that are true by virtue of how the words used in them are defined) to identify and synthesize the component predictions. For example, we typically model population change over time for a population defined over a large enough area to be closed to immigration and emigration (for a so-called "unit stock") by a simple balance relationship that we treat as a tautology: (population next year) = (survivors after harvesting this year) + (surviving new recruits). If there is no net migration (and no spontaneous creation of organisms), this balance relationship is a tautology because it re-expresses what we mean by (population next year) in terms of the component creatures that make up that population. In mathematical terms, the simplest way to express the balance relationship is

$$N_{t+1} = s_t(N_t - C_t) + R_{t+1},$$

where N_t = population size at the start of year t, s_t = survival rate, C_t = catch taken in year t, and R_{t+1} = surviving recruits. Note that "surviving recruits" generally refers to animals that graduate to an age or size class that are vulnerable to fishing gear. This statement tells us that to predict population change, (N_{t+1}), we need to have information on N_t and C_t, and we need to make some assumptions (called "functional relationships") about the survival and recruitment rates (s, R). That is, the balance relationship tells us that we can reduce the prediction problem to two "simpler" problems, predicting survival and recruitment rates, while accounting in the overall balance structure for two "known" temporal factors, N_t and C_t. In this approach, long-term predictions are constructed by applying the balance relationships recursively (repeatedly); by making a series of short-term, incremental predictions, we hope to be able to account for ecological feedback effects as expressed through possible changes in the s and R rates. For more complex situations, e.g., multiple stocks, we solve a list of such balance relationships in parallel, perhaps including terms that represent linkages among the variables (e.g., we might include predation effects as being either additions to the catch or effects on the survival rate s_t).

Several obvious things can, and regularly do, go wrong with this approach. We almost always leave out important variables, or equivalently fail to represent factors that cause change by treating some "parameters" such as the survival rate s_t as constant over time. We commonly use poor approximations for the forms of (and key variables that cause change in) functional relationships, particularly for the prediction of recruitment rates R_{t+1}. Solving the balance relationships recursively to obtain long-term predictions can lead to large, cumulative errors if the initial state and/or some key parameters are specified incorrectly. We will provide repeated examples and warnings of these and some other problems with mathematical modeling, and linking the models to data, throughout the book.

1.4 EXPERIMENTAL MANAGEMENT

The basic concept in this "actively adaptive" management approach (Walters and Hilborn 1976; Walters 1986) is not to pretend that a best policy option can be identified from experimental components modeling and analysis of historical data but, rather, to "embrace uncertainty" by using the modeling and analysis to identify a set of candidate policy options that are all defensible (and to screen out options that are likely inadequate to meet management objectives). Then these candidates are each given a "day in court" by applying them to the managed system as a set of experimental treatment options, either sequentially over time or on a set of hopefully similar experimental locations or units.

This approach has been successfully implemented in only a very few cases and has failed miserably in many, many others (Walters 1997). The failures have been caused by many factors. It has proven extremely difficult to obtain institutional support for programs that take a long time to produce results (sequential experiments may take decades to complete). There is a common management perception that experimentation is just too "risky" (see, e.g., Walters and Collie 1989; Parma and Deriso 1990b). Monitoring costs may be prohibitively large, especially for spatial experiments with a variety of experimental units and treatments. And perhaps worst of all, there is now quite a large community of scientists who are willing to sell modeling to managers as an alternative to hard, expensive experimentation, and this is too often an easy sell.

Theory versus Practice in Decision-Making under Uncertainty: Indecision as Rational Choice

For almost all important fisheries-management choices that have long-term ecological and economic consequences (e.g., the choice of target exploitation rate or stock size), we have to admit a wide range of uncertainty about those consequences. There are three basic reasons for uncertainty in long-term predictions, and only two of these can be reduced through an investment in measurement and modeling: (1) we do not know the current system state precisely (predictions must look forward from an uncertain starting point); (2) we do not know all of the "rules for change" (interactions, functional relationships) that will govern future dynamics; and (3) ecological dynamics are strongly influenced by environmental factors (physical and geochemical forcing, e.g., upwelling) that are not (as yet) predictable, especially in view of likely climate change. So even if we could measure ecosystem states (population sizes and such) much more accurately, and even if we knew and could model all of the ecological and economic interactions precisely, there would still be gross uncertainty about future change due to uncertainty about future environmental "forcing" patterns. This means that at best we can make only probabilistic statements about alternative futures, and much of the emphasis in stock-assessment research and modeling today is on how to do such probabilistic calculations more realistically (Patterson et al. 2001).

To objectively and quantitatively compare choices involving a range of possible outcomes, we need not only to place odds on those outcomes but also to combine the possible outcomes for each choice into some kind of overall utility measure for that choice (Raiffa 1982; Keeney and Raiffa 1976). There is no general standard or procedure for constructing utility functions to combine or weigh the possible outcomes in public decision-making that involves multiple stakeholders with varying interests and aversions to particular outcomes. The simplest or "expected value" utility measure would be to take an average of the outcomes, weighing each by its probability of occurrence. But this simple measure would not be acceptable to most fisheries stakeholders: people concerned with long-term conservation want to see a differentially low utility placed on poor long-term outcomes, while people concerned with immediate income and employment want to see low utilities placed on outcomes that would involve short-term economic hardship.

A common reaction from fishing stakeholders to uncertainty has been to demand that governments "prove" that there will be a problem before introducing more restrictive harvest regulations. In decision theoretic terms, this amounts to demanding that the utility function for combining and weighing alternative outcomes place very low utility on outcomes that cause immediate economic hardship and/or demanding that utilities for long-term outcomes be discounted at high rates. Most fisheries-management agencies have now been given a mandate to resist such demands, through the widespread adoption of the "precautionary principle" (FAO 1995; UN 1996; Dayton 1998). According to this principle, the "burden of proof" should be reversed, i.e., it should be up to fishing interests to demonstrate that the odds of long-term harm are low. This amounts to demanding that the overall utility function be exactly the opposite to what fishing interests would advocate, placing a very low utility on undesirable long-term outcomes.

It will probably not be that difficult to apply the precautionary principle to new and developing fisheries, so as to ensure that development proceeds with a relatively low risk of overfishing and economic hardship. Unfortunately, most important fisheries management choices today involve the opposite end of the development spectrum, at which choices that might reverse historical declines (and improve the odds of long-term sustainability) are ones that would create immediate economic hardship (loss of income and employment, social displacement) for relatively large, dependent communities of fishers. In such situations, there are strong (and what conservationists might call "perverse") incentives for fisheries managers to avoid making hard choices, i.e., the rational personal choice for them is to be indecisive (fig. 1.1). To understand figure 1.1, try to put yourself in the position of a senior fishery administrator or politician faced with scientists who have come forward with dire predictions of ecological collapse unless a fishery is cut back severely. You know that if you do follow their advice, you will be vilified by people who depend on the fishery for their livelihoods, and you know from much experience with fisheries scientists that their predictions are not exactly reliable (to say the least). On the other hand, if you delay action and keep the support of fishing interests, your experience tells you that there is at least some chance the problem will correct itself (will turn out to have been caused not by fishing but by some environmental "regime shift" that will eventually reverse). Even if the scientists are right, there is a good chance that you can move along or retire before the situation becomes so poor that no rational person could ignore it. But even if you are somehow legally required to adopt a precautionary principle in making your decision, you have a variety of options for clouding the issue (and making the easy choice) by appealing to evidence (which some scientists are sure to have) that

(Choice Point)   (Possible outcomes)

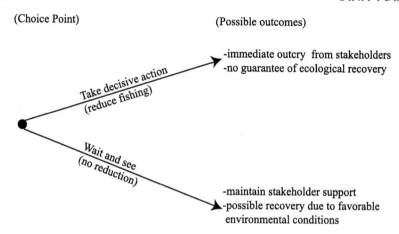

FIGURE 1.1: Indecision as a rational choice during a fishery decline. Viewed from a fishery manager's perspective in terms of a simple decision tree, it can be fully rational for the manager faced with either certain outcry from fishing stakeholders if decisive action is taken, or the possibility that nature will correct the situation without intervention even if no action is taken, to gamble on the situation's correcting itself (i.e., to be deliberately indecisive).

environmental factors have, indeed, been at least partly responsible for the "signals" that other scientists have interpreted as overfishing.

We warn readers of this book that there is no use pretending that the decision-making "pathology" (with respect to prudent, sustainable harvest management) shown in figure 1.1 can somehow be overcome through better scientific research and modeling. We cannot, even in principle, provide the certitude of predictions that would be required to demonstrate that gambling on inaction is wrong. Even more important, science cannot tell us what is right or wrong when there is a trade-off involving a hardship for people today versus a possible gain for people in the future.

1.5 THE ECOLOGICAL BASIS OF SUSTAINABLE HARVESTING

However fisheries managers might behave when faced with a conflict between fishing and conservation interests, there is broad public support for moving toward fisheries that are sustainable in the long term and for implementing policy options that avoid the decision-making pathology (figure 1.1) that develops during fishery declines. There is also broad support for sustainable policies that recognize other ecosystem "values" or "services" besides harvesting, such as protecting the capability of ecosystems to support diverse assemblages of creatures that are valued in their own right (e.g., marine mammals, birds).

To people who pretend (or have been led to believe through popular ecology literature) that natural ecosystems are finely tuned machines that are highly vulnerable to human disturbances, it might appear impossible to ever harvest various creatures on a "sustainable" basis without ultimately destroying the machinery. Never mind that people have been harvesting various creatures from most of the world's ecosystems for many thousands of years, so that we are hard put to even find a "natural" ecosystem. There is much fear that whatever early humans might have gotten away with, modern technology creates a destructive capability that is somehow unmanageable. It is easy to confuse two really different issues: what nature can produce, and what we can do to manage the activities of those who would capture that production. Presumably, we can do a better job with the management issue if we better understand the production issue.

Texts on fisheries science typically introduce the idea of sustainable harvesting in terms of "surplus production." A very simple logistic model, or any model with density-dependent rate processes, for population growth is used to argue that natural populations tend to "push back" against the impacts of harvesting by exhibiting positive population growth (surplus that can be harvested without further reducing the population) after being reduced in numbers by any harvest removal. Many ecologists are suspicious of this argument, not because of the ecological relationships behind it but because of the way the argument is built from a population model that we know is too simple to explain most of our field experience with how natural populations actually behave.

A more general way to understand the basic dynamics of, and basis for making predictions about, sustainable harvesting is to imagine going into a natural, unharvested ecosystem, picking a "target" population more or less at random, then starting to remove proportions N_t of that population over years t. If the ecosystem is not already undergoing some progressive development or recovery process (i.e., is not at an early successional stage), and if the study area is large enough so that numerical population changes are dominated by birth-death processes within the area (rather than dispersal to/from other areas), then we expect the chosen target population numbers N_t of animals at least one year old to satisfy the accounting balance relationship

$$N_{t+1} = s_{(a)t}N_t + s_{(j)t}f_t N_t = (s_{(a)t} + s_{(j)t}f_t)N_t = r_t N_t \qquad (1.1)$$

where $s_{(a)t}$ is the annual survival rate of animals more than one year old, $s_{(j)t}$ is the survival rate from egg/birth to age 1, f_t is the mean egg/birth production per animal present at time t, and $r_t = s_{(a)t} + s_{(j)t}f_t$ is the relative population growth rate from time t to time $t+1$. If the target population is a naturally sustainable part of the ecosystem, i.e., is not on its way to natural extinction or on its way to becoming a much more dominant part of the ecosystem, then we expect to find the average value $\bar{r}$ of its r_t to be $\bar{r} = 1.0$, i.e., on average $N_{t+1} = N_t$. At this point, the reader needs to be really careful about equation 1.1; most biologists would automatically assume that since it is a very simple equation, it must be based on very simple biological assumptions, e.g., that every animal has the same survival rate and fecundity. That is not

BOX 1.1
REPRESENTATION OF RATE PROCESSES AND STATE CHANGE
IN FISHERIES MODELS

Quantitative models for fisheries-policy analysis generally involve predictions of change in numbers and/or biomass over time. Typically, the predictions are made in a series of time steps. For each time step, discrete "inputs" or gains due to processes like recruitment that typically occur over short periods or seasons are usually treated as occurring at the start of each step, then loss processes are treated as occurring continuously over the step. Two apparently distinct types of equations are used to represent the loss processes:

1. discrete-time survival equations that predict net, proportional change, like the term in equation 1.1 for surviving older fish: (surviving older fish) $= s_{(a)t} N_t$.
2. instantaneous rate equations of the form $dN/dt = -ZN$, where Z is called the "instantaneous rate"; if Z is constant from t to $t + 1$, the solution of such equations is $N_{t+1} = e^{-Z} N_t$, which is exactly the same as the discrete survival prediction if we set $s_{(a)t} = e^{-Z}$.

Note that instantaneous loss rates Z can take any positive value, while survival rates like $s_{(a)t}$ are bounded between zero and 1.0. For example, $Z = 3$ implies $s_{(a)t} = e^{-3} = 0.0498$. A word of warning for biologists who are trained to think about complexity in visual terms: it is common to confuse complexity created by mathematical notation with complexity created by realistic assumptions. Instantaneous-rate formulations typically look more complex and realistic to naïve biologists, even if they make simplistic assumptions. For example, the equation $N_{t+1} = N_t e^{-Z}$ appears more complex than $N_{t+1} = N_t s$ but, in fact, says exactly the same thing (makes exactly the same prediction when s is set to e^{-Z}).

The more cumbersome instantaneous-rate formulation is used in most fisheries-assessment literature for two reasons. First, it provides a convenient way to deal with risk factors, such as predation and seasonal fishing, which involve very high rates and can cause rapid change over short periods. For example, purse seine fisheries for Pacific salmon off the coast of British Columbia can generate fishing mortality rates on the order of 500/year, i.e., they knock down the fish abundance at rates that would remove 500 times the number of fish initially present if those fish kept being replaced over a whole year so as to prevent changes in the number of fish present at any moment during the year. Second, these formulations make it simple to partition losses among mortality agents. So, e.g., if we predict the total mortality rate Z to be $Z = M_o + M_p + F$, where the component rates are defined by $M_o =$ natural loss rate due to factors other than predation, $M_p =$ loss rate due to predation,

(Continued)

(BOX 1.1 continued)

and F = loss rate due to fishing, then we can calculate the net loss of fish to each rate process as that rate over Z times the total deaths. Total deaths are predicted by $D_t = N_t(1 - e^{-Z})$ (numbers at t minus number of survivors to $t + 1$); e.g., loss to fishing (catch) is given by $D_t F/Z$, and total predator consumption by $D_t M_p/Z$. ∎

correct: there is no such simplifying assumption at all in equation 1.1; the individuals making up N_t can, and generally do, consist of a complex mixture of ages, sizes, sexes, home-range locations, etc. To say that these creatures produce total eggs $f_t N_t$, or survivors $s_{(a)t} N_t$, is not to say that every animal is the same but, rather, just that there is some rate value f_t or $s_{(a)t}$ such that multiplying this value by N_t gives the numbers of eggs or survivors for year t. That is, the parameters f_t, $s_{(a)t}$, and $s_{(j)t}$ represent per-capita averages over N_t, and one of the reasons that we need to think of them as time-varying (t subscripts) is that they are likely to change with changes in the composition of N_t—e.g., f_t is likely to be larger in years when more of the N_t individuals are large, highly fecund females. We discuss methods for making more or less precise numerical predictions about changes in N_t using composition information in chapter 5 (single-species assessment).

Harvesting proportions u_t of the 1-year-old and older animals from the target population will obviously change the balance relationship, to $N_{t+1} = s_{(a)t}(1 - u_t)N_t + s_{(j)t}f_t N_t$, and this will result initially at least in $r_t < 1.0$, i.e., in population decline. Now suppose that variations in $s_{(a)t}$, $s_{(j)t}$, and f_t over t are due solely to what ecologists call "density-independent factors," i.e., the variations are (statistically) unrelated to N_t. In that case, the mean value of r_t will be less than 1.0 for any $u_t > 0$, and the expected long-term population trajectory is a decline toward extinction. That is, the only possible long-term ("sustainable") outcome of harvesting given only density-independent variation in the specific rates is extinction. Thankfully, this outcome is not what has been observed in virtually every case in which populations have been monitored during harvest development, and it is hard to imagine any viable natural population that would still be around if it exhibited such lack of response to variation in natural factors that have had an impact comparable to u_t. What we have seen, in fact, is at least some "density-dependent" or "compensatory" change in at least one of the specific rates, leading to improved survival and/or fecundity in response to a reduction in N_t. For modest u_t, such compensatory change tends to return r_t to a mean of 1.0, i.e., to stop the decline. *Hence, compensatory change in survival rates and/or fecundity is the fundamental ecological basis of sustainable harvesting.* So if someone argues that a given population exhibits no density-dependent or compensatory rate changes, i.e., if someone makes an oxymoron assertion like "the population is regulated purely by density-independent factors," then that person is, in fact, asserting that the population is incapable of producing a sustainable yield (and is incapable of exhibiting any sort of stable average population size under natural conditions either).

When we have been able to estimate changes in the rate factors $s_{(a)t}$, $s_{(j)t}$, and f_t over the history of fishery development, e.g., figure 1.2 from Martell (2002), a quite consistent response pattern has been observed that is largely independent of the type of creature being harvested (vertebrate or invertebrate, benthic or pelagic, lower trophic level or top predator, etc.). Methods for obtaining such estimates or historical "reconstructions" of population change are discussed in chapter 5. The typical response pattern has the following main features:

1. Mean fecundity f_t either remains relatively stable (in semelparous species like Pacific salmon that die immediately after first reproduction) or declines due to a reduction in the proportion of older, more fecund individuals in N_t; i.e., there is generally not a strong compensatory response in f_t.
2. Natural survival rate of older animals $s_{(a)t}$ also remains relatively stable, seldom showing any consistent compensatory improvement with reductions in N_t and often showing relatively little change even with large changes in presumed predation mortality.
3. "Juvenile" survival rate $s_{(j)t}$ shows compensatory improvement that is sometimes remarkably strong, typically leading to the total recruitment $R_t = s_{(j)t}f_tN_t$ being nearly independent of N_t over a wide range of N_t, even despite considerable decreases in f_t.

We discuss the ecological basis for observations (2) and (3) in chapters 6 and 10. Methods for predicting changes in $s_{(a)}$ and $s_{(j)}$ due to trophic relationships (predation, competition) are discussed in chapters 11 and 12, with particular emphasis on the observation by Hollowed et al. (2000) that useful predictive models for ecosystem management may need to involve a careful analysis of stage- and scale-dependent interaction impacts. Problems in measuring compensatory responses in $s_{(j)t}$ are discussed in chapter 7. In various chapters, we point out things that can go wrong with response (3), in particular factors that can cause a "depensatory" decrease in juvenile survival rates at low population sizes, so as to cause extinction or the failure of population recovery efforts. Chapter 13 discusses what happens when people get greedy and try to supplement or enhance natural recruitment through artificial propagation programs, since this has been one of the main suggestions for trying to beat the apparent natural limits to harvesting implied by compensatory changes in rate processes.

Just how well are the three assertions of the previous paragraph supported by empirical evidence, rather than by arm waving about how "viable" natural populations must exhibit some compensatory response(s)? There is no question about how most fisheries cause declines in average size (age) of fishes and, hence, declines in the mean fecundity f_t because of strong size-fecundity linkages. There are dozens of long-time series of age-composition

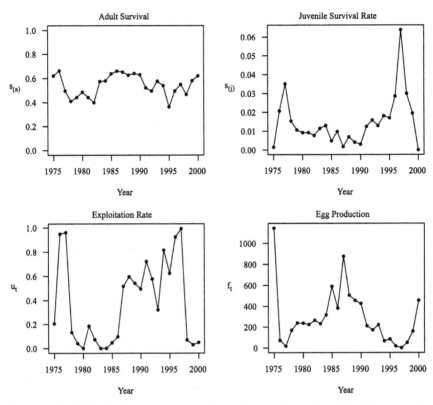

FIGURE 1.2: Variation in components of population change for pink shrimp on the west coast of Vancouver Island, B.C., from Martell (2002). These estimates were obtained by fitting an age-structured population-dynamics model to time-series data on relative-abundance, size-composition, and area-swept estimates of fishing mortality.

data that at least appear to support the assertion of stable adult survival $s_{(a)}$, though, in fact, what these data tell us is mainly about stability in the harvest-natural survival product $s_{(a)}(1-u_t)$. There are literally hundreds of data sets that demonstrate a lack of change in the recruitment rate with a change in spawner/egg abundance, i.e., density dependence in $s_{(j)}$. These recruitment data sets are readily available thanks to the painstaking efforts of Ransom Myers and his colleagues (*http://www.mscs.dal.ca/~myers/welcome.html*), and we strongly urge the reader to scan through them to see the variety of patterns that fish recruitments have exhibited.

It is not typical in fisheries texts to introduce the idea of surplus production in terms only of the numbers balance in equation 1.1. Most fisheries are measured and valued in terms of biomass (numbers x body weight), and growth in body weight is typically represented as an important component of biomass production. Equations such as production = growth + recruitment − mortality invite us to think of growth and recruitment as additive and to imagine that growth might be more "important" than recruitment or might occur as an additive effect on production independent of what might happen to recruitment. Such equations are misleading. The biomass yield Y_t from a fishery can be represented as $Y_t = u_t N_t w_t$, where $u_t N_t$ is the numbers yield and w_t is the average body size of the harvested fish. Much of the classic "theory of fishing" (Beverton and Holt 1957) was concerned with how to adjust u_t and the size of fish harvested so as maximize the average Y_t, but subject to the assumption of a strong compensatory improvement in $s_{(j)}$. In fact, the average body size w_t typically decreases with increases in the harvest rate u_t, as does the average abundance N_t, even if there are (relatively uncommon) compensatory improvements in fish body-growth rates as abundance decreases. When predicting surplus production and average yield Y_t, it is not helpful to point out that biomass production due to body growth tends to increase on a per-capita (per N_t) basis due to shifting the composition of N_t toward younger, faster-growing individuals. Numbers sustainability, i.e., $s_{(a)t}(1 - u_t) + s_{(j)t} f_t = 1$ on average, remains a basic requirement no matter what might happen to per-capita body-growth rates, and no matter what units we might use to measure yield or value.

Unfortunately, the qualitative knowledge that a given species is likely to exhibit compensatory responses in $s_{(j)}$ is not a sufficient basis for designing sustainable harvest policies. Even if there is no concern about the impact of harvesting the species on other ecosystem functions and species, we must still deal with two difficult (and quantitative) issues: (1) how to vary the strategic harvest-rate goals u_t over the long term in response to uncontrolled natural changes in $s_{(a)}, s_{(j)}$, and f; and (2) how to limit u_t in the short term by using various harvest-regulation "tactics" such as closed areas. We discuss the first of these, the so-called "harvest strategy" problem, in chapter 3. We discuss the second in three chapters: chapter 4 discusses broad options for limiting u_t, while chapters 8 and 9 discuss spatial models that are needed for the evaluation of closed-area policies and models for the evaluation of the effects of unregulated fisher behavior (fishing-effort dynamics) on the efficacy of regulation schemes.

Further, modern fisheries management involves making predictions about far more complex trade-offs than those involved in single-species abundances, survival rates, and body sizes (chapter 2). In particular, it is no longer acceptable in many management settings to ignore the possible ecosystem effects of harvesting each species. The mortality losses $(1 - s_{(a)t})N_t$ and $(1 - s_{(j)t}) f_t N_t$ are not just disappearances from ecosystems; rather, at least part of these losses represents "trophic support" provided by a species to higher trophic levels, i.e., part of the food supply of the species' predators. In single-species management, the historical tradition was either to treat such support functions as having no economic or social value, or to pretend that there is an ample supply of other food organisms to take up the slack when the production of any given species has been appropriated by fishing. Further, we have largely ignored the other side of the trophic coin, namely the responses of other organisms when the "demand" on a given species's food supply (on "lower trophic levels") has been reduced through fishing on that species. For example, other organisms might use that food supply to prosper and become replacement food sources for predators, hence reducing the net effect on predators of taking away some of their usual prey. Chapters 10, 11, and 12 discuss our emerging ability to make useful management predictions about such food-web interaction effects.

There is a critical point for readers to keep in mind about the complex biology and modeling introduced in chapters 10–12 for making predictions about the effects of food-web interactions. We are not introducing this material as a substitute for single-species population-dynamics modeling or management, or on the pretense that including trophic interaction effects in predictions of $s_{(a)t}$, $s_{(j)t}$, and f_t will somehow lead to better, more precise predictions about how each species is likely to respond to harvesting. In fact, from a single-species management perspective, trying to model all of the interactions that lead to variation in survival rates, especially of juvenile fish, can easily result in an "overparameterized" calculation, subject to a larger average prediction error than could be achieved with more precise estimates for fewer parameters. Rather, our aim in introducing these models is to provide a capability for fisheries scientists to respond to a broader set of policy questions and predictive demands than can single-species analysis. These questions lead to a much broader set of options for future ecosystem management than might ever be imagined by thinking only of species populations one at a time (chapter 14).

CHAPTER 2

Trade-Offs in Fisheries Management

TEXTS ON FISHERIES and ecosystem management typically begin with discussions about the objectives of management. It is easy to say that the objectives of modern fisheries management should be to ensure sustainable harvests, viable fishing communities, and healthy ecosystems, i.e., to sustain a mix of production, economic, and ecological values. Unfortunately, such platitudes are of little value in guiding either management or science, because there can be conflicts among the component objectives.

A more useful statement of objectives for scientists and managers would be to say that the central objective of modern fisheries science should be to clearly expose trade-offs among conflicting objectives, and the central objective of modern fisheries management should be to develop effective ways to decide where to operate along the trade-offs, and how to operate successfully (Mangel 2000).

Two kinds of things can go very wrong when people fail to acknowledge that trade-offs are, in fact, the central problem of fisheries decision-making:

1. The trade-offs end up happening "by default," either in the form of depleted fish stocks, unhealthy fishing communities, and severely degraded habitats when no one accepts responsibility for explicit trade-offs between short- versus long-term values, or in the form of extreme and destructive policy choices based on the interests and political power of particular stakeholders. The balance of political and economic bargaining power is rapidly shifting from fishers to broader communities of public interest, particularly toward conservation interests and groups; this has resulted in the promotion of some quite extreme positions about values (i.e., fisheries should not be allowed to accidentally kill even one sea turtle), and the very real possibility of economically destructive closures of some of the world's largest fisheries.

2. Scientists end up being forced, encouraged, or allowed to make value judgments about matters for which they have no particular competence or wisdom, particularly in relation to trade-offs involving risk and existence (nonconsumptive) values. That is, scientists end up being asked how fisheries *should* be managed rather than how they *could* be managed. Science is about what can and cannot happen in the world around us, and a scientist who steps beyond this to advocate particular policies or objectives is proposing that the public should adopt his or her personal preferences in making trade-off choices. That is an unethical thing for a scientist to do because it is an abuse of the respect that the public may afford the scientist for his or her knowledge and, more importantly, because in most cases "talk is cheap": scientists are generally well insulated from the economic consequences of their recommendations (it is easy to advocate

fisheries closures, protected areas, and such when it is not you who will bear the social and economic costs of such policies).

Indeed, one of the main reasons we have written this chapter is to encourage young scientists to scrupulously avoid mixing their personal values with their analyses *as scientists* of how the world functions. This means avoiding even subtle advocacy that can undermine the credibility of scientists as honest providers of information, such as occurs in the following statement from a recent National Research Council report on marine reserves (NRC 2001, p.2):

> Based on evidence from existing marine area closures in both temperate and tropical regions, marine reserves and protected areas will be effective tools for addressing conservation needs as part of integrated coastal and marine area management.

The scientists who wrote this statement were asked for their opinions as scientists about the efficacy of marine reserves as management tools; they were not asked to advocate reserves (the phrase "will be effective" is an implicit advocacy statement), and they certainly were not asked whether the public "needs" conservation. In using the word "needs" rather than some weaker term such as "preferences," these scientists invited readers to predict some dire consequence of not sharing or adopting their personal preferences concerning the "healthy" ecosystems that they enjoy observing and studying. They crossed the line between science and advocacy and in doing so should make readers of the report suspicious about whether the rest of the report is an honest and balanced presentation of the scientific evidence concerning marine reserves. Walters and Martell happen to be strong proponents of marine reserves, because we are conservationists as well as scientists. But we utterly fail to see why people should pay more attention to our arguments about the values of conservation just because we happen to be scientists.

Here is a short list of specific trade-offs that fishery managers must face on a regular or routine basis:

1. abundance of target species versus fishing effort (less for each fisher when there are more fishers);
2. harvest today versus harvest in the future (more for each fisher today can mean less for future fishers, especially when stocks have already been severely fished);
3. abundance of unproductive stocks and species versus harvest of more productive stocks, when nonselective fishing gear takes them all;
4. profit versus employment (providing an excellent income for a few fishers, or lower incomes for many fishers);
5. public expenditure on fisheries (monitoring, research, management, and subsidies to fishers) versus expenditure on other public services such as health care or education;
6. harvest of valued species versus abundance of other species that depend on the valued species as food;
7. inexpensive fishing practices that can have severe bycatch and habitat impacts versus selective fishing practices.

There are also some severe trade-offs in how financial and human resources are deployed for fisheries science:

1. basic research on ecological processes versus routine monitoring for assessment;
2. development of new monitoring and assessment methods versus gathering data to support existing methods;
3. investment in research and assessment versus other management functions such as enforcement and habitat restoration;
4. research that interests other scientists versus research on how to manage better even if that research is mundane.

All of these trade-offs involve hard choices, and in most cases they also represent points of conflict among fisheries stakeholders. When conflicting objectives involve the interests and values of different people, it is important not to think of "solving" trade-off problems as involving a "sole owner" making choices to maximize some personal utility, value, or preference function. That is, we cannot solve most trade-off problems just by resorting to the elegant theory of helping people make personal decisions by using "multi-attribute decision analysis" (Keeney and Raiffa 1976).

This chapter begins with a general look at ways to visualize and explain trade-off relationships and at the form of objective functions that might be used to make formal "optimization" choices about where to operate along some trade-off relationship. Then we examine several of the most important trade-offs in fisheries more closely, to see what forms of relationships are most likely to arise from scientific prediction models and how alternative objective functions or values or political bargaining processes can affect choices for these particular trade-offs.

2.1 Trade-Off Relationships and Policy Choices

Trade-offs between two conflicting values or measures of management performance can generally be represented as graphs showing how much of one measure can be achieved as a function of how much is achieved of the other measure (fig. 2.1). We "sketch" these relationships by defining a set of policy choices, then plotting the combinations of measures achieved under different choices. While the shapes of such relationships can be quite complex, we most often predict them to be of one of three possible general shapes:

Convex set: one value can be increased considerably without affecting the ability to achieve the other value (these are the easy trade-offs to deal with in conflict situations since a "balanced" or compromise policy can be found in which both values are high);

Concave set: small increases in one value measure require or force a disproportionate drop in the highest achievable value of the other

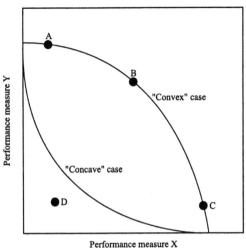

FIGURE 2.1: Trade-off relationships can often be usefully viewed as graphs of two value measures against one another. There is a trade-off when an increase in measure X leads to a predicted decrease in measure Y. Such relationships should generally be calculated as "constraints" or maximum achievable combinations of X and Y, to aid in the identification of "dominated" policies in which both measures could be improved. A, B, C, and D are outcomes of example policy choices: A is a policy that gives high Y but low X; B is a policy that "balances" X and Y; C is a policy that gives high X but low Y; and D is a "dominated policy" that gives less of both X and Y than could be achieved by one of the choices along the "frontier" or "constraint set."

measure (these are difficult trade-offs to manage because adding much of either value causes a severe decrease in how much of the other value can be had);

Linear set: increases in one value lead to a proportionate drop in the other value (these mainly arise in simple economic situations, in which there is a simple constraint on total expenditure, and spending more on one thing means spending just that much less on another).

For some trade-off relationships, the only combinations of values that can be achieved are ones along the trade-off curve or "frontier" itself. In other cases, there are policy options that would achieve less of both values than the relationship predicts would be possible. We say that such options are "dominated" by alternative choices that would give results along the trade-off frontier, or that movement away from such choices to a choice that does give values on the frontier would be a "win-win" policy choice. When we see a fishery that is apparently being operated with a dominated (interior, suboptimal) policy, we should think about three possibilities for why this unhappy state of affairs exists:

1. we have left something out in the trade-off calculation, and the values being achieved are, in fact, already ones along the trade-off frontier;
2. there are "hidden objectives," so that trade-offs have been made with respect to some third value measure not visible in the two-dimensional graph;
3. somebody goofed, and the system just hasn't been managed as well as it could be.

We can usually think about broad trade-off relationships and whether existing policies might be dominated ones without being very precise about exactly how each value or performance axis is measured. We need to choose specific performance measures when we use mathematical models to predict the trade-off relationships, but often there are several different measures that are about equally as good as indices of value.

In some cases, we can "solve" trade-off problems by exposing the trade-off relationship clearly to stakeholders (e.g., plotting it), then asking them to be "reasonable" by making a choice that represents some fair balance of performance measures. But in other cases it may be considered preferable or more fair to ask stakeholders to agree on some overall system-scale objective function that contains their values as contributing components, then formally seek a policy choice that maximizes this function. Fisheries debates typically involve arguments that imply two possible forms for such an objective function, linear and logarithmic. A linear objective function is an equation of the form

$$V = w_1 X + w_2 Y \tag{2.1}$$

where V is the total value, X and Y are (conflicting) value measures, and w_1 and w_2 are "value weights" determined by socio-economic or political bargaining-power considerations. A logarithmic objective function is of the form

$$V = X^{w_1} Y^{w_2} \quad \text{or} \quad V = w_1 \, ln(X) + w_2 \, ln(Y) \tag{2.2}$$

Linear objective functions usually arise in settings in which stakeholders (or managers or politicians) are willing to trade-off freely between X and Y, without prejudice about extreme values of either choice. Logarithmic functions arise when extreme values of either X or Y are seen as unfair, destructive, or perhaps even illegal. The log utility form comes from arguments about performance measures such as X = income; generally, people view positive increments in income as important when their income is small to start with but do not view similar increments as being so important when income is large to start with.

A useful way to visualize the application of formal objective functions to trade-off problems is to plot the $X - Y$ combinations associated with particular overall values V, then vary V so as to find the V that has the highest

Linear Objective Function

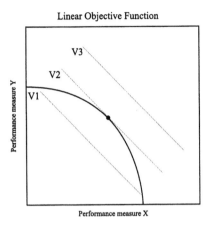

Logarithmic Objective Function

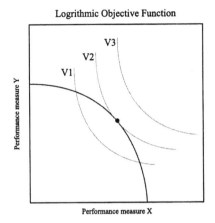

FIGURE 2.2: The formal optimization of a linear or logarithmic objective function like equation 2.1 or equation 2.2 can be done graphically by drawing lines representing increasing total values, then seeing which line (V value) is just barely feasible (gives the highest V). The lines labeled V1 represent objective function values that can be improved upon, the lines V2 represent maximum value, and the lines V3 represent value that cannot be achieved. The black dots show the "optimum" achievable X, Y performance combinations.

feasible $X - Y$ combination. For example, equation 2.1 can be written as $Y = (V/w_2) - (w_1/w_2)X$, i.e., the equation of a straight line $Y = a - bX$, where the intercept a depends on V. If we plot a set of such straight lines representing progressively higher V on a graph showing a trade-off relationship, we can easily "read off" the $X - Y$ combination that gives the highest feasible V, as shown in figure 2.2. The same basic idea can be applied to logarithmic objective functions, except that the $X - Y$ combinations for a given V lie along a curve of $1/X$ shape. A very important distinction between linear and logarithmic objective functions becomes obvious when we do this: if the trade-off relationship is concave or linear, a linear objective function will always tell us to manage so as to produce either all X or all Y; if the relationship is logarithmic, the function will tell us always to try to produce at least a little of both X and Y. As we will see in the next section, this distinction is particularly critical in deciding how to deal with trade-offs over time, that is, short- versus long-term harvest values.

2.2   SHORT-TERM VERSUS LONG-TERM VALUES

Easily the single most difficult and pervasive trade-off issue in fisheries management is between catching fish now versus leaving them in the water to produce surplus for harvesting in the future. There has been much research on how to optimize the long-term value of fisheries (see, e.g., Clark 1976, 1985; Walters 1986), mainly under the hidden assumption that society is best

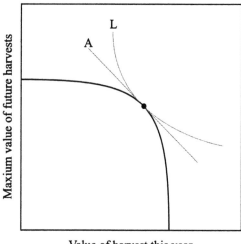

FIGURE 2.3: Trade-off relationship between the immediate and long-term harvest values for a fishery where fish abundance is high ("healthy") at the time of decision. Note that whether the fishers' objective function is additive (A) or logarithmic (L), the management recommendation will still be to take a moderate catch that does not jeopardize the long-term harvest value. For harvests "this year" below the optimum point, there is little impact of this year's harvest on the maximum that can be achieved from next year on.

served by managing fisheries as would a sole owner who took a relatively long view of future benefits and costs, i.e., a sole owner who would calculate the present value of harvest decisions by adding up all the future values resulting from present and future decisions, with future values discounted at relatively low rates to reflect the interests of "future generations." Formal optimization methods generally tell us that such a sole owner would follow a relatively simple "feedback rule" in response to changes in stock size, varying harvests so as to keep the stock near the level at which it produces the largest average "surplus" (potential population growth) each year.

The sole owner's feedback rule is basically calculated by examining the trade-off relationship shown in figure 2.3. For a set of choices of harvest this year (trade-off relationship X axis), calculate the best harvest that could be achieved from next year forward in time (trade-off Y axis) as a function of how many fish would be left after this year's harvest choice (that is, apply Bellman's 1957 famous "principle of optimality," the basic building block of much optimization theory for dynamic systems). Such calculations can be done with a variety of population dynamics models, and they generally produce a trade-off pattern similar to that of figure 2.3: if the stock is at a healthy level this year, increasing this year's harvest up to some point has little impact on potential future value. But if this year's harvest is too high, so too few fish are left to produce growth and recruitment for future years, the

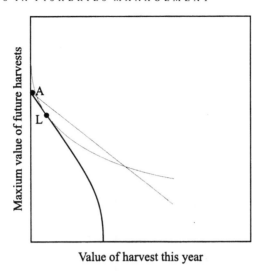

Value of harvest this year

FIGURE 2.4: The same trade-off relationship between the immediate and long-term harvest as in figure 2.3, but for a time when fish abundance is low ("unhealthy") at the time of decision, due to historical mismanagement or the impacts of unpredicted environmental factors. Note that very different recommendations follow from assuming an additive objective function (do not fish at all, point A) versus a logarithmic one (try to take enough fish to at least keep the industry from bankruptcy, point L).

long-term harvest will be severely impaired. For both sole owners and sensible fishers in a competitive fleet, this problem has the same simple, equitable answer (and recommendation to public management agencies): constrain fishing this year so as to provide a reasonable harvest, without reducing the stock below that level that will maximize its average surplus production in future years.

Figures 2.3 and 2.4 tell us that managers face different decision problems when stocks are still healthy than after they have been overfished. In particular, there is likely to be controversy over whether to view the value of immediate catches as having logarithmic utility for fishers, on the grounds that complete fishery closure will force them into bankruptcy or other employment (with the attendant disruption in their lives). Further, there develops a separation of interests for the fishers (box 2.1): if a fishery is closed entirely to allow the stock to rebuild as rapidly as possible, it is quite likely that the benefits from this decision will accrue not to the present community of fishers but to other people entirely (and not necessarily even the fishers' children). But from a public perspective, any fishing at all can delay stock recovery and the reestablishment of a healthy fishing community. Fishers often do not appreciate just how severe an impact "just a little fishing" can have on stock rebuilding. An example of how severe this impact can be is the set of predictions made shortly after the Newfoundland cod collapse in the early 1990s (fig. 2.5); at that time, 20- to 40-year recovery times were predicted,

Recovery of Newfoundland's nothern cod stock

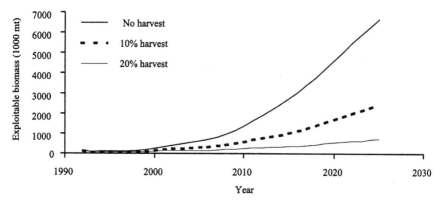

FIGURE 2.5: The effect of complete closures versus low annual harvest rates on the potential recovery of the Newfoundland cod stock. Note that even very low harvest rates are predicted to delay the recovery and realization of the long-term potential of the stock to produce sustainable harvests. The recovery pattern for a 20% harvest rate corresponds roughly to harvesting at the MSY fishing-mortality rate.

even if the fishery were closed entirely, increasing to perhaps as much as a century if even modest (10%) harvest rates were allowed. Ten years later, we now fear that the reality may be even worse than we predicted, since the stock has not yet shown even the modest recovery that we predicted for the 1990s. Various hypotheses have been advanced to explain this failure, ranging from continued fishing to predation effects (Hutchings 2000; Shelton and Healey 1999; Anderson and Rose 2001). In April 2003 the Canadian government closed the Newfoundland cod fishery for an indefinite period.

What is the optimum regulatory response or policy in these increasingly common situations, when we must look forward in time from a severely depleted ecological basis for production? It is pointless to recriminate by pointing out that the best answer is not to get into such situations in the first place. In our judgment, there is no good answer to this question, and the best that managers can do is:

1. develop incentives for fishers to cooperate in rebuilding (box 2.1);
2. be scrupulously honest and open about predicting how long the stock recovery will likely take, and who will eventually benefit from it;
3. do not pretend that stakeholders will reach a "reasonable" compromise solution, especially considering that one of the most important stakeholder groups is not even at the table (future fishers).

There is some hope that we can "buy" our way out of some depleted situations by using stock-enhancement methods, but such investments have their own serious risks (see chapter 12).

BOX 2.1

WHY FISHERMEN OPPOSE STOCK-REBUILDING PLANS, AND WHAT CAN BE
DONE TO OBTAIN THEIR SUPPORT

The following is a very long quotation that essentially reprints the whole
of Hilborn (2000). It is a collection of insightful comments about values,
trade-offs, and incentives.

It is often a mystery to managers and scientists why fishermen oppose rebuilding plans such as in most New England groundfish stocks, where the
evidence is that catch restrictions would result in long term benefits. In reality
rebuilding plans are often not in the economic interest of the fishermen. Superficially this defies the basic dynamics of populations and economics, that
the stocks could produce higher yields if allowed to rebuild, and the fishermen
would enjoy the "long term gain for short term pain." However, this is often
not the case.

- Reason 1: Fishermen who make the reduction in catch will not enjoy all
  the benefits of the rebuilding. There are often inactive permits or lightly
  fished permits that would reenter the fishery if the stock were rebuilt to
  higher catches, so those who pay the "short term pain" would not enjoy all
  the "long term gain." Simple economics shows that the fishermen may be
  better off accepting lower catches now than reducing catches for benefits
  they may never see.
- Solution 1: Some form of property right needs to be instituted while the
  stock is depleted so that those who suffer the short term pain enjoy all the
  long term gain.
- Reason 2: In today's political climate fishermen have no assurance that
  catches will be allowed to increase after the stocks are rebuilt because of
  uncertainty regarding implementation of the precautionary approach and
  ecosystem management. There would then be no long term gain for the
  fishermen.
- Solution 2: Regulators and fishermen need to agree on firm decision rules
  specifying how catches will change in relation to the data collected on
  the fishery, so that if the stock rebuilds, based on data the fishermen can
  understand such as surveys, catches will increase according to those rules.
- Reason 3: The price of rebuilding in multispecies fisheries is often reduced
  catches of more abundant species, and in such cases it is not economically
  optimal to rebuild the depleted stocks. It is inherent in multispecies management that when a fishery is economically optimally managed, some
  stocks may be overfished.
- Solution 3: In multispecies fisheries, regulators must distinguish between
  stocks that are truly threatened or endangered and those that are fished

(Continued)

(*BOX 2.1 continued*)

harder than would be optimal on a single species basis. In many U.S. fisheries the current system does not encourage fishermen to avoid depleted stocks. Other systems provide such encouragement. In the Canadian west coast groundfish fishery, a vessel cannot fish in an area if it does not have quota for every major species that may be caught. This is combined with 100% observer coverage to prevent discarding. The result is that fishermen expend great effort finding ways to fish without catching species for which there is little allowable catch.

- Reason 4: The increase in potential yield may be small in relation to the cost of rebuilding. In New Zealand the government was concerned about the status of what is known as the snapper 1 fishery, one of the most important fisheries in the country. The stock biomass was estimated to be at a level that would support MSY, which would be considered depleted in the U.S. The government proposed a 40% reduction in commercial catch for an estimated 20-year rebuilding period. At the end of that period the stock assessment suggested that catches could rise to 8% above what they had been before the rebuilding plan. Most sensible people would oppose a 40% pay reduction for 20 years in order to get a 8% pay increase 20 years later!
- Solution 4: Management plans need to balance expected benefits against potential costs and it may be that for some stocks the benefits of rebuilding are low. If there are noneconomic reasons to encourage rebuilding, it might be appropriate to provide financial compensation to fishermen.
- Reason 5: The stock assessments on which the classification as depleted and the benefits of rebuilding are based may be wrong. In 1997 the "new" assessment for sablefish on the west coast suggested the quota should be reduced to 2,200 tons from 9,000 tons. But by 1998 the next assessment "corrected" the previous low estimate of biomass, and the quota was increased again to 7,200 tons.
- Solution 5: The more fishermen are able to participate in stock assessment and provide data input into the process, the better the assessments will be and the more confidence fishermen will have in them. Considerable progress is being made at industry participation in assessment, and this progress needs to be continued.

The five reasons above are some of the reasons why fishermen may quite rightly oppose rebuilding plans, while in most cases rebuilding is likely beneficial to society as a whole. The fishermen's opposition can be overcome by making sure that the system is structured in such a way that it is in their and society's best interest to support the rebuilding. ∎

One symptom of how difficult it is to trade off between immediate and long-term values, and one of the ways that stock depletion can occur in the first place, is a decision pathology that we call "indecision as rational choice" as introduced in chapter 1 (fig. 1.1). It is worth saying a bit more about this pathology, in the terminology of trade-off analysis. Try to imagine being a fishery manager or politician who has just been approached by scientists warning of an imminent collapse and recommending harsh, expensive action (such as a fishery closure). You have basically two choices: (1) you can heed their advice, in which case you know that you will face immediate, harsh criticism from fishers and may even risk losing your job, or (2) you can make some excuse (e.g., "the problem needs further study") to delay action, in hopes that either the scientists were wrong and the apparent problem will correct itself or you can move to another position and leave the decision to your successor. There is a trade-off between the immediate pain of facing irate fishers versus the risk of having to face even worse pain in the future. But as you reflect on this trade-off, it will doubtless occur to you that fisheries scientists are habitual "fear-mongers" and have a rather poor track record with predictions in general (so you know that you cannot trust their appraisal of the long-term risk). So in the end, you will likely see the decision as a no-brainer: you will opt for the inaction extreme along the trade-off relationship every time, *if you are not personally accountable should the scientists' prediction prove correct*. In fact, government fisheries managers and politicians are seldom directly accountable for much of anything. It is fine for scientists to argue that some vaguely defined "we" should "reverse the burden of proof" in such situations and opt for safer if more immediately painful choices. Bluntly, such arguments are meaningless: what we should be arguing for is not a "precautionary approach" but, rather, the development of law and regulation to *make inaction more costly and painful than action, by making managers personally accountable for myopic decisions.* In Ray Hilborn's words, the problem is not lack of good intentions, it is lack of good incentives (box 2.1).

2.3   BIOLOGICAL DIVERSITY VERSUS PRODUCTIVITY

While short-term versus long-term harvest trade-offs have been the most serious decision problem in traditional (single-species) fisheries management, an equally difficult problem that we must face in moving toward "ecosystem management" is the impact of fishing on natural biodiversity. Fisheries impact nontarget species through a variety of mechanisms, ranging from direct bycatch mortality to habitat damage by gear to the appropriation of production that would otherwise be taken by natural predators. It is foolish to pretend that there is some ecosystem principle stating that ecosystems are most productive of harvestable surpluses when they are maintained in the most natural and diverse possible state. An aquatic ecosystem in its natural state, in fact, exhibits no net production at all; the various organisms that prosper in this state have already appropriated all surplus production,

otherwise they would be growing in abundance and/or would likely have
already been subject to invasions by other species that could use them as resources. Further, long-term data sets on places such as the North Sea do not
even support the reasonable proposition that the simplification of ecosystems
through fishing might cause increased variability, vulnerability to invasion
by exotic species, or even "dynamic instability" among the abundances of
interacting species (Hall 1999; Pimm and Hyman 1987).

So we cannot convincingly argue that the maintenance of a natural community structure is a win-win option for everyone, including fishers as well
as people who value various creatures (and diversity itself) for other reasons
(or who feel that other creatures have some intrinsic right to existence and
protection). Producing catch is damaging to other ecosystem values, and we
have to face this trade-off more and more often today as people demand consideration of those other values. Fishers and fishery managers who ignore
these demands (and the trade-offs implied by them) invite sanctions ranging from marketing boycotts (e.g., dolphin-free tuna) to closures driven by
an insistence on the strict application of legislation (e.g., endangered-species
acts).

Almost all fisheries catch more than one species of fish, and many catch
multiple genetic races or stocks of the same species (e.g., salmon ocean fishing). Among the species and stocks captured by any fishery, there is always
a variation in vulnerability (the mortality rate caused by a given amount
of fishing) and sensitivity (the mortality rate that the species can sustain or
tolerate). So there is a distribution over the ecosystem of the impacts on
species/stocks, some suffering little impact and some a lot. Ecological interactions (competition and predation relationships among the species) may
mitigate or exaggerate the effects of fishing but do not compensate for the
effects of varying vulnerability and sensitivity. These observations mean that
for almost every fishery, there is a fundamental trade-off relationship between
the intensity or size of the fishery and the long-term, sustainable abundance
and diversity of species/stocks. We cannot think of a fishery as either destructive or not destructive, sustainable or not sustainable; diversity impacts
are not a yes-no proposition. To make matters worse, the trade-off axes
involve fundamentally different public-interest groups. Biodiversity and existence values mainly represent the interests of urban communities who do
not depend on fishing for their livelihoods, and for whom it is personally
not expensive to advocate low levels of fishing. Consumptive values (retail
cost, sport catch, etc.) mainly represent the interests of the small minority
of people who actually do depend for fun and sustenance on the fisheries.

The predicted trade-off relationships between diversity and consumptive
values are strongly dependent on what management controls are recognized
or considered practical to implement. Recall from the previous section that
in plots of long-term versus short-term harvest values, we generally try to
plot the maximum long-term value that can be achieved for a given short-
term value choice. Similarly, when the Y-axis of a trade-off relationship is

some measure of biological diversity or "ecosystem health," and the X-axis is some measure of consumptive value (productivity or yield or economic value or employment), we need to ask what the best achievable level of biodiversity is for each possible level of consumptive output. If we assume that only very crude fishery controls are available or practical, such as catch quotas or large space/time closures, we will conclude that the trade-off is much more severe than would be the case if it were possible to use more selective fishing practices and local space/time controls. Conservationists are sometimes not even aware of (or are mistrustful of) the rich variety of options actually available to fishers and fishery managers to avoid damaging ecological impacts and, hence, end up recommending blunt policies such as large protected areas that are in fact "dominated" (below a potential trade-off line, so that all values could be bettered, as with option D in fig. 2.1) by more complex regulatory options. Of course, more complex options may have much higher costs to implement (e.g., monitoring, enforcement, gear changes) than public agencies or fishing industries are willing or able to endure.

There is a global move toward changes in fishing practices and gears to avoid obvious, direct (bycatch) impacts on nontarget species, and it is turning out that the apparent economic trade-offs associated with such changes (less efficient and more costly practices) are typically much less severe than fishers have claimed they would be. Avoiding dolphin catch in tuna purse seines does have some time/labor cost, but it certainly has not shut down the tuna industry. BRDs and TEDs (bycatch reduction and turtle exclusion devices) do reduce catch rates in shrimp trawl fisheries but also save time and labor in sorting catches. Devices (line shooters) to keep pelagic birds from taking longline baits as the lines enter the water may even turn out to be a win-win change in practices, by reducing bait losses.

There is an interesting point about how more selective practices are being developed: most of them are being invented by fishers themselves because there is a threat of fishery closure. That is, selective practices are not coming out of scientific research by management agencies but, rather, from the much simpler tactic of saying, "All right, you are going to be closed down unless you find a way to be less destructive; go find it"; fishers have displayed remarkable inventiveness when faced with such threats.

Unfortunately, bycatch mortalities might not be the biggest causes of diversity impacts in the first place. The two most severe diversity/productivity trade-off issues today are (1) the appropriation by fisheries of production that would otherwise support natural predators that we value, and (2) the erosion of within-species diversity in the stock structure by "mixed-stock" fisheries.

Ecosystem models such as Ecosim (see chapters 10 and 11), and basic thermodynamic considerations, warn us to expect concave trade-off relationships between a sustainable abundance of natural predators and the catches that we take of their prey (fig. 2.6) when we harvest widely over the set of

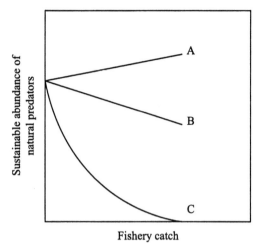

Fishery catch

FIGURE 2.6: The possible trade-off relationships between a sustainable abundance of natural predators (birds, mammals) and fishery catches. A-fishery targets species that are competitors/predators of preferred food species for the natural predators; B-fishery targets a small subset of prey species; C-fishery is the food-chain case in which the fishery directly appropriates production that would otherwise support the natural predators.

natural prey species. That is, it is not true, as some proponents of single-species management have claimed, that we can expect to take reasonable harvests of fish without impairing the capacity of these fish to also support birds, marine mammals, and other natural predators. No quantitative model supports such assertions, except in cases in which a very particular food-web organization leads to the enhancement of some prey species when others are fished (Yodzis 2001), or in which the fishery impacts only a few prey species from a large natural array. We do not always see large, immediate declines in birds and mammals as fisheries develop (e.g., Shima et al. 2000), presumably because most mammalian and avian predators are flexible about switching to alternative prey as fisheries impact particular prey species, but such options can disappear rapidly as other fisheries develop and/or switch to other species themselves. That is, the prey-switching argument applies just when little of the total fish production is being appropriated by fisheries. Considering the possible complexity of both natural food-web interactions and possible "novel" behavioral responses like prey switching, we generally have to treat the impacts of fishing as grossly uncertain (we might see A or B or C in fig. 2.6), subject to understanding only through hard field experience or adaptive management experiments.

Formal optimization methods applied to ecosystem models suggest an even more worrisome trade-off possibility than case C in figure 2.6. If we set up multiple fisheries in ecosystem models, then search for an "optimum" mix of fishing efforts to maximize some simple economic objective like total profit from the ecosystem (the sum of catches multiplied by prices minus fishing costs, over all fisheries and species), we find that this optimum mix often involves an ecosystem-scale "farming" strategy (Pitcher and Cochrane 2002). That is, the optimization tells us to use some of the fisheries to deliberately overexploit (control, cull) the competitors/predators of a few most valued species, so as to direct as much ecosystem production as possible through these species. This is, of course, exactly what people do in modern farming, through cultivation and pest-control practices. Almost every fisheries scientist who has seen such results has reacted with disgust and has turned to alternative objective functions that represent both broader values (existence as well as catch values) and also the "risk-spreading" value of maintaining a diversified "portfolio" of productive populations. But the simple fact is that we do not have strong, objective evidence from past fisheries experience to demonstrate that it would ultimately be destructive to engage in ecosystem-scale farming, i.e., to move a long ways along a trade-off relationship between natural diversity and the total economic value of fish production.

Trade-offs between diversity and productivity are also a major concern in single-species management, when harvesting takes multiple substocks in "mixed fishery" areas. For example, most coho salmon (*Oncorhynchus kisutch*) harvested in British Columbia are taken in some eight large fishing areas, where fish from several thousand small spawning streams mix for feeding before they begin their spawning migrations. For some 900 of these substocks, we have 10–50 years of simple trend data on spawning abundance, from crude surveys conducted by fisheries officers. We can estimate the intrinsic rates of population growth, i.e., the maximum sustainable exploitation rates (stock-recruitment slope parameters) for these populations, using stock-recruitment fitting methods that correct for some biases (discussed in chapter 7). This results in an estimated distribution of productivities over the substocks (fig. 2.7), where substocks in the lower tail of the distribution are those that have declined greatly over the last 50 years (presumably due to relatively poor freshwater survival rates and/or high relative vulnerability to fishing) and substocks in the upper tail are those that have prospered. (The distribution would be even wider if we included productive hatchery stocks in the mix.) If we then treat the average ocean fishing-mortality rate as a basic policy variable, and plot both the expected total catch over all stocks and the proportion of stocks that should tend toward extinction as functions of this average rate (fig. 2.8), we see a clear, convex trade-off relationship. Fishing can be increased up to some point with little loss to the substock structure, but further increases to capture the last-catch increments up to the mixed-stock MSY are likely to cause (and in fact already have caused) a fairly high proportion of the stocks to be driven toward extinction. There has been fierce debate in British Columbia about whether to "write off" the less-productive coho stocks or to try to mitigate the overharvesting effects on them through habitat improvements and/or hatchery rearing. In Washington state, another way of trying to deal with this problem has been to

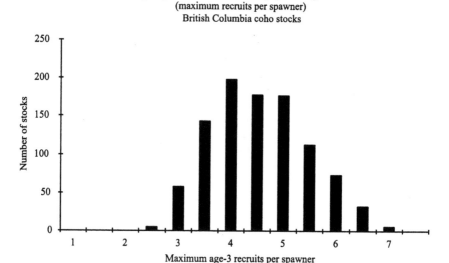

FIGURE 2.7: Distribution of intrinsic rates of population growth for 943 British Columbia coho salmon substocks, estimated from 1950 to 1999 escapement trend data. The rate of increase for each stock in the distribution is estimated by fitting the Beverton-Holt stock-recruitment model (rate of increase $= e^a$ in the relationship $ln(S_{t+3}) = ln(S_t) + a - F_t - ln(1 + bS_t)$, where F_t = fishing-mortality rate, S_t = spawners, b = carrying capacity parameter) with various bias corrections. The relative shortage of low-productivity stocks (skew in distribution) likely represents the loss of less-productive stocks due to fishing before 1950.

make the fisheries more complex by allowing only relatively low harvests in the mixed-stock areas and placing "terminal" fisheries in locations where stocks can be taken more selectively (Wright 1981, 1993).

These examples point out that it is a complex matter to try to predict trade-offs involving diversity/existence versus productivity/consumption values. In many cases there is not even a clearly defined forum in which trade-off relationships can be presented to all who consider themselves stakeholders, so that debate might lead to some consensus about the best balance. Rather, these trade-off issues are debated in multiple jurisdictions, using multiple routes to access political decision-makers, and with scientific exaggeration and misinformation used routinely as debating tools.

Effect of increasing coastwide exploitation rate on B.C. coho stocks

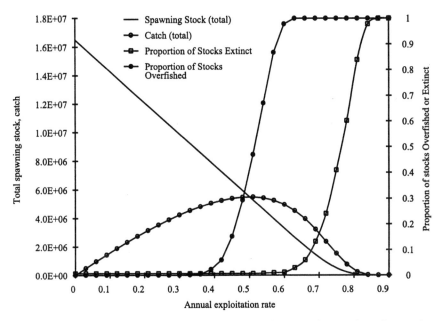

FIGURE 2.8: The predicted mean annual total yield summed over the coho stocks in figure 2.7, and the proportion of stocks expected to trend toward extinction, as functions of the mean coastwide exploitation rate along the British Columbia coast. The historical mean exploitation rate since 1950 has mainly been in the range of 0.7–0.8.

2.4   ECONOMIC EFFICIENCY VERSUS DIVERSITY OF EMPLOYMENT OPPORTUNITIES

During the 1960s and 1970s, many fisheries that had not yet been biologically overfished came under scrutiny by economists, who pointed out that any profits these fisheries might have generated for fishers or the public (as economic "rent") had been dissipated: profitable fishing had attracted unregulated investment and the entry of new fishers, until available catches were distributed over so many vessels that most fishers were barely meeting their operating costs (Scott 1979). It was noted that the average incomes at such "bionomic equilibrium" were commonly lower than wage levels for most employment; this prompted the suggestion that many fishers are apparently willing to "pay" quite a bit (in foregone income) to keep fishing, and hence must prize the lifestyle despite the work being visibly brutal (long hours, backbreaking, etc.) to outsiders. Economists then went on to point out that such industries do not make the "best" (most efficient) possible use of scarce resources (fuel, labor, capital equipment), and that the economy (public) as a whole would be better served by cutting back fleet sizes (and improving technology) so as to maximize the total industry-scale profit. It was argued that the public would benefit not only from the "efficient" deployment of resources but also from the direct capture of some of the profits (rents) as taxes. It was argued that instead of all these low-life shirkers running around wasting gas and their labor, we should have much smaller, high-technology, profitable fishing fleets.

In some nations (e.g., Canada), the economists found sympathetic ears in government, and their arguments drove the development of major programs to "buy back" and retire fishing licenses and to encourage the development of large, efficient fishing vessels. Reductions in fleet sizes led to immediate improvements in profitability, and this did not go unnoticed: vessel ownership began to shift from individual "owner-operators" to entrepreneurs who started to see fishing as another profitable investment. Many fishers ended up being more productively employed in places such as pulp mills, a few got rich, and some people who were already wealthy saw some increases in the values of their investment portfolios. Fishery regulators faced smaller and a bit less vicious communities of fishers in public meetings, but had new headaches in the form of highly efficient vessels that could overfish the stocks much more quickly. The economists were quite proud of themselves and went on to develop still more concepts for improving efficiency, like ITQs (individual transferable quotas) that would reduce competition among fishers and give them more flexibility to plan how to fish in the most profitable ways (see, e.g., Davis 1996; Anderson 2000; Eythorsson 2000).

So are we (the public of nations such as Canada) better off now that this trade-off from employment to efficiency has been largely completed? Certainly we have lost something in terms of diversity of employment opportunities; very few people can now choose to give up the urban life in favor of the amenities and hardships of living in small fishing communities. Labor (and gasoline, and boat repairs, and...) is not noticeably cheaper now that it is not being "wasted," nor is our overall economic productivity noticeably higher than we would expect from the impacts of other factors like technology. We cannot buy cheaper fish in the market. But have we at least been well served through the capture of rent in the form of taxes and expenditures by the remaining fishers? With higher earnings, many fishers are not even spending much of their incomes at home in their coastal communities, wisely spending it instead in warm places abroad. Their taxes, which the economists assured us could go toward important things such as education and health care, now go largely to pay the salaries of civil servants who mostly produce nothing of visible value to anyone.

The point here is that employment versus efficiency trade-offs, which have had a huge impact on the lives of many fishers and others in fishing communities, have been promoted on the basis of simplistic and misleading economic models. We have hurt a lot of people, without much real gain to the public welfare. You can safely bet that none of the economists who promoted these changes is working in a pulp mill. This is another example of our admonition that those who are not accountable or responsible for the consequences of their recommendations should perhaps not be so zealous in promoting them.

Similar trade-offs between employment and economic efficiency are being made around the world, with the substitution of "industrial" for "artisanal" fisheries. Artisanal fishers typically have a diversified portfolio of employment activities, including various kinds of fishing (e.g., nets, lines, etc.) as well as nonfishing activities. When policy advisors recommend replacing such fisheries with efficient, industrialized, high-risk fleets, they are making much the same mistake as a stock market investment advisor who tells clients to buy particular, high-reward/high-risk stocks or mutual funds instead of a low-reward, diversified portfolio. A competent and ethical market advisor never makes such recommendations but, instead, offers a range of portfolio options (points along a risk/reward trade-off relationship) and assistance to clients in clarifying their personal objectives and attitudes toward risk.

2.5 ALLOCATION OF MANAGEMENT-AGENCY RESOURCES

Driven by threats of funding cuts and new public demands, there has been much soul searching in public fisheries-management agencies about the best allocation of funds and personnel among basic agency functions: monitoring, enforcement, research, fish production (hatcheries), and habitat protection. Agency employees who are proponents of (and whose jobs depend upon) particular activities have sometimes blatantly appealed to the public for more support for their activities, so that agency allocations sometimes end up bearing little relationship to any real needs for sustainable management. In particular, in some agencies there has tended to be a reallocation of resources away from basic harvest-management functions (monitoring, stock assessment, enforcement) toward "supply side" activities such as hatchery enhancement and habitat protection/restoration. Some newer and very expensive activities, such as genetic stock identification and the extensive GIS mapping of aquatic habitats, have produced some apparently useful results (pretty maps) but suspiciously little contribution to the actual practice of management. Added to all this is a warning from economists that public expenditures on fisheries management are often far larger than what the value of the fisheries would appear to justify (Cochrane 2000; FAO 2001).

While there is much bickering over broad trade-off issues involving general management responsibilities and approaches, there are a few very specific trade-off issues in stock assessment and harvest management that should be getting more attention than they have. As will be discussed in chapters 3 and 5, some of our most cherished stock-assessment approaches involve the analysis of abundance trend (survey) and composition (age, size) data that are very expensive to collect. But these methods can fail during fishery declines and rebuilding and, in any case, will probably not be helpful in monitoring the effectiveness of some "new" regulatory approaches such as marine protected areas, which will require better (fine-scale) spatial information on fish, habitat, and fishing interactions. There may be ways to avoid such failures, and possibly even to reduce assessment costs considerably, by investing in two alternative assessment approaches: (1) the direct measurement of fishing mortality rates using tagging, and (2) direct assessments of stock size using a local density estimation combined with spatial habitat mapping (to expand the local density estimates). However, any development of these methods will be contingent on considerable research investments in technologies for in situ tagging, tagging systems that permit the automated detection and counting of tags in catches, and technologies for more effective direct-density estimation. Such research investments are risky and do not provide exciting research topics for biologists. The major trade-off here is between "business as usual" monitoring and research versus investment in innovative monitoring methods, when there is little incentive for at least one major stakeholder group (fisheries scientists) to participate in the investment. There is also a trade-off in relation to risk management: it may easily be seen as less risky and more cost effective to try to "shore up" existing monitoring methods (e.g., better survey designs and statistical analysis methods) than to gamble that we can solve some of the difficult problems of tagging studies and direct fish counting.

Another major trade-off issue that deserves much more careful attention is the allocation of agency resources to enforcement. There have been sharp declines in enforcement budgets in some fisheries agencies in favor of other activities such as habitat management. Yet many regulatory schemes have failed because the regulations are not enforced at all. This will certainly become a major issue in the implementation of marine protected areas, which are likely to attract "poachers" and, hence, may even end up with fishing mortalities at least as high as those of open areas (for an example of this effect, see comments by Russ [2002] on fishing effects on the Great Barrier Reef). The "persecution" (shooting, seal-bombing, etc.) of marine mammals by fishers continues largely unabated in some North American waters where there are important marine-mammal conservation concerns, despite marine-mammal protection laws and the threat of stiff penalties for such activities. Discarding and high-grading practices continue to cause unnecessary waste and unmonitored impacts on nontarget populations, because there is no one around to prevent them. Very likely there should be the same investments in the development of new monitoring/enforcement technologies as are needed for stock assessment, but research in this area is even less interesting for biologists than research on tagging/counting methods. So it probably will not happen unless there are strategic decisions to force a reallocation of agency resources toward the problem.

PART TWO

Elementary Concepts in Population

Dynamics and Harvest Regulation

Strategic Requirements for Sustainable Fisheries

MANY LONG-TERM DATA sets are now available to show us that fish populations fluctuate in remarkably complex patterns over time. Most fisheries texts show various examples of these patterns, and Caddy and Gulland (1983) note that the patterns range from "spasmodic" (boom and bust) to quite stable or "regular" to strongly "cyclic"; a more complex classification of patterns is proposed by Spencer and Collie (1997). Complex patterns are also evident when we plot historical changes in fishing mortality rates (measured as the ratio of catch to abundance); examples of these are shown in figure 3.1, with many stocks subject as we would expect to growing mortality over time and some suffering "runaway" increases in mortality followed by stock collapse. R.J.H. Beverton (Anderson 2002) provides an eloquent analysis of a variety of such collapses, showing how they involve the interaction of environment, stock dynamics, and changes in the efficiency of fishing. For a variety of reasons that we will discuss in chapter 5, such runaway cases have typically involved large errors in stock assessments, with the assessment procedures failing to detect and warn of collapse in a timely way.

Some students (and naïve biologists) intuitively suppose that complexity in population behavior over time implies a need for corresponding complexity in the "rules" used to monitor and manage harvests, i.e., in the "strategies" needed for predicting and coping with natural variation. They take the relatively stable fishing-mortality-rate patterns that are common in figure 3.1 as evidence of poor management performance and/or lack of "active" management responses to natural variability. It is routine to hear questions such as "How can you possibly manage a fish population successfully in the face of all that variation, without understanding what is causing it and hence without being able to predict future variation?" Part of the answer to such questions is to point out that we have been able to manage some highly variable populations quite successfully (e.g., many Pacific salmon stocks) by finding strategies to cope with the variation despite a pretty much complete lack of understanding of what causes it. Another answer is that we had better find strategies for management that do not depend on accurate prediction, even if we do come to understand the causes of variation, because at least some of that variation is almost certainly due to "environmental factors" (climate changes, oceanographic regime shifts, etc.) for which there is little prospect that we will be able to successfully predict future changes. About all that we do understand today is that most of the variation is concentrated in the early life-history stages of fish, i.e., in the juvenile survival (recruitment) process; in so far as we can tell from size-age composition sampling, most fish populations have apparently shown relatively stable natural mortality and growth rates.

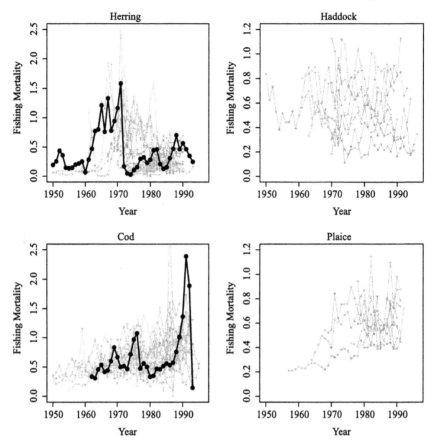

FIGURE 3.1: Several examples of changes in fishing mortality rates for some major fish stocks in the North Atlantic. Fishing mortality is measured as the ratio of catch to abundance, where abundance is estimated using various stock-assessment methods that are summarized in the Myers stock-recruitment database (*www.mscs.dal.ca/~myers/*).

A rather surprising result has emerged from the large number of simulation and optimization studies that have been done to identify strategies for coping with unpredictable (and largely uncontrollable) natural variation. The optimum way to cope with variation may, in fact, be a quite simple "feedback policy" or "decision rule" specifying how much to harvest each year as a function of the population size that nature (and historical management) presents us with that year, and this rule may be to take the same proportion of the stock every year (i.e., to maintain a constant fishing mortality or exploitation rate over time, as has apparently happened in some of the examples in fig. 3.1). This chapter first reviews the findings of those studies, then goes on to show a simple way to construct feedback policies for complex population dynamics by first finding the "open loop" optimum sequence of harvest decisions that an "omniscient manager" (with perfect knowledge of the stock dynamics and future variation) would make. It then points out that the performance of such policies is critically dependent upon how the strategy is actually implemented (stock-assessment and regulatory methods used), and it reviews modeling techniques for evaluating "closed loop" management performance. Such feedback-policy design techniques are mainly valuable for cases in which we have enough historical data to make some reasonable predictions about the possible future patterns and extent of variation. Absent such experience, we then point out that the optimum management strategy may involve so-called "actively adaptive" or deliberate experimental variation in harvest rates so as to gain more information about compensatory responses to harvesting.

For the harvest strategies discussed in this chapter to work, there must, of course, also be a responsible and accountable management authority to ensure that they are followed. On our doorstep is the Georgia Strait, a lovely marine area that is home to a remarkable diversity of marine life. It is also home to one of the highest densities of fisheries biologists on the planet, with two major government research laboratories, fisheries-management offices, and three academic institutions with educational programs in natural-resource management and fisheries. To most professionals, the Strait is a precious place; we watch it, dive in it, and fish in it regularly. One of the most common types of fish in the Strait used to be rockfishes of the genus *Sebastes*. About 25 years ago it was discovered that rockfish grow very slowly and can live to remarkable ages: 60, 100, perhaps even 200+ years (Munk 2001). This means that the abundance that we used to see represented an accumulation of fish from many cohorts or year classes, similar to the situation in a slow-growing forest. Unfortunately, longevity and wariness about fishers are not well correlated in fish, and rockfishes readily accept almost anything with a hook in it. So we have watched them virtually disappear from the Strait, to the point at which we would now call them "commercially extinct" in most places. Today we ask ourselves how such a thing could have happened, right under our noses, and in a country that has loudly proclaimed its commitment to and public investment in sustainable fisheries management for many years. The basic answer appears to be that no one was watching; rockfish were historically not very valuable, so not much was spent on stock assessments. They are mainly taken in "mixed stock" recreational and small-scale commercial fisheries that are expensive to monitor and regulate. They are spatially concentrated (mainly on rocky bottoms) and so can be targeted by fishers who continue to see reasonable catch per effort until the stock size is very low, so simple fishery statistics did not reveal the decline in a convincing way as it proceeded. Most important, there was no public commitment at any point to a real management strategy, in the sense of a planned response to changes in abundance.

3.1 HARVEST OPTIMIZATION MODELS

Beginning in the late 1950s, scientists working on creatures ranging from whales to Pacific salmon suggested that one way to deal with changes in stock size caused by both fishing and natural variation, and to reduce acrimonious debates among stakeholders about what to do each year in the face of such variation, would be to get everyone to agree to follow a simple year-independent rule relating the harvest goal for any year to the stock size estimated to be available that year (fig. 3.2). Simulation studies compared various simple rules, e.g., taking the surplus of stock size above a fixed base-stock or "escapement" goal, or taking the same fixed proportion of the stock every year (Ricker 1958; Allen 1973); these studies demonstrated that fixed base-stock rules seemed to produce about the highest possible average harvests over many years, while fixed exploitation rules produced more stable harvests over time. It was noted that fixed quota rules (taking the same number or biomass of fish every year independent of population size) are liable to produce a catastrophic spiral leading to collapse, by causing progressively more severe "depensatory" increases in fishing impact should a stock "accidentally" fall below the level that could (on average) sustain the quota removal. So it was suggested that a good feedback rule is one that reduces catch when stock size declines, so as to help natural compensatory mechanisms (improved survival and growth at low stock size) push the stock back toward a more productive level. Modern variations on simple feedback rules are the quite complex relationships between fishing and mortality rate that have been used by the U.S. National Marine Fishery Service and other agencies to define targets and reference points for management (Mace 1994; Restrepo et al. 1998).

Whether or not they can be justified as optimal with respect to maximizing the average annual yield or other particular performance measures, the stock-size–dependent harvest rules have been a substantial improvement on "seat of the pants" decision rules that some fishery managers have at least claimed to follow, e.g., by monitoring trends in catch per effort and adjusting harvest policies "as needed." Stock size–dependent rules can be constructed so as to make the most careful possible use of historical data on productivity and variability to set base stock and exploitation-rate goals (whereas seat-of-the-pants rules are usually advocated by people who would not have a clue about how to use the data most effectively and who would, instead, resort to arguments about how nature is just too complex to model). Further, clear rules leave no room for hedging and procrastination in the application of needed harvest restrictions, which managers have often pretended are needed in order to meet the so-called "complex objectives" related to the economic stability of fishing (but which, in fact, are mainly just an excuse for inaction in the face of pressure from stakeholders).

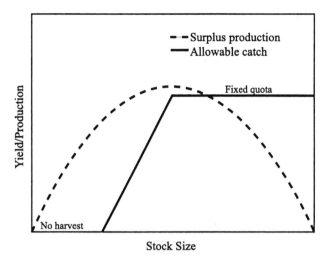

FIGURE 3.2: A decision rule suggested by International Whaling Commission scientists for setting annual harvests of a whale population as a function of its estimated abundance (see Donovan 1989 for a review of such rules). Note that the rule calls for not harvesting when the population is small, and increasing the harvest to some safe limit in the event of population recovery to a productive level.

During the 1970s and 1980s, intuitions regarding the performance of stock size–dependent rules received strong support in the form of results from formal optimization studies, mainly involving the application of methods from control-systems theory known as "stochastic dynamic programming" (see reviews in Hilborn and Walters 1992; Walters 1986. For early work mainly on semelparous—i.e., stock-recruitment relationship only—species, see Mangel 1985, Walters 1986, and Reed 1979. For more recent examples, see Parma and Deriso 1990b; Walters and Parma 1996; Spencer 1997). Dynamic programming is essentially just a (messy) numerical way to solve the short- versus long-term value trade-off relationships discussed in chapter 2. Early studies used (1) quite restrictive ecological assumptions, e.g., that stock size is known exactly at the start of each fishing season and that the future probability distribution of natural variations is stationary (the mean production versus stock size relationship is stable), with independent variation each year; (2) quite simple measures of performance, such as the average annual catch; and (3) simple population-dynamics models. These restrictions, particularly the use of population-dynamics models with few state variables, were imposed because of a technical problem with dynamic programming, known as the "curse of dimensionality" (the method cannot be applied to models with many state variables). Despite such restrictions, the early optimization models repeatedly made two apparently robust predictions, for a wide variety of species and for a wide variety of assumptions about the statistical distributions of variation (the same form of decision rule whether the optimization was for a salmon or a shark or a whale or ...):

1. If management performance is measured by the average annual (or total long-term) catch, the best harvest policy is indeed a "stationary" (time-independent) rule relating harvest to stock size, and this rule is to aim for the same, fixed base stock or escapement each year (i.e., harvest the surplus above this base stock level, do not fish at all in years when the stock is below the optimum base level). If there is slow (regime shift, decadal scale) variation in the mean relationship between stock size and productivity, so that the stock "carrying capacity" changes slowly over time, adjust the target base stock so that the stock is kept near its most productive level for the current carrying capacity.

2. If management performance is measured by a risk-averse utility function such as the sum over time of log(catch), so stakeholders are highly averse to low catches at any time, then the best policy is very close to (or exactly at, for some simple models) a fixed exploitation rate rule: harvest the same percentage of the population each year. If there is a slow variation in the mean production relationship, either (a) do not even change the exploitation rate if only the stock carrying capacity is changing, or (b) vary the exploitation rate to track changes in the population's intrinsic rate of increase "r" (or changes in the mean productivity at low stock sizes).

Of particular interest are the findings about how to respond to slow variation or "nonstationarity" (Walters 1987) in ecological production relationships (Walters and Parma 1996). The optimization calculations tell us that we can sometimes deal effectively with the very complex ecological dynamics of long-term environmental changes by applying simple decision rules, provided we can "track" the long-term changes so as to make appropriate adjustments in the "parameters" (target stock size, exploitation rate) of the decision rules.

Implementation of a fixed-exploitation-rate harvest strategy appears to place a strong onus on assessment scientists to estimate the optimum rate precisely, and we seldom have enough data to make such a claim. The relationship between average yield and exploitation rate is usually expected to be asymmetrical, as shown in figure 3.3, with exploitation rates that are too high causing big decreases in yield. Ricker (1963) aptly termed this effect "big effects from small causes." However, note that yield is the product of exploitation rate (u) times stock size (B), i.e., yield $= uB$. Suppose we recommend a u that is somewhat lower than we think is optimum, to hedge against the risk of falling off Ricker's cliff edge (overfishing). It might seem that this recommendation would proportionally reduce yield. That is wrong, because the average biomass B will generally be higher when a lower exploitation rate is used. This gain in biomass tends to compensate for a

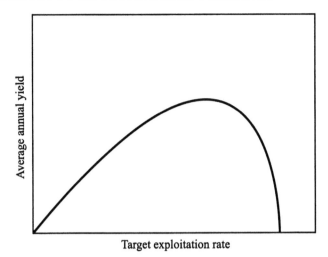

FIGURE 3.3: Population-dynamics models with realistic recruitment relationships typically predict the relationship between average long-term yield and exploitation rate to be asymmetric: small overestimates of the optimum rate can cause catastrophic decline, Ricker's (1963) "big effects from small causes." But note that the average yield drops off slowly as the exploitation rate is reduced from the one giving the highest average yield, because increases in the average stock size at lower exploitation rates tend to compensate for immediate losses due to the lower rates.

decrease in the exploitation rate, meaning that there is a relatively small long-term cost (as lost yield) of ensuring against overfishing by using conservative exploitation-rate estimates.

Numerical techniques of stochastic dynamic programming are no longer widely used, since the basic results are usually so simple. Mainly, these techniques are used by analysts who want to examine particular issues related to more complex objectives or to the relaxation of assumptions about stationarity and stock size being known at the time of each decision. Uncertainty and stochastic variation in stock-size estimates are certainly major issues in fisheries policy design, but we believe that the best way to deal with these issues is in the design of "implementation tactics" (next chapter) rather than in the form or parameters of harvest strategy rules. In short, there are ways to implement a feedback-policy rule without knowing absolute stock size, so as to achieve the exploitation-rate goal implied by the policy even if the current stock size is wildly uncertain.

3.2   CONSTRUCTING FEEDBACK POLICIES

Another good reason not to bother learning dynamic programming methods is that we now have a method for constructing harvest feedback policies when there is good information on the likely patterns of unpredictable variation, and anyone with a basic knowledge of personal computers and spreadsheet techniques can use this method. This method also allows biologists to use much more realistic, age-size structured population-dynamics models, so as not to be left with nagging doubts about whether the optimization results might somehow be overly simplistic because they are based on simple population models. The method involves the following basic steps:

1. *Construct a realistic spreadsheet model of long term population dynamics and harvest, including both "random" and persistent environmental effects on juvenile survival and recruitment.*

Spreadsheets are a convenient tool for constructing complex, age-structured population models. For example, the barramundi (*Lates calcifer*) supports important sport and commercial fisheries in northern Australia. Its life history is a fishery biologist's worst nightmare—it is a catadramous protandric hermaphrodite: adults spawn in estuaries; juveniles migrate upstream into freshwater nursery areas for 2–3 years, then mature as males and migrate back downstream to rear and breed with females, changing to female themselves at ages 3–6. These life-history characteristics result in age schedules of size, fecundity, and vulnerability to commercial harvests (in estuaries):

TABLE 1: An example of an age schedule for barramundi, where the grid represents rows and columns in a spreadsheet.

Age	1	2	3	4	5
Weight	0.77	3.83	8.24	12.83	16.92
Fecundity	0	0	0	0.1	0.5
Vulnerability	0	0.06	0.5	0.88	0.97

Age	6	7	8	9	10
Weight	20.27	22.88	24.85	26.31	27.36
Fecundity	1	1.3	1.5	1.8	2.2
Vulnerability	0.99	1	1	1	1

Such schedules can be used to predict the annual vulnerable biomass and egg production in a spreadsheet simulation that keeps track of numbers-at-age ($N_{a,t}$) over ages and years.

A key starting point for age-structure models is to specify a relationship between recruitment and the factors affecting it, including total egg production and habitat changes. Freshwater rearing area and rearing conditions for barramundi can be quite variable from year to year, and it is reasonable to assume an age-1 recruitment relationship of the Beverton-Holt form (see chapters 6 and 7):

$$N_{1,t} = \alpha s_t \epsilon_{t-1}/(1 + b\epsilon_{t-1}/c_t) \qquad (3.1)$$

Where $N_{1,t}$ is the age-1 recruits in year t, ϵ_{t-1} is the total eggs produced in year $t-1$, α and b are long-term mean survival and carrying capacity parameters, and s_t and c_t are time-varying relative survival and carrying-capacity parameters scaled so as to average 1.0. Complexity in the recruitment pattern over time is represented in equation 3.1 by changes in s_t and c_t: these are likely to vary randomly from year to year, and there is likely to be long-term (regime shift) change in at least the rearing-capacity parameter c_t. To "challenge" an omniscient manager, define an arbitrary but reasonably complex pattern of long-term changes in s_t and c_t, and drive the spreadsheet recruitment rates with this pattern. For example, panel A in figure 3.4 shows a test pattern in which there is a relatively modest variation in the maximum juvenile survival at a low stock size (s_t being relatively stable over time), but a threefold variation in carrying capacity c_t organized as a set of regime periods. More complex patterns can be used, but these tend to obscure unnecessarily the underlying prediction/response patterns that an omniscient manager would use.

The harvesting effects over time can be included in spreadsheet models by calculating changes in numbers-at-age following recruitment with an accounting equation of the form

$$N_{a+1,t+1} = S(1 - v_a u_t)N_{a,t} \qquad (3.2)$$

where S is a natural survival rate (around 0.6 for barramundi; can be made age-dependent if you wish), v_a is the age-specific vulnerability to harvest (see the schedule above for barramundi), and u_t is an annual, overall harvest rate or management-strategy goal for year t. Note that in the spreadsheet setting, u_t will be represented by a vector of "year effects" just like s_t and c_t. Note also that in the barramundi age schedules above, fish become vulnerable to harvest well before they begin to produce eggs; this means that cumulative harvesting effects (on numbers at age) can have quite severe effects on the total annual egg production (the sum of numbers-at-age times fecundities-at-age) even if the annual harvest rate is quite low. For each model year, the total biomass of fish vulnerable to harvest can be calculated as $B_t = \sum_a v_a w_a N_{a,t}$, where w_a is the body weight at age a, and the predicted total yield (biomass harvest) can be calculated as just this vulnerable biomass times the exploitation rate, i.e., $C_t = u_t B_t$.

2. Use the spreadsheet optimization procedure (Solver in Excel, or Optimizer in Quattro Pro) to estimate the optimum harvest rate for every simulated year, under alternative objective functions for measuring management performance.

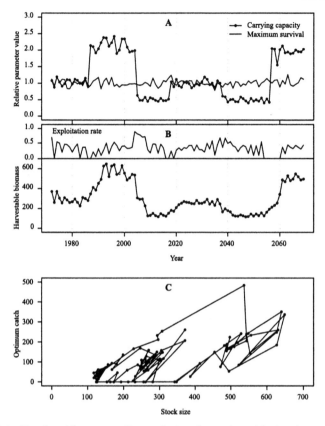

FIGURE 3.4: Simulated barramundi population dynamics with the changes in environmental effects shown in panel A, and with the exploitation-rate sequence shown in the top of panel B that the spreadsheet optimization procedure Solver tells us would be used by an omniscient manager to maximize the total long-term yield. Panel C plots the annual sequence of optimum catches $C_t = u_t B_t$ against vulnerable biomasses B_t.

Suppose next that a particular spreadsheet cell is set to some overall management-performance measure V, such as the total long-term catch (V= $\Sigma_t C_t$) or the log risk-averse utility measure V= $\Sigma_t ln(C_t)$ (or, if you prefer, use discounted measures such as V= $\Sigma_t \lambda^t C_t$, where λ is 1 minus the discount rate). Note that if V is calculated from the simulated catches C_t, it will depend upon the annual exploitation rates u_t. In the control systems literature, any particular setting for the u_t time vector is called an "open loop policy." Now, spreadsheets offer numerical optimization procedures (e.g., Solver in Excel) for trying to find the maximum or minimum value of any "target" cell by varying the "inputs" represented by a set of other cells. That is, when an optimization procedure is invoked, it systematically varies the values in the input cells, recalculating the whole spreadsheet model each time it changes an input number, to move toward a combination of input cell values that maximizes or minimizes the target cell value. For an age-structured population model that simulates changes in numbers and harvests over 10 ages for 100 years, Solver can find the 100 annual harvest rates u_t that maximize either of the value measures V above, in less than 50 iterative search steps. This takes only 10–30 seconds on the typical notebook computers available in 2003. Hence, finding the optimum open-loop policy is a surprisingly trivial computational problem, even considering that the search algorithm should be restarted several times with different initial u_t estimates, to determine whether there is, in fact, a single global maximum for V and whether the u_t that give this maximum are unique (this is called "multiple shooting").

Once such a numerical search has been completed, the resulting u_t estimates represent the behavior that an omniscient manager would be expected to follow. That is, each time the search procedure has varied a particular future annual harvest rate u_t to see how such a change would affect the total value V, it has been able to "see" (through the linked spreadsheet calculations) the full future effects of that variation on population dynamics and future yields, while fully accounting also for future disturbance effects represented in the recruitment relationship by the s_t and c_t sequence.

Examples of optimum open-loop u_t sequences for maximizing long-term catch and sum of $ln(C_t)$ are shown for the barramundi example in figures 3.4 and 3.5. A key feature of the exploitation-rate sequence that the omniscient manager would follow is "anticipation" (Parma and Deriso 1990b). A few years before the omniscient manager "knows" that a persistent change in recruitment carrying capacity or juvenile survival will occur, the manager either (1) increases the exploitation rate if a period of poor recruitment potential is coming, so as to avoid "wasting" a large spawning stock to produce the relatively few juveniles that will soon be supportable, or (2) reduces the exploitation rate if a period of good recruitment potential is coming, so as to ensure a large enough spawning stock to make full use of that potential. These anticipatory behaviors are a bizarre reversal of our usual prescriptions about how to hedge against uncertain future changes in productivity and would almost certainly never be acceptable as management recommendations in practical decision settings. Imagine telling a group of scientists and conservationists that fishing should be increased because a natural stock decline is in the offing, or telling a group of fishers that they should reduce fishing now because good production will be coming in a few years.

An interesting effect can happen when fish are fully vulnerable to harvest at all ages (e.g., if we set all vulnerabilities to 1.0 in the barramundi example), as shown in figure 3.6. In such cases, annual harvesting can cause "growth overfishing," a waste of potential biomass harvest by taking fish while they are still growing rapidly. The optimum harvest regime then becomes "pulse fishing" (Walters 1969): the fish are harvested hard only once every few years and are left alone to grow without harvesting for a few years between harvest pulses. However, pulse-fishing yields can generally be bettered by

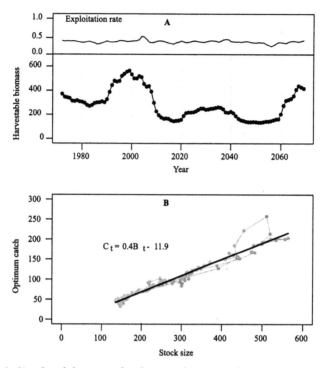

FIGURE 3.5: Simulated dynamics for the same barramundi population (and environmental "forcing pattern") as in figure 3.4, but here the exploitation rate sequence in
panel A is for maximizing the sum of ln(C_t) over time. Panel B shows the optimum
catches plotted against biomasses; the good fit to a straight line with a small intercept
implies that the omniscient manager would follow close to a fixed-exploitation-rate
policy, taking close to 40% of the vulnerable stock each year.

changing the age-vulnerability relationship so as to avoid killing small fish
in the first place.

    3. *Plot the estimated optimum catch as a function of simulated stock
size, to define a harvest policy rule.*

    Suppose we now treat the sequence of catches $C_t = u_t B_t$ and biomasses B_t
that an omniscient manager would generate over time as "data," and plot
C_t versus B_t to see if there is an average feedback relationship that could
be implemented in practice without any pretense of anticipatory behavior.
Even if we do not bother to filter out the "nonrepresentative" data points
for years when the manager would make strong anticipatory responses to
abrupt and persistent recruitment changes, such graphs typically show that
the omniscient manager would follow either a fixed escapement rule (not fish
at low B_t, vary C_t along a line with slope 1.0 for higher B_t) to maximize the
average catch, or a fixed-exploitation-rate rule to maximize ln(catch) utility
(figures 3.4 panel C, 3.5 panel B). In figure 3.4, note that the omniscient man-

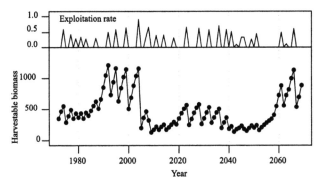

FIGURE 3.6: The barramundi example yet again, but with the age-specific vulnerabilities set to 1.0 for all ages to force the harvest of sub-optimally small fish every year. Note that the optimum exploitation-rate sequence is a regular series of "pulse harvests," with no fishing at all in the majority of years.

ager would actually come close to following three different fixed escapement policies, corresponding to the three periods of low, medium, and high juvenile carrying capacity in panel A, figure 3.4. Further, the "average" of these three policies (the single line through all the C_t and B_t data points) would be quite close to the fixed-exploitation-rate policy that would be followed under the ln(catch) objective, but this is an artifact of the roughly balanced sampling from the three recruitment-capacity regimes.

The basic message from such regression exercises is quite clear: we have not been led astray in past optimization studies by ignoring complex age-structure effects (lags, the cumulative effects of fishing on fecundity) and complex changes in recruitment relationships due to environmental factors. In fact, a simple equilibrium analysis based on deterministic population-dynamics models (box 3.1) typically gives quite good estimates of the average exploitation rates and base stock sizes that an omniscient manager would follow.

Further, we can easily calculate the "cost" of not being omniscient by replacing the optimum u_t sequence estimated by Solver with the u_t sequence implied by following a simple feedback rule based on the regression analysis. Such exercises show that we should generally be able to obtain at least 80% of the maximum possible performance given omniscience, and most often better than 90% of that performance (Walters and Parma 1996). This is comforting and has also led to questions about the value of expensive research aimed at trying to understand and forecast any recruitment variation (Walters 1989). Such forecasts would be valuable mainly for fishing industry and short-term regulatory planning and are certainly not a precondition for successful, sustainable harvest management. The critical precondition for sustainability is some ability to estimate and implement a feedback rule to strongly adjust harvests in response to uncontrollable changes in abundance.

BOX 3.1
EQUILIBRIUM ANALYSIS OF AGE STRUCTURED POPULATION MODELS TO
ESTIMATE OPTIMUM EXPLOITATION RATE AND BASE STOCK SIZE

Botsford and Wickham (1979) and Botsford (1981a, 1981b) developed a relatively simple and clever way to do many of the important equilibrium calculations needed for an age-structured analysis of fish-population responses to harvesting, integrating "per recruit" methods with predictions about stock-recruitment effects. This approach involves using survivorship-to-age calculations in conjunction with age schedules of size, vulnerability, and fecundity to calculate equilibrium "incidence functions"; each incidence function represents a sum over ages of some quantity like fecundity, weighted by survivorship (which is the net probability of surviving to each age, from age 1). In standard fisheries-population notation, the survivorship to age a in a fish population is given recursively by:

$$lx_1 = 1, lx_a = lx_{a-1}e^{-M}(1 - uv_{a-1}) \quad \text{for} \quad a > 1$$

This recursion captures the cumulative effect of fishing and natural mortality on animals, as they age. Here M is called the instantaneous natural mortality rate, u is the annual exploitation rate for fully vulnerable animals, and v_a is the relative vulnerability of an average age-a animal to harvesting. A population at equilibrium that has R age-1 recruits should on average have Rlx_2 two-year-olds, Rlx_3 three-year-olds, etc. The annual egg production ϵ from this population should average $\epsilon = Rf_1 + Rlx_2f_2 + Rlx_3f_3 + \ldots$, where f_a is the average fecundity of an age-a fish. Note that any such sum over ages can be written as $\epsilon = R\{f_1 + lx_2f_2 + lx_3f_3 + \ldots\} = R\phi_\epsilon$, where ϕ_ϵ is the "fecundity incidence function" $\phi_\epsilon = \{f_1 + lx_2f_2 + lx_3f_3 + \ldots\}$. We can obviously write several other useful incidence functions, e.g., the vulnerable biomass $B = \Sigma_a v_a N_a w_a$ (w_a = body weight at age a, v_a = vulnerability) can be written as $R\phi_{BV}$, where the "vulnerable biomass incidence function" $\phi_{BV} = w_1v_1 + lx_2w_2v_2 + lx_3w_3v_3 + \ldots$. The equilibrium yield per recruit can be predicted as just $u\phi_{BV}$, and the equilibrium yield Y as $Y = uR\phi_{BV}$ (note how ϕ_{BV} depends on u through the effect of u on the survivorships lx_a). A key point about each of these incidence functions is that they depend on age schedules of survival and per-capita characteristics but not on the total population size in any way; that means we can first calculate them without reference to the total population size, then include the total population size in the analysis through a separate prediction of the average recruitment R.

Now, here is Botsford's most useful trick. Suppose we want to assume that the average R depends on the average egg production ϵ, and that the dependence can be modeled with a stock-recruitment function of the form $R = \alpha\epsilon/(1 + b\epsilon)$ (Beverton-Holt) or $R = \alpha\epsilon e^{-b\epsilon}$ (Ricker), where we have estimates

(Continued)

(BOX 3.1 continued)

of α and b from the analysis of historical data. Since at equilibrium ϵ can be written as $\epsilon = R\phi_\epsilon$, we can write these dependencies as $R = \alpha R\phi_\epsilon/(1 + bR\phi_\epsilon)$ and/or $R = \alpha R\phi_\epsilon e^{-bR\phi_\epsilon}$. Each of these equations contains only the equilibrium R as an "unknown," and we can solve them for this unknown to predict equilibrium recruitment as a function of the factors (especially u) that affect ϕ_ϵ. These solutions are:

$$R = \frac{\alpha\phi_\epsilon - 1}{b\phi_\epsilon} \quad \text{(Beverton-Holt)}$$

$$R = \frac{ln(\alpha\phi_\epsilon)}{b\phi_\epsilon} \quad \text{(Ricker)}$$

(Negative solutions to these equations, i.e., for high u that cause low ϕ_ϵ, imply that the u would drive the stock toward extinction). Another powerful way to write these predictions is obtained by noting that α can be expressed as $\alpha = \kappa R_0/\epsilon_0 = \kappa/\phi_{\epsilon_0}$, where R_0 and ϵ_0 are estimates of the average recruitment and egg production for the unharvested population, κ is the "compensation ratio" (Goodyear 1977, 1980) representing the maximum possible relative increase in juvenile survival as a stock is reduced from unfished to near zero, and ϕ_{ϵ_0} is the fecundity-incidence function evaluated at $u = 0$ (unfished stock). Myers et al. (1999) present estimates of κ from many data sets as their $\widehat{\alpha}$.

Using these relationships, it is very easy to set up spreadsheets to calculate equilibrium abundance and harvest predictions as functions of the proposed exploitation rates u, in the following sequence:

$$u \rightarrow \phi_\epsilon, \phi_{BV} \rightarrow R \rightarrow B, Y.$$

(Each of these arrows means to apply the information at the left, along with age-specific values of v_a, f_a, and w_a as needed, to calculate the function at the right end of the arrow). Then the u that maximizes Y (and is usually close to the optimum u found by more complex optimization procedures) can be found by a numerical search (e.g., Solver in Excel) or "Table function" procedures. It is also easy to initialize age-structured simulations of changes in numbers at age $N_{a,t}$, where the recruitment-compensation parameter κ and unfished average recruitment R_0 are treated as important "leading parameters" for policy analysis and model fitting. κ and R_0 along with the unfished eggs-per-recruit (ϕ_{ϵ_0}) determine the stock-recruitment parameters α and b, as $\alpha = \kappa/\phi_{\epsilon_0}$ and $b = (\alpha\phi_{\epsilon_0} - 1)/(R_0\phi_{\epsilon_0})$ (Beverton-Holt) or $b = ln(\alpha\phi_{\epsilon_0})/(R_0\phi_{\epsilon_0})$ (Ricker). Then setting $N_{a,0} = R_0 lx_a$ for all a, where lx_a is calculated with $u = 0$, allows a simulation of how the stock's age composition likely changed as fishing developed. ∎

The open-loop optimization approach is currently being used to help design ecosystem-scale feedback policies, for cases in which single-species assessment models may be misleading for various ecological and stakeholder value reasons (see chapter 11). For instance, the Ecopath with Ecosim software for ecosystem modeling has essentially the same search procedures as Solver and allows users to seek optimum exploitation-rate sequences over multiple fishing fleets as well as species. So far, we have been seeing the same basic results as presented here: omniscient managers would show dangerous anticipatory behaviors in response to ecosystem-scale changes in productivity but would otherwise follow relatively simple feedback policies (Pitcher and Cochrane 2002).

3.3   FEEDBACK POLICY IMPLEMENTATION

The steps outlined in the previous section make it look simple to design effective feedback policies for dealing with natural variability. However, those calculations lie at the end of a complex inference chain that begins with raw (and often misleading) data:

> Biology and fishery data $\Longrightarrow$ reconstruction of past population changes $\Longrightarrow$ population dynamics parameters $\Longrightarrow$ policy parameters.

Obviously, a lot can go wrong in the above chain, from a misinterpretation of the raw data to an inappropriate choice of a population-dynamics model to poor parameter estimates caused by the lack of data at informative stock sizes (e.g., there is often not enough data at a low stock size to give a good estimate of the stock-recruitment slope "α" parameter, which is critical in calculations of the optimum exploitation rate). Most of the tools of "modern" fish stock assessment are aimed at doing better at the first step in this chain, i.e., obtaining the best possible estimates of the historical changes in the recruitment/growth/mortality rates and the resulting abundances. Some assessment methods attempt to "leap" directly from data to mean population-response parameters (e.g., the α and b parameters of the stock-recruitment relationship in the last section) by fitting models containing these parameters directly to the historical data.

But in most cases, stock-assessment methods are recommended and evaluated largely on the basis of their expected statistical performance in the first two steps in the inference chain, and this does not necessarily mean that they give the best performance at estimating the best average exploitation rate or other key policy parameters. For example, Hilborn (1979) and Ludwig and Walters (1985) have shown (by simulating the management of age-structured populations using various stock-assessment procedures) that better management performance can often be expected from deliberately using an oversimplified model (that is known to be "wrong") for the assessment, because the oversimplified model may give a better statistical performance in estimating the key policy parameters. In the control systems literature, such modeling exercises that include a simulation of the full management cycle (fig. 3.7) of

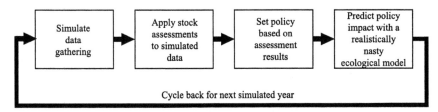

FIGURE 3.7: A closed-loop or management-procedure evaluation involves simulating not just ecological dynamics but also the full decision loop from data gathering to policy implementation.

data gathering, analysis, and policy setting are called "closed loop" analyses (e.g., Walters 1998a; Martell and Walters 2002). In recent years, fisheries scientists have more often called them "management strategy evaluations" (Sainsbury et al. 2000) or "management procedure evaluations" (Butterworth et al. 1997; McAllister and Kirkwood 1998; Punt 1993; Punt and Smith 1999a; Punt and Smith 1999b). These exercises have been extraordinarily helpful in detecting problems in the stock-assessment methods, evaluating alternative investments in data gathering, and solving problems that can arise when assessment procedures are "linked" to practical management.

To make matters still more complicated, the "best" strategy for proceeding through the inference chain depends on how policy is to be implemented, i.e., on the tactics of management to be discussed in the next chapter. There are two basic ways to implement a feedback harvest policy:

1. Output control approach: set a quota or allowable catch each year based on assessments of the absolute abundance and expected pattern of recruitment. In this case, the stock-assessment data analysis must try to estimate the total abundance and changes in recruitment rates. Complex assessment models may be required, and with such models it is easy for scientists to get sidetracked in trying to estimate and explain variability in the data.

2. u-control approach: aim regulations to directly control the exploitation rate (u) itself, independent of the absolute abundance. In this case, the inference-chain worries are about growth, vulnerability, M, and the slope of the stock-recruitment relationship; good estimates of absolute population sizes and recruitments are not required.

Notice that the information requirements for the output approach are much more rigorous than for u-control; in particular, the assessment methods must be at least reasonably good at providing estimates of the total stock size. As we shall see in later chapters, such requirements may be impractical or uneconomical, which argues in favor of trying to control u directly whenever possible.

Hilborn and Sibert (1998) point out that the full inference chain from data to policy can produce precise estimates of the optimum exploitation rate and stock size only in cases for which the data include observations of the recruitment response at a very low stock size, i.e., cases for which there has actually been historical overfishing or collapse caused by other factors. Fortunately, enough such cases have now been documented (Myers et al. 1999) to demonstrate that there are consistencies over many types of fishes in at least recruitment performance at low stock sizes, so we can set prudent exploitation-rate goals without having to deliberately advocate or permit overfishing. If we do see a future shift away from the output-control approaches in favor of methods for directly limiting exploitation rates, we may be able to radically shorten the inference chain leading to policy parameter estimates and hence greatly reduce the risk of something going wrong in complex stock-assessment calculations (and data). There are at least two ways to shorten the inference chain by using historical experience with stocks that have already been overfished: the $u = XM$ approach and the SPR approach.

A very old suggestion for managing stocks with limited historical data was to set an initial estimate of MSY to be $MSY = XMB_o$, where M is an estimate of the natural mortality rate for a stock, B_o is an estimate of the unfished stock size, and X is an empirical factor that some early models and management experience suggested should be around 0.5. To avoid the requirement for an estimate of unfished stock size B_o, we have (Walters 1998a; Walters and Martell 2002) recommended using inference chains of the form

$$(\text{growth parameters}) \rightarrow M \rightarrow u = XM.$$

The idea here is that we can usually obtain good estimates of the von Berta-lanffy body-growth parameters K and L_∞. These in turn are reasonably good predictors of the natural mortality rate M (Pauly 1980, but see Beverton 1992; Pascual and Iribarne 1993; Charnov 1996). Much simulation experience and some limited metanalyses of historical population trends tell us that the exploitation rate u should be set to XM, where X is not large. Patterson (1992) has shown that X should be around 0.6 for small pelagic fishes in order to avoid stock collapses. Our own experience with a variety of demersal fish species has suggested that $X = 0.8$ may commonly give a u near u_{MSY}. Hilborn (pers. comm.) notes that X depends importantly on the age when fish become vulnerable to fishing in relation to the age at maturity; low X values (0.6 or less) should be used for stocks that are fished well before the age at maturity, while $X > 1$ may be perfectly safe for stocks that spawn at least once before reaching harvestable sizes. The $F_{0.1}$ policy based on a yield-per-recruit analysis commonly results in the recommendation of $X = 1$ (Deriso 1987), but this value is often too optimistic, particularly for stocks that are harvested before maturity.

Another simple calculation that is becoming popular as a way to estimate prudent exploitation-rate goals and to evaluate past (and proposed) management performance is the "spawning potential ratio," SPR (Goodyear 1989; Mace and Sissenwine 1993). This ratio is $SPR = $ (annual egg production under harvest)/(annual egg production for an unfished stock), i.e., the proportion of natural spawning that would be expected under a given harvest policy. For harvest policies that are not expected to reduce (or have not reduced) spawning stock enough to cause a measurable decrease in the average recruitment, *SPR* can be calculated as the per-recruit ratio $SPR = \Sigma_a lx_a^{(fished)} f_a / \Sigma_a lx_a^{(unfished)} f_a$, where a = age, lx_a = survivorship to age a, and f_a = mean fecundity at age a. The survivorship $lx_a^{(policy)}$ for a given historical or proposed exploitation-rate policy u can be calculated in a spreadsheet using M and a proposed or estimated vulnerability schedule v_a, using the recursive relationship: $lx_1 = 1$, $lx_a = lx_{a-1} e^{-M}(1 - uv_{a-1})$ for $a > 1$ (u is set to 0 to calculate $lx_a^{(unfished)}$). This calculation is obviously much simpler than the full equilibrium analysis shown in box 3.1, and does not require the assessment of stock-recruitment parameters. An inspection of historical stock-recruitment data tells us that the risk of seeing recruitment overfishing starts to increase considerably for $SPR < 0.3$; Walters and Kitchell (2001) warn that $SPR < 0.5$ may invite long-term changes in the community structure that can lead to apparent depensatory declines in recruitment. Very low *SPR* estimates (< 0.1) have been used as a central argument for curtailing or closing a variety of fisheries, e.g., the Nassau grouper and Goliath grouper fisheries of the American southeast and Caribbean.

3.4 FEEDBACK POLICIES FOR INCREMENTAL QUOTA CHANGE

Most assessments of optimum feedback policies and decision rules have been based on the naïve assumption that harvests can be varied freely over time in response to changes in the assessed abundance. But the politics of setting harvest limits, particularly for fisheries that are managed through allowable-catch or quota-output controls, almost always results in an "incremental" approach to policy change, such that large year-to-year changes in catch limits are avoided because of their economic and social impacts. This approach to decision making can be represented in dynamic-optimization frameworks by treating last year's allowable catch or quota as another "state variable" in the system dynamics, and treating the annual increment in quota as the policy "control variable" (Walters 1978).

Treating the annual quota increment rather than the quota itself as a key management variable results in optimum feedback policies (calculated by dynamic programming methods) that are strikingly different from the typical pictures of limit and target harvest rates as functions of stock size (fig. 3.8). When both the quota increase and decrease per year are severely limited, the optimum average quota is somewhat reduced, and the optimum decision rule involves domains of current quota and current population size within which it is optimum to decrease or increase the quota as rapidly as possible. There is a line or set of current quota/current stock-size combinations

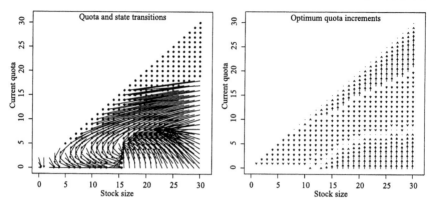

FIGURE 3.8: An example of optimum quota increments, and the resulting expected dynamic "flow" of quota and stock size over time, resulting from dynamic programming aimed at finding the maximum long-term average yield from a stock with logistic-surplus-production dynamics, but for which the annual harvest "policy" is specified as quota increments rather than an annual quota.

along which it is best to leave the quota at its current value (zero increment); when the current quota is above this line, it should be decreased as quickly as possible, and when the current quota is below the line, it should be increased quickly. There is also a "no-return" domain of quota and stock combinations, representing situations in which the stock will be severely depleted even if the quota is reduced as rapidly as possible; if the system enters this domain, there is a very low probability that natural production processes will overcome the depensatory effects of the quota before the stock is collapsed. Dynamic programming methods "see" the risk of entering this domain by accident (through stochastic production variations and/or errors in stock-size assessments) and adjust the "target" stock size (such that the quota removal balances production on average) downward so as to reduce the risk.

Optimization results such as those shown in figure 3.8 warn us that incremental approaches to setting harvest quotas can be both extremely dangerous and extremely costly (in terms of the lost average yield at the "safe" average quota). The answer to such situations is not more elaborate optimization or population models but, rather, a broadening of the options for management so as to include more rapid harvest change when needed, especially harvest reductions during periods of low natural productivity. This flexibility can be "bought" through a variety of management tactics, ranging from alternative fishing options to insurance programs to help fishers through tough times without continuing to fish.

3.5 ACTIVELY ADAPTIVE POLICIES

Most stock assessments today are based on the analysis of population and harvest data from a limited range of years. An increased commitment to data gathering for assessment since the mid-1980s has often led scientists to do detailed assessments with only a decade or so of the most recent data considered "reliable" or defensible enough to use for estimating abundance. This scientific approach promotes the "shifting baseline syndrome" (Pauly 1995): if abundances appear relatively stable over the short period of analysis, readers of complex stock-assessment reports can easily conclude that these abundances represent reasonable goals for long-term management, despite "anecdotal" evidence that stock sizes were once much higher. Further, statistical "errors in variables" and "time-series bias" effects in such assessments (see chapter 7) can easily produce the impression that recruitment is independent of stock size, i.e., that it would not be worthwhile to attempt to rebuild a stock size toward historical levels.

Hence while we can use much historical experience to avoid overfishing by recommending "prudent" exploitation rates as discussed in the previous section, we very often cannot provide sound justification based on detailed stock assessments for recommending stock-rebuilding plans. It was precisely such a situation, with sockeye salmon of the Fraser River, British Columbia, that led Walters and Hilborn (1976) to propose the idea of "actively adaptive management" or "adaptive control" of fishing systems. At that time, we began to use a variety of extremely complex stochastic optimization methods (reviewed in Walters 1986) to demonstrate that the best management policy given a very limited range of historical stock sizes can, in fact, be a "probing experiment" to deliberately vary the exploitation rates so as to gain information about stock responses. Such experiments involve a severe trade-off between policy choices that are informative and policies that are "tried and true," with the value of informative policies accruing to future managers and stakeholders as a "legacy of uncertainty." We warned that the successful implementation of a fixed-base-stock or fixed-exploitation-rate feedback policy, without any deliberate probing component, could even stabilize stock sizes so as to prevent any informative variation that future stakeholders might use to do better assessments.

There has been a lively debate about whether experimental management policies can be avoided through better data analysis and modeling techniques. Some scientists have argued that "extrapolation" beyond the range of historical data using models is just too dangerous (Castleberry et al. 1996), while others have argued that we have much better models these days and that it is often experimentation that is too dangerous and costly (Van Winkle et al. 1997). It remains to be seen whether modern modeling techniques can, in fact, produce more reliable policy predictions; at present, there is little empirical reason to trust the claims of modelers, and the main impact of such claims has been to provide managers with an excuse for inaction (spend money on modeling rather than bite the bullet and take decisive management action). Probably the best way at present to view our modeling methods is as a "screening tool" to help avoid silly decisions and to help define clear alternative hypotheses (about a policy response) that can be used to better structure stakeholder debates about whether to proceed with risky and costly policy changes.

Adaptive probing or experimental policies have now been implemented for fisheries ranging from salmon in British Columbia (Walters et al. 1993) to the line fisheries of the Great Barrier Reef in Australia (Campbell et al. 2001; Punt et al. 2001), generally in the face of much opposition from stakeholders who see such policies as too risky or beneficial to someone else (e.g., future stakeholders but not them). A broader application of experimental policies to issues beyond harvest management (e.g., habitat restoration) has also been proposed in many cases, but successfully implemented in very few of these (Walters 1997). Harvest-management experiments have been plagued by at least three serious problems (see also chapter 7):

1. Inadequate or extremely costly monitoring programs to measure responses. "We cannot see any clear trend in the population data."
2. Confounding of "treatment" effects with the effects on ongoing environmental changes. "We see a trend, but it could well be due to oceanographic changes rather than the management change."
3. Implementation of inappropriate or inadequate management treatments. "The policy you recommend is too controversial; we will try a smaller change and see if it works."

In hindsight, many of these problems could probably have been anticipated and, hopefully, avoided by more judicious and critical use of closed-loop simulation (fig. 3.6) and management gaming (Walters 1995) techniques to "troubleshoot" possible experimental management outcomes, particularly more realistic (even pessimistic) "observation models" for predicting future data. But one thing is very clear from adaptive-harvest management experience to date: the biggest need today is not for more modeling but, rather, for major innovations in our measurement and monitoring methods, especially for spatially structured experiments in which responses need to be measured in several or many places. As we will see in chapter 5 (stock-assessment methods), this admonition applies to fisheries stock assessment in general, for which we have repeatedly seen assessment failures caused not by the population models that biologists worry about a lot but, rather, by uninformative data and poor assumptions about the data.

CHAPTER 4

Tactics for Effective Harvest Regulation

IN FISHERIES POLICY DEVELOPMENT and implementation, there has traditionally been a strong separation between the "scientific" functions of stock-assessment/harvest-strategy analysis as discussed in the previous chapter, and the development of regulatory "implementation" tactics to meet the strategic goals related to stock sizes and exploitation rates. Most scientific assessments stop short of providing advice about the "mundane" issues of how to regulate harvests safely and, instead, provide only broad advice about quantities such as total sustainable yield. On the other side of the split, there has been a growing tendency in recent years for management agencies to populate their management positions with people chosen for administrative (coordination, communication, facilitation) rather than analytic skills. Bluntly, the typical fisheries manager today knows very little about ecological dynamics and harvesting strategies and would have no idea about how to evaluate whether some particular regulatory proposal or method would likely meet strategic objectives.

When both scientific and managerial responsibilities are defined too narrowly, the result can easily be a "tactical hiatus" such that no one takes responsibility for the often complex analyses that are needed to design and implement effective practices for regulating harvest rates and, in particular, for developing regulatory schemes that are robust to inevitable uncertainties about stock size and productivity. This hiatus has encouraged, or at least condoned, a broad movement in management toward "output controls" in the form of relatively simple quotas or ITQs (individual transferrable quotas) without a clear recognition of the dangers of such an approach. Further, it has fueled the development of mistrust between scientists and managers: neither has good reason to trust the other. To make matters still worse, it is very easy for both scientists and managers to forget (or never to have been told in the first place) that fisheries management agencies are first and primarily regulatory agencies; people who do not understand (or care) that their primary role is a regulatory one tend to become promoters of particular fishing interests rather than the interests of their primary employer, the public (this phenomenon is sometimes called "industry capture").

Two examples from the coasts of Canada illustrate why mistrust between scientists and managers is so common today. On the east coast, the collapse of the northern cod stock off Newfoundland in the early 1990s was one of the largest social and economic disasters in Canada's history, costing the public billions of dollars in lost incomes and social assistance programs. This disaster was at least due partly to an overestimation of the stock size and productivity after Canada took control of the fishery in the late 1970s

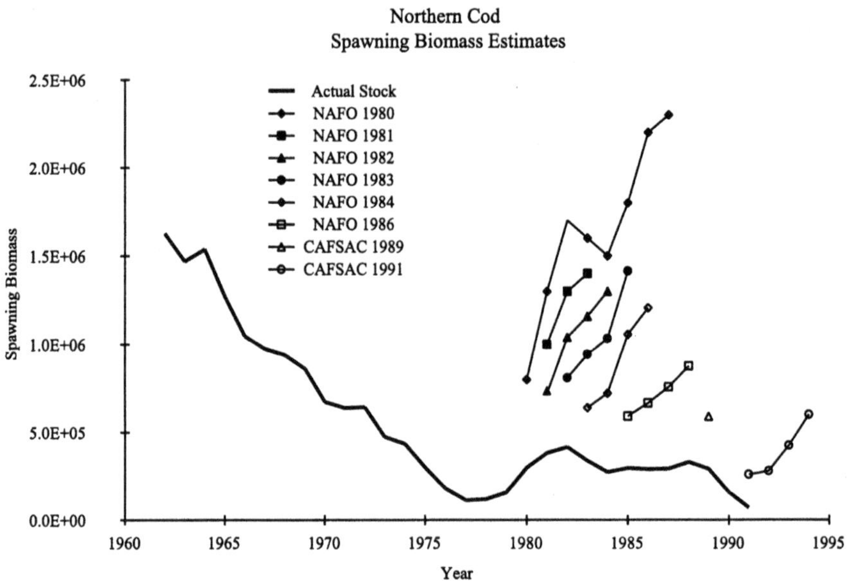

Northern Cod
Spawning Biomass Estimates

FIGURE 4.1: A comparison of the best current estimates of the historical stock size of the Northern cod stock off Newfoundland to estimates and projections made by scientists in the late 1970s and 1980s. Each year, scientists overestimated the stock size and predicted a rapid stock recovery. Redrawn from Walters and Maguire 1996.

(with extended jurisdiction). For more than a decade leading up to the collapse, assessment scientists had overestimated the stock size by as much as 300% (and underestimated exploitation rates) and confidently predicted a stock recovery (fig. 4.1; Walters and Maguire 1996). So in this case, the scientific assessment advice was taken seriously but was misleading. Meanwhile, a similar disaster was developing on the west coast, with the collapse of a valuable sport and commercial fishery for chinook salmon in the Georgia Strait (fig. 4.2). In that case, scientists were warning about the decline well before it started and carried out a variety of detailed modeling exercises to evaluate regulatory tactics for reducing exploitation rates (Argue et al. 1983). This scientific advice was largely ignored; under strong political pressure from sport-fishing interests, the management agency (Department of Fisheries and Oceans) handed responsibility for the development of regulatory tactics largely to a "sport fishery advisory board" consisting mainly of avid sport-fishing interests. This board predictably recommended ineffective regulations ranging from modest decreases in bag limits (that were not being filled anyway) to "spot closures" that were intended to reduce harvest rates in fishing "hotspots" (but acted only to delay the harvest in such areas for a few weeks each year).

These examples warn us that it is every bit as important to have effective regulatory methods or tactics as it is to have good strategic goals (stock size, exploitation rate) based on long-term population information. This chapter discusses some of the basic options that are available for limiting exploitation

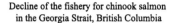

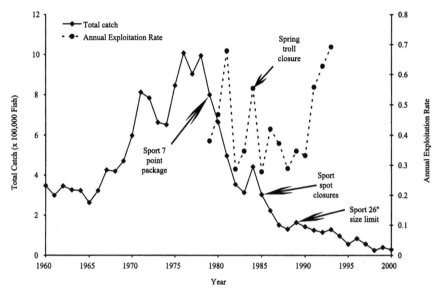

FIGURE 4.2: The collapse of the Georgia Strait chinook salmon fishery was accompanied by a variety of sport-fishing regulations and a commercial troll-fishing closure, but none of these regulations was adequate to stop the decline or even to reduce the exploitation rates (the exploitation rates being estimated from returns of coded-wire tags applied mainly to hatchery fish).

rates and, in particular, for preventing exploitation rates from increasing and causing more severe damage during stock declines and recoveries.

4.1 TACTICAL OPTIONS FOR LIMITING EXPLOITATION RATES

Suppose that harvest-strategy analyses of the sort discussed in the previous chapter, along with discussions with stakeholders about short- versus long-term harvest values and stability of harvests, indicate that a good exploitation-rate goal would be, say, 15% of the stock size each year. As shown in figure 4.3, there are basically two possible approaches to regulating the fishery so as to meet this strategic goal. The first, and most popular, approach today for commercial fisheries is "output control": obtain an estimate of the current stock size, multiply this by 0.15 to obtain a total harvest quota or total allowable catch (TAC) for the year, and regulate the fishery in some way so as to prevent the catch from exceeding the quota/TAC. Some naïve observers of fishery management appear to suppose that this is, in fact, the only way to manage fisheries and ask, "How can you possibly manage the fishery if you don't know how many fish there are?"

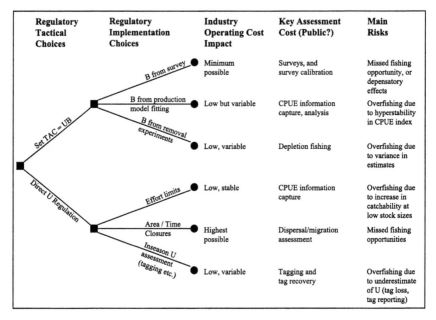

Regulatory Tactical Choices	Regulatory Implementation Choices	Industry Operating Cost Impact	Key Assessment Cost (Public?)	Main Risks
	B from survey	Minimum possible	Surveys, and survey calibration	Missed fishing opportunity, or depensatory effects
Set TAC = UB	B from production model fitting	Low but variable	CPUE information capture, analysis	Overfishing due to hyperstability in CPUE index
	B from removal experiments	Low, variable	Depletion fishing	Overfishing due to variance in estimates
Direct U Regulation	Effort limits	Low, stable	CPUE information capture	Overfishing due to increase in catchability at low stock sizes
	Area / Time Closures	Highest possible	Dispersal/migration assessment	Missed fishing opportunities
	Inseason U assessment (tagging etc.)	Low, variable	Tagging and tag recovery	Overfishing due to underestimate of U (tag loss, tag reporting)

FIGURE 4.3: A decision tree of tactical choices for trying to meet an exploitation-rate goal. Modified from a scheme suggested in Perry et al. 1999.

The second or "input control" approach is to use regulations aimed at directly limiting the exploitation rate, whatever the stock size might turn out to be; such regulations are called input controls because historically they have aimed to limit fisher activity in various ways. A better term for this approach would be "direct exploitation rate control," or *u*-control, since it might involve monitoring a fishery output (catch of tagged or marked fish) and trying to regulate this specific output "indicator" of exploitation rate directly.

Output controls are obviously simpler for fisheries managers to explain and implement, and place the entire burden of conservation planning on the stock-assessment system that provides estimates of annual stock sizes. Output controls can be implemented either as simple limits on the total catch allowed by a fleet or as individual vessel quotas (IVQs or ITQs). Total catch limits, without any allocation to individual fishers, typically result in "derby fishing," in which there is strong incentive for individuals to race to catch as much of the quota as possible before the fishery is closed. Under open access (no limit on the number of fishers), this typically results not only in a "pathological" investment in technologies for getting to the fish more quickly, but also in the progressive shortening of fishing seasons until fishing becomes dangerous and precarious (see, e.g., the history of halibut fishery openings, figure 4.3 on page 109 in Hilborn and Walters 1992). Under the IVQ approach, fishers can plan their activities so as to minimize their costs and obtain the best prices.

The traditional idea behind the input control of the exploitation rate was to assume that the fishing-mortality rate F is proportional to the fishing effort, i.e., $F = qE$, where q = the catchability coefficient estimated using stock-assessment models. This approach fell into disfavor when it was discovered that the catchability coefficient is seldom stable; in particular, (1) q tends to grow over time due to technological innovation, and this growth can be difficult to measure, and (2) q is most often inversely density-dependent, i.e., q increases as stock size decreases, due to nonrandom fisher search patterns and to decreases in the range area occupied by the fish. Recognizing these risks (of underestimating the fishing impact by using an "outdated" q in predictions of allowable effort), managers have sometimes turned to time-area closures (if you don't want to catch more than 15% of the fish, do not open more than 15% of the grounds to fishing) and, more recently, to the direct assessment of the exploitation rate by tagging studies.

In the following sections, we first examine ways to make output control safer as a tactical option, then look in more detail at the options for direct control of the exploitation rates on the assumption that in most fisheries it will never be economical to gather the accurate stock-size information needed to make output control truly safe or economical. We then examine in-season management systems for using fine-scale (time, space) information to limit the exploitation rates. Finally, we review the information-gathering priorities for effective regulation.

4.2   MANAGING THE RISK OF DEPENSATORY EFFECTS UNDER OUTPUT CONTROL

The basic strategic problem with output-control approaches is that most modern fisheries are technically capable of generating very high annual exploitation rates, so that overestimates of stock abundance can lead to high quotas that in turn cause u = (quota)/(actual stock) to be dangerously depensatory (i.e., to increase as the actual stock declines). This is basically what happened in the last years of the Newfoundland cod fishery (fig. 4.1). Unfortunately, retrospective and simulation studies of the stock-assessment methods typically used for annual stock assessments indicate that these methods are particularly prone to give overestimates of stock size during periods of decline (NRC 1998), which can then result in a "vicious circle" in which an increasing exploitation rate then drives an even more rapid decline and an even more severe underestimation of the decline.

There are at least three tactical options for reducing the risk of depensatory effects caused by an overestimation of the stock size during periods of decline:

1. time-area closures and other measures that provide bounds on the exploitation rate independent of the annual stock-size estimate;
2. calculation of an annual quota using a low stock-size estimate, near the lower tail of a probability distribution for the current stock size;

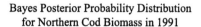

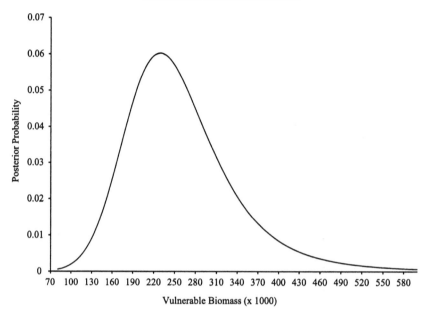

FIGURE 4.4: The Bayes posterior probability distribution for the Northern cod vulnerable biomass (*t*) in 1991. The mode of the distribution corresponds to the most credible hypothesis, and the width of the distribution reflects uncertainty.

3. calculation of an annual quota using only a "proven production potential" equal to the target exploitation rate times a minimum stock size that the industry can demonstrate unequivocally to be present by using direct stock-size assessment/counting methods.

The first of these approaches is basically just to combine input- and output-control approaches and can involve any of the tactics discussed in the next section. The other two involve adjustments in the stock-size estimate used each year in setting the quota.

Modern stock-assessment procedures (chapter 5) typically result in a Bayes posterior probability distribution for the current stock size (fig. 4.4), which can be used in decision analysis to provide some weighted or expected value outcome for each of a series of policy choices (see box 4.1). The width of the posterior distribution reflects the uncertainty arising from both variability in the historical data and uncertainty about the assumptions used in deriving estimates from the data (e.g., about whether historical abundance indices have, in fact, been proportional to historical stock sizes). A reader of modern stock-assessment papers and texts might imagine that we are getting better at assessment and that better models along with more accumulated data have resulted in narrower probability distributions for stock size (more accurate

BOX 4.1
DECISION ANALYSIS AND BAYES STATISTICS

A powerful method for organizing information about policy choices and possible outcomes in decision situations involving uncertainty is to construct a decision table (Hilborn 2001):

	Possible "states of nature"		
Policy Options	*Model A*	*Model B*	*Model C*
(1) Treat A as true	$V_{(1),A}$	$V_{(1),B}$	$V_{(1),C}$
(2) Treat B as true	$V_{(2),A}$	$V_{(2),B}$	$V_{(2),C}$
(3) Treat C as true	$V_{(3),A}$	$V_{(3),B}$	$V_{(3),C}$
(4) Take precaution	$V_{(4),A}$	$V_{(4),B}$	$V_{(4),C}$
(5) Experiment, then decide ...	$V_{(5),A}$	$V_{(5),B}$	$V_{(5),C}$

Such tables expose uncertainty as alternative hypotheses or models (or parameter values of models) represented by the columns, and identify alternative choices or policy options as table rows. Typically, the policy options can be organized as choices that would be optimum given each of the models, along with additional choices representing hopefully "robust" choices that might be good no matter what model is correct and/or "actively adaptive" choices aimed at experimentally revealing the correct model before deciding on a long-term policy. The table entries $V_{policy,model}$ represent measures of how well the policy in that table row would perform if the world turns out to be close to the model in the column. By looking across each row, we display (and a policy maker can obtain) a quick look at the range of possible consequences of the policy choice for that row, allowing a quick identification of the robust choices (rows that have relatively good V values for most/all states of nature).

Decision tables can provide much useful information about uncertainty, possible policy impacts, and ways of measuring policy performance whether or not they are used in any formal decision analysis aimed at identifying a particular best policy choice. But when they are used in attempts to identify such a choice, it is generally necessary to say something about the relative odds that should be placed on each possible model given the available data, i.e., to place probabilities P(model) on each of the states of nature. Absent such probabilities, we may be able to identify policies that are not robust (i.e., that would perform poorly if the state of nature for which they were designed turns out not to be correct), but we cannot say whether the chances of bad performance are high enough to warrant avoiding such choices (we cannot rely upon "min-max" or "maximum regret" criteria for screening

(Continued)

(BOX 4.1 continued)

out policy choices that might perform poorly under some states of nature, without reference to whether such states of nature are at all probable).

There is a potential problem with the idea of placing probabilities P(model) on alternative hypotheses or states of nature. Classical or frequentist statistics defines probabilities strictly as measures of the frequency of occurrence of events over sample spaces, and under this definition it is meaningless to speak about P(model), i.e., a model is or is not the one from which observations were collected, period. Under the frequentist definition, we can speak about likelihoods, i.e., P(data|model), but not about what Bayesian statisticians call the "posterior probabilities" P(model|data) that might represent the best assessments of the odds to place on each model in the light of available data. Bayes's theorem says that there is a way to calculate P(model|data), but only if we are willing to place some prior odds on the models, P_o(model): P(model|data) = P(data|model)P_o(model)/P(data), where P(data) is the integral probability of getting the available data, over all models that could have given rise to those data. In practice we do not need to calculate P(data), since it is the same for all models, and we can place relative odds on alternative models just by examining the relative values of P(data|model)P_o(model), i.e., the relative values of likelihoods × priors.

The classical statistical objection to using the Bayes posterior probabilities P(model|data) for weighing alternative outcomes in decision analysis is that these values are necessarily contaminated by subjective or judgmental choices for the prior probabilities P_o(model). That is, the Bayes calculations open the door for analysts to include personal, possibly unscrupulous beliefs and judgments that may favor particular values or decision choices. Proponents of the Bayes methods answer this objection in several ways: (1) it is nonsense to claim that there is no subjectivity in frequentist analysis—all sorts of judgment calls are made in model definitions, choice of likelihood functions, and decisions about whether even to report anomalous results; (2) the Bayes methods expose the judgment calls to scrutiny, i.e., they attempt to make hidden assumptions more transparent to review and criticism, which is generally a healthy thing for scientists (and everyone else); and (3) prior probability distributions can often be constructed quite objectively from metanalyses of historical case experience with hopefully similar systems, allowing the quantitative use of that experience in the analysis of each new case or situation, and, indeed, it might even be considered scientifically irresponsible to ignore such experience. ■

and precise stock-size estimates). But in fact, just the opposite has been occurring: the widths of probability distributions for stock size have tended to widen in recent years, particularly as we have uncovered more problems with our assumptions about historical data.

So the simplest risk-management approach under output control is to calculate the annual quota as the target exploitation rate times a conservative
stock-size estimate near the lower tail of the stock-size probability distribution. Some simulation studies of the expected long-term management
performance (e.g., Walters 1998a; Walters and Bonfil 1999) suggest that a
good choice is the 15% or 20% cumulative probability point on the distribution (i.e., the stock size for which there is less than a 15% or 20%
chance that the current stock is actually smaller). Because the distributions
generally have wide tails, choosing to use even lower probability points (further reducing the risk of setting too high a quota) is quite costly in terms of
the immediate quota value but does not substantially improve the long-term
management performance. Unfortunately, such simulation studies do not
resolve the basic trade-off problem of how risk averse to be (how big an
insurance payment to make in terms of lost immediate quota) in the face
of possible short-term overfishing caused by an overestimation of the stock
size. A very conservative "investor" or stakeholder might reasonably argue
for using a stock size as low as the 1% probability point, especially in view
of the tendency for assessment scientists to be "overconfident" (to underestimate uncertainty) for various reasons. A more short-sighted or risk-prone
stakeholder might make an equally reasonable argument that the 20% estimate is more than adequate, especially considering that even if it does result
in overfishing (exceeding the annual target exploitation rate) for a few years,
this error is likely to be detected and corrected before there is a catastrophic
stock collapse (you will not hear any Newfoundland cod fishers making such
an argument anymore, but perhaps they are a unique case). The point is,
we do not yet have any sound, objective way to resolve such arguments and
to set reasonable probability standards, so the whole approach is very likely
to fail in practice. Further, we have a discouraging track record of being far
too optimistic about assessment assumptions, leading to probability distributions that are far too narrow (implying considerable risk that even a very
"precautionary" stock-size estimate—e.g., the 5% probability point on the
stock-size distribution—may be far too high).

The "proven production potential" (PPP) concept takes the idea of setting
quotas based on conservative stock estimates a step further, while creating
incentive and responsibility for fishers to invest in stock assessment (Pearse
and Walters 1992). The concept is based on treating fish stocks as a "public" resource (like oil and gas fields on public lands) from which particular
industries might profit, and where it is seen as the industry's (not the public's)
responsibility to invest in (bear the cost of) any information gathering needed
to demonstrate the economic potential ("proven reserves") of those resources
as part of the industry's economic development. For a fishery, the concept would work like this: (1) government agencies would estimate an optimum sustainable exploitation rate U, using the optimization methods discussed in chapter 3; (2) government and industry would jointly invest, with
some reasonable cost-sharing arrangement reflecting the public's interest

in future fishery values, in an annual stock assessment to "prove" that the current stock size B_t is at least B_{proven} tons; and (3) the quota for the year Q_t would be set to the proven production potential $PPP_t = UB_{proven}$.

The information "investment" problem in the PPP approach then becomes how and how much to spend on a determination of B_{proven}. Very likely the investment choice by the fishing industry would not involve spending money on gathering traditional fisheries stock-assessment data (relative abundance, size-age composition data) nor on traditional stock-assessment methods. Rather, it would involve some more reliable (and understandable to fishers) direct assessment approach involving a recognition that B_{proven} can be estimated as the product of two numbers: (density of fish per unit area) × (area over which the density per area estimate applies), i.e., $B_{proven} = b_{proven}A_{proven}$. Using this relationship, it is easy to see that fishers can invest in both improving the accuracy and precision of the estimates of average fish density b_{proven}, and in demonstrating that the estimate applies to a larger area A_{proven}. Because of the range-contraction changes that typically accompany stock declines, historical data on spatial distributions would not be acceptable in the estimation of A_{proven}; any surveys aimed at improving the estimates of B_{proven} would have to demonstrate b_{proven} over just the particular area A_{proven} that is surveyed in a fully representative way (random or systematic arrangements of b_{proven} estimation points) before the current fishing season or year. Fishers might "gamble" on demonstrating that the stock is abundant over a larger area by spreading b_{proven} sample points over that larger area, but at the risk of finding no fish and/or an increasing variance of the sample density observations (a higher variance means a lower "proven" b, if b_{proven} is taken to be a lower bound or confidence limit on the estimate of mean density). There are many ways to obtain reliable minimum estimates of the local fish densities b_{proven}, ranging from trawl surveys to visual and acoustic counts to localized depletion and mark-recapture experiments. With an economic incentive to improve (and increase, e.g., trawl efficiency) these, fishers would probably invest much more in innovative technologies and methods for a local density estimation than have most management agencies. Indeed, some of the most exciting research on innovative stock-assessment methods today is being done in the Antipodes, where there has already been a shift in the financial responsibility for stock assessment from government to industry.

4.3   TACTICS FOR DIRECT CONTROL OF EXPLOITATION RATES

In many fisheries we can avoid the difficult problem of estimating total abundance completely, by using management tactics that directly limit the exploitation rate no matter what the stock size turns out to be each year. Direct control of the exploitation rate has been particularly important in management "success stories" such as Pacific salmon fisheries, where time-area closures have long been used as the main tool to limit the proportion of maturing, migrant adult fish exposed to very intensive, efficient fisheries (see chapter 8, fig. 8.6 on page 197 and fig. 8.7 on page 198). The basic concept in these early systems for controlling the exploitation rate has been to close most of the ocean for most of the time, and to make time-area fishery openings small enough so that only a small proportion of the total stock (rarely >20%) is exposed to harvest in each opening, while assuming that virtually all exposed fish will be harvested during that opening. When a gillnet or purse seine opening for salmon occurs somewhere in British Columbia or Alaska today, we expect that these very efficient gears will remove most (90% or more) of the fish present in the open area, usually within the first 12–24 hours of fishing.

Besides the elaborate time-area closures used for the last century in salmon management, the oldest approach to direct control of the exploitation rates has been to control the fishing effort. The basic theory of effort control is quite simple. If a typical unit of fishing effort "sweeps" the fish from an area a_s, if the stock is distributed over an area A_s, and if the swept areas a_s are randomly distributed over A_s (whether or not the fish are distributed at random), then the average proportion of the stock taken by each unit of effort should be the "catchability" $q = a_s/A_s$. Then a total effort of E swept area units should generate a total instantaneous risk of harvest $F = qE$, and (accounting for overlap in areas swept when F is high) the annual exploitation rate should vary as $U = 1 - e^{-qE}$. A few fisheries probably do operate about this way, especially in cases in which historical fishing has greatly reduced the fish density over large areas so that fishers cannot do better than to sweep the areas more or less at random. But in most fisheries, it has turned out that neither a_s nor the effective stock area A_s are stable over time. Gear improvements cause a_s to increase, sometimes quite rapidly. More important, fishing is rarely randomly distributed in space, and the area A_s most often changes with changes in the total fish abundance. Fishers most often "see" the fish as distributed patchily over an area A'_s that is much smaller than the A_s that biologists might define as the area over which the fish can be found in random or systematic surveys, and both A_s and A'_s are likely to shrink during stock declines. Since $q = a_s/A'_s$, q is likely to increase with both technological change and stock range "erosion." Such density dependence in the catchability coefficient has been documented for many fisheries (Harley et al. 2001) and means that traditional effort control is likely to result in depensatory exploitation rates unless there is some way to track the changes in a_s/A'_s as they occur.

Some shrimp and lobster fisheries in Australia and North America have been managed quite successfully using simple effort limits implemented by limiting the numbers of licenses, gear (e.g., on the number of lobster pots per license, net sizes), and seasonal closures. In these fisheries, a key goal of much stock-assessment work is to "track" what have turned out (luckily) to be relatively slow changes in catchability q caused by gear improvements and changes in fish distributions, and to make incremental adjustments to the effort limits in response to these changes in q. Methods for tracking the changes in q have ranged from the back calculation of the total fishing rates in past years from post-fishing estimates of total catch and stock size (using $F = $ catch/mean stock size), to the use of tagging experiments to directly estimate U as the proportion of tagged fish captured in the first fishing season after release. Martell and Walters (2002) describe methods for updating estimates of q based on tagging and other assessment data, using variations on the idea of Kalman filtering.

Unfortunately, for many fisheries, especially for shoaling/schooling species, range contractions during stock declines can cause A_s to decrease very rapidly and q to increase correspondingly. In these cases, predictions of F as qE can be completely misleading over time scales of even one year (q can easily double in just one year, if the stock has dropped by 50% due to either fishing or natural factors and if A_s has also dropped by 50%), and direct control of the exploitation rate must involve some other method. Two main alternatives have been tried or suggested:

1. the "salmon approach" of conducting fisheries in small time-area openings only, with mapping of fish distributions preceding each fishery so as to define opening areas where only a limited proportion of the fish are exposed to fishing;
2. the direct estimation of the exploitation rate as fishing proceeds, from in-season returns of tags from known numbers of fish tagged before the start of the fishing season.

The first of these approaches has been used mainly in herring fisheries, and it has turned out to be very expensive to do the extensive surveys (most often with acoustic methods) needed to map the year's fish distributions accurately enough to ensure safe fishery opening locations. To our knowledge, the second has not been used yet in management but is potentially much cheaper, provided the tagging can be done so as to (1) involve negligible tag loss and tag mortality; (2) tag fish representatively over the population (each exploitable fish being equally likely to be tagged); and (3) recover all or a known proportion of those tagged fish actually harvested (Martell and Walters 2002). So far, the limiting factor has been requirement (3): when tag returns will be used for regulation/closure of the fishery, there is a strong incentive for fishers not to report tags, and there are myriad ways for them to avoid reporting their tag catches. Martell and Walters (2002) suggest developing "blind" tagging programs, in which fishers cannot tell the difference between marked and unmarked fish and port sampling programs or remote sensing technologies can be developed to recover marked fish. Internal tags, such as PIT tags (PIT = Passive Integrated Transponder) would be appropriate for solving two potential problems—tag loss and nonreporting—because random samples from the landings are examined for the presence of PIT tags.

A very interesting variation on the idea of directly estimating (and controlling) the exploitation rates from short-term tagging experiments is the radio-tagging study reported by Hightower et al. (2001). In this study, a relatively large number ($N = 60$) of striped bass were radio-tagged in a reservoir from which relatively few fish were expected to emigrate during the fishing season. Then the study area was swept repeatedly in search of the radio signals, so as to provide estimates of the times of disappearance (last radio contact) for fish that were removed (or that emigrated) from the system, and the times of death of fish that died naturally (indicated by unchanging radio positions in successive sample sweeps). This study revealed considerably higher exploitation rates, and lower natural mortality rates, than might have been expected for this species in a recreational fishery. We suspect that similar patterns (lower natural mortality, higher fishing impact) will be revealed by this methodology for a variety of fish species. Tracking the fates of radio or acoustic tagged fish can obviously help get around problems with tag reporting and tag loss and provides the added bonus of valuable information on fish movements and tag mixing relative to fishing areas. Tagging and tracking fish with dozens or even hundreds of transmitting tags may seem like an expensive way to measure exploitation rates, but it may, in fact, be grossly less expensive (and less risky) than collecting and analyzing the survey and size/age composition data needed for traditional stock-assessment methods (which we usually cannot trust even when multimillion-dollar data-collection systems provide apparently quite reliable data).

An important benefit from any regulatory tactics that provide direct estimates of the exploitation rate (e.g., tagging) is that these tactics also "solve" the larger stock-assessment problem of estimating the total stock size. That is, if the direct estimate of the exploitation rate is U, and if the measured total catch is C, then we can estimate the vulnerable stock biomass as $B = C/U$ (since $C = UB$). The precision of this estimate depends upon the precision of the catch and U estimates, and in tagging studies the precision of the direct U estimate depends, in turn, on the number of fish tagged and on the exploitation rate. Typically, we warn against using mark-recapture estimates of the population size from small tagging studies (in which only a small proportion of the total stock is actually tagged), but that is partly because in such studies we typically examine only a relatively small recapture sample of fish for marks. If the entire catch C is somehow examined for marks so as to directly estimate the proportion of tagged fish caught, the mark-recapture estimator C/U can be much more precise than for typical mark-recapture studies.

4.4 REGULATION OF EXPLOITATION RATES IN RECREATIONAL FISHERIES

Recreational fisheries have been dominant users of freshwater fish in many regions of the world, and they are now taking over as the dominant fisheries in many coastal areas as well. The traditional assumption in managing these fisheries has been that recreational fishers are unlikely to cause recruitment overfishing because they show strong effort responses (fishers go away when the fish get scarce; see chapter 9). It has further been assumed that simple regulations involving mainly size and bag limits should be sufficient to ensure sustainability and equitable allocation of fish among individual anglers. But there is growing evidence that these assumptions are very wrong, at least for more accessible, heavily used fishing areas (Post et al. 2002). The balance point at which fish abundance is driven low enough so as not to attract additional fishing effort may well be at an abundance low enough to cause reduced recruitment. Further, in multispecies situations such as the reef fisheries off the coast of Florida (about one-third of the global expenditure of around $25 billion per year on sports fishing occurs in Florida), the bionomic equilibrium with respect to the targeted fish species may be at efforts high enough to drive some "incidental" or nontarget species toward biological extinction (Ault et al. 1998).

The simple regulation of sports fishing via size and bag limits has been called "passive" management. The extreme of this approach is "catch and release" fishing, which has been used for a long time for some game fishes such as tarpon and is enjoying growing popularity among the prosperous, urban-based anglers that dominate many modern sport fisheries. But there are at least three reasons to question the efficacy of these passive regulations: (1) they do not deal effectively with the quality versus quantity problem of recreational fishing, i.e., whether to provide good fishing (high catch per effort, fish size) for fewer anglers or poor fishing for as many anglers as possible; (2) there is at least some probability of mortality for each fish released, and large size limits may result in smaller fish being released many times (and hence suffering high cumulative mortality) before contributing to the spawning population; and (3) taken together, size and bag limits may actually generate conflicting effects, i.e., may work against each other (when a low bag limit is used, more fish can often be "saved," or at least released, by removing all size limits so as to fill as many bag limits as possible with smaller fish). Further, it can be extremely difficult to measure the efficacy of changes in bag and size limits (Allen and Pine 2000).

In view of these problems, there has been a growing call for "active" management of the total effort in recreational fisheries (Post et al. 2002; Walters and Cox 1999), based on some lottery-access system similar to those now used for "limited entry" management of big-game harvests. Should lottery-access systems prove as popular as we suspect, there will be a need for much improved analyses of the relationships involving recreational fishing effort, in particular the estimation and prediction of exploitation rates as a function of effort, and the determination of whether recreational fishing is likely to involve density-dependence in catchability coefficients (so that even a constant, regulated effort can result in a depensatory increase in the exploitation rate during a stock-size decline, as suggested by Post et al. 2002 for Ontario lakes). Density dependence in catchability coefficients in sports fishing can occur not only due to nonrandom search processes by fisheries combined with fish range changes (the usual mechanism causing density-dependent q in commercial fisheries), but also due to "effort sorting": as fish abundance declines, less capable anglers tend to give up fishing first, so that the anglers still going out are those with the highest individual catchabilities. Since recreational fisheries have not generally been monitored carefully in the past, there is little information about the dynamics of recreational catchability. This means that active, direct-effort regulation systems will generally need to be accompanied by other monitoring/regulation methods, such as tagging programs, that provide more direct information (and capability of control) on the exploitation rates.

4.5 In-Season Adaptive Management Systems

Some wildly variable and unpredictable populations, particularly of Pacific salmon and herring, have been managed quite successfully over most of this century by using complicated "in-season" assessment and regulatory procedures. These procedures involve monitoring catches and abundance indices on very short time scales (days) during fishing seasons, and updating the abundance estimates and fishery openings in response to the data so as to meet spawning abundance- and exploitation-rate goals (Cave and Gazey 1994; Claytor 1996; Hilborn and Luedke 1987; Link and Peterman 1998; Mackinson 2001; Sprout and Kadawaki 1987; Su and Adkinson 2002). Indeed, a wide misunderstanding about salmon is that they are easy to manage because they are easy to count (as catches and as spawners moving up rivers); this is nonsense: salmon managers are grossly uncertain about abundance at the times when they must make key harvest-management decisions, which is well before the fish reach their spawning rivers. They have learned through much hard experience, i.e., many costly mistakes over the years because of poor forecasts and/or misleading data during fishing seasons, to proceed with considerable caution in managing annual harvests: fishing areas are small, fishery openings are short, and at least the first few openings each year are carefully monitored to provide indices of abundance that have been demonstrated (by statistical regression analysis of historical data) to be well correlated with total abundance.

In-season adaptive management systems critically depend upon building a history of experience about when fish will be available and about how much impact a given fishery opening or other regulation is likely to have. But even given many years of such experience and data, fish can still do surprising things that cause a risk of overfishing. For example, the salmon manager's worst fear is the "little run coming early." What happens in such cases is that the salmon arrive at the coast early, causing high indices of abundance in early fisheries (because early fisheries that should be seeing the "front edge" of the run are, in fact, seeing the bulk of that run). Given this misleading signal, large fishery openings are then declared on the assumption that the stock will be large. But then the fish do not arrive, and the large opening decimates the run that does arrive. Another big fear is that the fish will not migrate at normal speeds/patterns through the fishing areas and, in particular, that they will "stall" in some area and will unexpectedly be exposed to more than one fishery opening (see fig. 8.6 on page 197 and fig. 8.7 on page 198). There are various ways to guard against these fears by improved monitoring programs (e.g., acoustic-survey grids, carefully placed test fisheries), but management agencies have usually seen such investments as uneconomical or unnecessary. A less expensive, simpler tactical option has been to hedge against overfishing risks by setting more conservative exploitation-rate goals in the first place, recognizing that this will result in some loss of potential catch in "average" years as an insurance against bigger losses of future production due to overfishing in occasional years. Unfortunately, there have been few attempts to carefully quantify the trade-off between a loss in the average catch due to precautionary harvest goals and the direct costs of obtaining better data in the first place.

There are strong incentives for fishers to cooperate with scientists and managers in gathering information for improved in-season management (Claytor and Allard 2001; Lane and Stephenson 1998), particularly in situations in which a precautionary reduction in the exploitation targets would otherwise be used to reduce the risk of overfishing. An excellent example of just how far this cooperation can go is in the Spencer's Gulf prawn fishery in South Australia (Carrick 1988). In that fishery, the fleet carries out a preseason trawl survey to map the distribution, abundance, and size structure of the prawns. This information is used to set an initial fishing plan for a first short (few weeks) fishery opening, including protected ("nursery") areas for smaller prawns. After this opening, catch-rate information from the opening and from a second survey fishing period are used to decide whether/how to proceed with further openings. This system has worked very well, and the fishery is healthy by both ecological and economic standards. In sharp contrast, the nearby St. Vincent's Gulf, South Australia, prawn fishery, where fishers have largely refused to cooperate with management, has been threatened with collapse and is a classic management nightmare in which scientific assessments and advice have been hotly contested by fishers.

4.6 MONITORING OPTIONS AND PRIORITIES

We have noted in previous sections that the two basic approaches to harvest-rate regulation (input vs. output control) have fundamentally different implications in relation to what needs to be measured. Output-control approaches require the estimation of absolute abundance, either by direct observation (density × area) or by the use of stock-assessment or state-reconstruction models. Input-control approaches require either the direct assessment of the exploitation rate U or information that will allow for the prediction of U from measurements of fishing activity (area swept by gear, distribution of fish).

The traditional priorities for information gathering in fisheries (Hilborn and Walters 1992) have emphasized the data needed for long-term assessments of abundance and productivity: catch and effort statistics first (to permit the estimation of total removal and trends in relative abundance as evidenced by catch-per-effort), then composition information (size/age structure) to permit more detailed population modeling, then perhaps information on the space/time details of the stock structure and fishing. In fact, these priorities were developed before much thought had been given to alternative tactics for achieving sustainable exploitation rates, and before comparative data were available to demonstrate that optimum rates can often be predicted from quite "simple" information on growth, longevity, and size/age selectivity of harvesting. Given what we know today about the problems with stock-assessment methods and about the prediction of optimum fishing mortality rate, a new set of priorities might well be:

1. Measurement of the growth curve parameter K: implies a reasonable estimate of the natural mortality rate M (Pauly 1980), where M is a good predictor of the optimum fishing mortality rate.
2. Spatial structure information on both fish and fishing: the fish life-history trajectory (see chapter 8), how fishing is distributed relative to this trajectory (a key determinant of overfishing risk, size-age selection), and where there are spatial opportunities for limiting the exploitation rate via closed areas. Note that this priority would imply spatially referenced, detailed logbook systems that are often not required until fisheries are "fully developed."
3. Ecosystem linkage information: diet compositions, diet information on the main predators that might be impacted by fishing, estimates of bycatch and discards of nontarget species.
4. Direct estimates of the fishing rate: swept area methods in conjunction with spatial mapping (see 2, above), routine tagging.
5. Surveys for multispecies abundance trend assessment.

Such a revised set of priorities would in many cases better reflect the need early in the development of modern fisheries for the data needed to control very rapidly changing exploitation rates, and to respond to concerns from nonfishing interests about the impact of development on nontarget organisms and ecosystem function. Unfortunately, every single element in this list is typically seen as one of those "would be nice to have" kinds of information that management agencies rarely get around to collecting until serious concerns about overfishing and ecological impacts have already been raised.

So there is a fundamental monitoring choice and trade-off (because the resources for monitoring are limited in every management agency) between investments in the assessment of abundance trends via surveys and the analysis of harvest statistics, versus investments in gathering data that directly measure the fishing mortality rate (compositional data, tagging). As we will see in chapter 5, compositional data have proven difficult to interpret, due to the confounding of recruitment, size-age selectivity, and time trends in the mortality rate. This means the real investment choices are between gathering information for shoring up "traditional" assessment procedures, and gathering direct information on the exploitation rate via tagging methods. There is the devil that we know, and the one with which we have much less practical experience. Just about every experienced biologist has a personal story to tell about a failed "tagging experiment" aimed at estimating the abundance or exploitation rate, due to tag loss and tag mortality, the nonrandom distribution of tags, changes in fish behavior due to tagging, or the nonreporting of tags. All of these bias sources result in an underestimation of F, a dangerous bias indeed. But these stories are about single tagging experiments, not about the use of tagging as a routine procedure. Perhaps there is a common ground, where information from tagging is combined with traditional assessment information for cross-validation and improvement of estimation performance (Martell and Walters 2002).

One of the most important questions we need to ask today about monitoring for harvest regulation is whether it will ultimately be cheaper to correct the bias problems with tagging and direct fish census methods, or to correct known problems with traditional assessment data. New, in situ tagging methods are appearing regularly, as are innovative systems for detecting tagged fish in the field and in catches. But the research needed to develop such systems is largely coming from a few highly visible research programs on fisheries such as salmon in the Columbia River. Few management agencies are making significant research investments in innovative monitoring methods, since their financial resources are largely tied up in keeping existing monitoring programs afloat.

Also, agency resources are more and more often being diverted to highly visible, intuitively appealing, but ultimately questionable investments in large-scale, detailed fish habitat assessment and mapping. Seldom have the massive data sets from such mapping exercises been used in stock assessment and regulatory design, and in fact, existing Geographic Information Systems (GIS) data make such uses painful and costly for biologists; the data are used mostly to produce pretty pictures of coastal habitats and not much else. Potentially, the map information could be very valuable for the stratification of survey programs, designing spatial allocation patterns for tagging and expanding local density estimates by total areas.

On top of all the trade-offs related specifically to the assessment of single-species harvesting impacts, what should we do to monitor more broadly for ecosystem management (Sainsbury 1998)? Options range from the assessment of community-structure indices using existing surveys (e.g., Tyler et al. 1982 concept of assembly monitoring for fish), to the monitoring of incidental catch/bycatch information on nontarget species for both a trend and absolute impact assessment, to the development of new technologies for broader ecosystem monitoring (e.g., acoustics, satellite information, and ecosystem-scale tagging using chemicals). The emergence of these broader monitoring methods as requirements for ecosystem management will make it even more difficult to find the financial resources to improve exploitation-rate monitoring.

4.7 MAINTAINING GENETIC DIVERSITY AND STRUCTURE IN HARVESTED POPULATIONS

Most natural "stocks" of fish are complex mixtures of quite variable genotypes with different susceptibilities to harvesting. This means that harvesting almost inevitably "erodes" natural genetic diversity by selecting for genotypes that are better able to withstand the impacts. Two types of genetic change have been particularly worrisome: erosion of spatial structure, and selection for changes in body size and maturity.

Many management units that are called stocks consist, in fact, of a variety of more or less distinct spatial "subpopulations" of individuals adapted to local spatial circumstances. Local spatial adaptation is particularly spectacular in Pacific salmon and other anadramous species, among which literally thousands of subpopulations return to particular spawning and rearing locations and exhibit a remarkable diversity of adaptations to these specific locations (spawn timing, migration and juvenile habitat use patterns, etc.). But even if local adaptation is not so easy for scientists to document, fishers can almost always provide anecdotal information about the existence of localized subpopulations that exhibit distinct behaviors (and often distinctive appearances to the experienced eye) and can describe how this diversity has been eroded over time. It is rarely practical or economical to regulate fishing so precisely as to selectively harvest every local subpopulation at its "best" rate, so most harvesting can be viewed as "mixed stock" fishing. This means that less productive (or more vulnerable by virtue of their proximity to fishing ports) subpopulations tend to be eliminated over time, and fisheries come to depend on fewer and fewer productive locations.

Almost all fish harvesting is deliberately size-selective, for both economic and biological reasons. We typically prefer to harvest larger fish, in order to (1) avoid growth overfishing (taking fish that would produce more by being allowed to stay in the water to grow rapidly); (2) reduce the harvesting and processing costs; (3) meet the market demands for larger individuals; and/or (4) reduce the risk of recruitment overfishing by allowing fish at least some opportunity to spawn before becoming vulnerable to fishing. Further, even if we wanted to harvest fish nonselectively with respect to size, there is virtually no way to do so; almost every fishing method (short of dynamite and poisons) is highly size-selective, whether it involves nets or hooks or traps or whatever. Indeed, this selectivity haunts us as scientists whenever we try to obtain a "representative" or random sample of the size-age composition

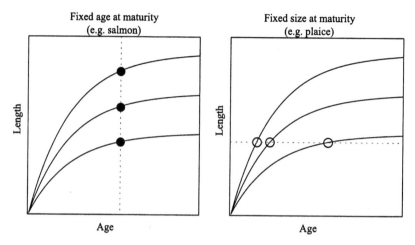

FIGURE 4.5: Fish typically show considerable individual variation in growth rates, with individuals tending to be either slower or faster growing throughout life. There are two possible extreme maturation "rules" for these individuals: they may mature at particular fixed age(s), at highly variable sizes (closed circles), or they may mature at particular size(s) over a range of ages (open circles).

of a population. But there is excellent evidence from selective harvesting experiments and experience with size-selective mating in aquaculture that there is considerable heritable variation in growth and maturation patterns (Conover and Munch 2002; Law and Grey 1989; see reviews by Smith 1999; Law 2000). So there is a risk that size-selective harvest regulations aimed at making fisheries both safer and more valuable may "backfire" by causing selection for undesirable changes in growth patterns.

Intuition might lead us to be concerned that the selective removal of larger fish will result in a selection for smaller body sizes and reduced growth, ultimately leading to reduced productivity. This is an excellent example of why ecologists should not trust intuition. Both simulation models that include genetic structure (e.g., Law and Grey 1989) and a considerable body of field evidence tell us that the selective harvesting of larger fish can lead to selection for either faster or slower growth, depending on how growth, maturation, and fecundity are linked. There are two qualitative, extreme ways that individual variation in growth and maturity might be linked (fig. 4.5) "age-linked" fish mature at particular ages, independent of size, while "size-linked" fish mature at a particular size so that faster-growing individuals mature at younger ages. Most semelparous fish such as Pacific salmon show mainly age-linked maturity, while iteroparous species such as cod and plaice show at least approximately size-linked maturation. What models and the data tell us is that age-linked fish are most likely to show declines in mean body size at age (in growth rates) under selective fishing, and this has been well documented for Pacific salmon (see review in Law 2000). In contrast, size-linked fish are more likely to display evidence of selection for increased growth rates (despite selection against large body size) because the early maturation associated with faster growth provides a strong selective advantage when the mortality rates of older, larger fish are higher. Law (2000) reviews a range of examples involving demersal fishes for which such increases in growth rates have in fact been observed, noting that it has been extremely difficult to separate the genetic selection effects of fishing from the effects of other factors that might also have contributed to increased growth (density-dependent effects, temperature warming effects).

Because of this selection for increased growth rates and earlier maturity, models for the long-term genetic effects of size-selective fishing produce ambiguous predictions about whether it would be better to avoid selectively harvesting larger fish. This ambiguity is particularly important in view of the other conservation benefits of avoiding harvesting fish until they have had at least some chance to spawn.

PART THREE

Use and Abuse of Single-Species

Assessment Models

An Overview of Single-Species Assessment Models

WHETHER THE AIM OF FISHERIES MODELING and analysis is to provide information to manage each species better while pretending it lives in isolation, or to provide historical reconstructions to help in assessing the parameters for ecosystem models, there will always be a critical need to analyze single-species historical data. There are now a number of useful texts on the methods and pitfalls of single-species assessment (Ricker 1975; Seber 1982; Hilborn and Walters 1992; Hilborn and Mangel 1997; Quinn and Deriso 1999). These texts describe a bewildering variety of methods, not because of the complexity of population-dynamics processes but because we must generally base assessments on a variety of largely fragmentary measurements that have been gathered over the years by biologists hoping to interpret what is going on in natural populations without having the tools or financial resources to directly measure the components of dynamic change in abundance and composition. Further, there is often much confusion about what the aims of an assessment should be in the first place, and about what criteria should be used in comparing alternative assessment techniques for the same data (e.g., statistical criteria versus criteria that measure performance at providing management advice).

In this chapter, we begin by asking the reader to think carefully about the alternative and sometimes conflicting objectives of single-species data analysis. We then review the state-observation dynamics structure that is rapidly becoming a standard way of thinking about assessment methods and of unifying apparently disparate methods and data. Then we examine two of the most difficult decision problems faced by stock-assessment scientists, namely, whether to use simple or complex models and what statistical criteria to use in the parameter estimation and assessment of uncertainty. We review the two main approaches to the reconstruction of population-dynamics histories from long-term data on population composition and relative abundance, virtual population analysis and stock synthesis, and provide advice on which of these to use under various circumstances. Finally, we discuss approaches to dealing with persistent change or "nonstationarity" in production dynamics.

5.1 OBJECTIVES OF SINGLE-SPECIES ASSESSMENT

Single-species assessment methods can be aimed at answering at least three quite distinct questions. The methods for answering each of these can be quite different, and confusion about this point has led to many published assessments that fail to address the question(s) for which they were funded by management agencies, to much time wasted by scientists trying to answer questions that no one asked or even needed to ask, and even to assessment results that are actively misleading due to statistical problems related to "overparameterization." The main questions or objectives that we might try to answer with single-species assessment are these:

1. **The Status Question:** What is the current level of harvesting impact on the stock, measured by the fishing-mortality or exploitation rate $F =$ (Catch)/(Abundance), in relation to goals or standards that may come from an analysis of the data from the stock or from other stocks for which there is historical experience about sustainable exploitation rates? Note that catch alone is never a meaningful measure of current fishing impact, since the catch cannot be sensibly interpreted without any information about how large it is relative to the stock size (10,000 tons may sound large, but it would likely be insignificant if taken from a stock of 1,000,000 tons). Often the most efficient way to answer this question would be to abandon the existing machinery for the analysis of historical data and to invest instead in direct stock-size and exploitation-rate estimation methods such as tagging experiments. However, there is typically much resistance to such proposals due to fears about investing in new, unproven field methods.

2. **The Mean Productivity Question:** What is the average long-term relationship between the fishing-mortality rate, stock size, and yield (how does the average net production rate of the stock vary with abundance and the harvest rate?), assuming that there is a "stationary" relationship between the average production rate and stock-size or exploitation rate? This question raises the main concerns about model complexity and overparameterization, since to estimate average response parameters it can often be best to deliberately use an oversimplified, "biased" model that sacrifices realism (in its representation of the sources of variation) for precision in its estimation of the key mean response parameters.

3. **The Stock Reconstruction Question:** How have the components of net production (growth, recruitment, natural mortality rate) varied over time, and what alternative hypotheses about the impact of stock size, fishing, and habitat/environmental changes could explain these variations? The most elaborate, complex stock-assessment models in use today are aimed at this question and use a variety of information sources about changes in relative abundance and size/age composition to say as much as possible about historical total abundances and, particularly, changes in recruitment rates.

The biggest single mistake that is being made routinely by today's inexperienced stock-assessment scientists is to assume that answering the third question is equivalent to, or the best route to, answering the first two questions. Complex stock-reconstruction models generally have 10s to 100s of unknown parameters to be estimated from the data, and many biologists appear to assume that having lots of parameters means that a model can both explain (fit) the data better and also filter out (correct for) more causes of variation so as to make the mean response parameters more visible. This intuition is often wrong: explaining the noise generally means admitting more, not less, uncertainty about the underlying average responses; i.e., the complex models admit that there are many more ways to explain the observations—meaning that there is more uncertainty about which particular way (if any) is the correct one.

One traditional division of approaches to this analysis involves a hierarchy of increasing detail as more information becomes available about a particular fishery, starting with a simple surplus-production-analysis (the net balance of recruitment + growth − natural mortality) and moving later to an estimation of the age-size and spatial components of these processes. This division has come with the presumption that methods should be used in progression over time, with more detailed methods being obviously "better" as more information becomes available. However, a more important distinction in practice is about whether composition information (size/age structure data) will be used in the analysis. We can always model or explain the overall biomass/numbers dynamics in terms of recruitment, growth, and natural mortality components, whether or not these component effects on past production can be separated by using more detailed data. In such modeling "beyond the data," we should not claim that a "better" model structure is needed to explain what data are available. Rather, we should admit that there are policy concerns and options that cannot be addressed without a more detailed calculation (e.g., the effects of changing size selectivity), and we should provide the capability to at least perform the calculations without pretending that the data justify them. That is, we should not pretend that the limitations of the available data should dictate or limit the range of policy concerns that we attempt to evaluate (policy concerns and options should drive the choice of models, not the other way around!).

5.2 STATE-OBSERVATION COMPONENTS

There has been a major change over the last 20 years in how we analyze historical fisheries data in order to estimate the past impacts of fishing and other factors. If you examine one of the older books on stock assessment and the analysis of fisheries statistics, e.g., Ricker (1975) or even Hilborn and Walters (1992), you will see mainly a collection of disjointed "recipes" for the analysis of bits of data such as catch curves, single tagging experiments, and time series of catch and CPUE data. These recipes were developed mainly by starting out with particular, usually simple population models and assumptions about how observations are related to population characteristics, then trying to interpret the predicted observations as having come from some standard statistical model such as a linear regression. That is, we tried to turn our data-analysis problems into familiar statistical problems. Examples include regression estimates of abundance from depletion experiments and estimates of the total mortality rate from catch curves.

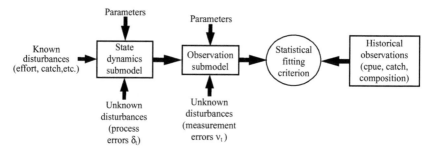

FIGURE 5.1: Most stock-assessment models are built with the general systems notion that to analyze a set of incomplete data on any dynamic system, we must hypothesize (assume) at least two distinct types of "submodels": one for the (generally unobserved) dynamics of system state in nature, and one for the relationship between system state and the available measurements. Arrows show the flow of calculations or dependencies (e.g., predicted observations depend on predicted states, plus observation submodel parameters). Note that catches are sometimes treated as input disturbances (estimation "conditioned on catch") or predicted outputs (estimation "conditioned on effort").

This changed dramatically in the 1980s, with the idea that we could build quite general models incorporating structural knowledge about population processes, use these to predict a rich variety of observations all at once, and use more complex statistical tools and criteria (nonlinear estimation, likelihood functions) to see how much could be said about all the data at once (fig. 5.1). At first, this change in approach was presented in an almost apologetic way, with titles such as Deriso et al. (1985), "catch-age analysis with auxiliary information."

A key computer innovation that helped promote the change was the spreadsheet. Using a spreadsheet, it is easy to lay out a block of cells representing known historical "drivers" of change (fishing effort, catches, hatchery stocking rates, etc.), a block of cells representing a possibly quite complex population model for response to these drivers, another block of cells representing predictions of a variety of observations, and blocks of cells with actual data. Obvious spreadsheet functions can then be used to calculate the statistical agreement between observations and predictions and to search for parameter values that give better agreement. Ray Hilborn was the first to show how general and powerful this approach could be, in his work on Hudson River fishes. For that work, he not only "drove" a set of spreadsheet population models with fishing disturbance patterns over time, but also estimated the historical mortality rates of larval fish due to entrainment in power plants. He predicted not only patterns of catch and age composition for commercially harvested species, but also relative-abundance trends for larval and juvenile index sampling. His models were set up so as to include a rich variety of parameters for the effects that had been hypothesized, e.g., whether there is compensatory improvement in the survival of young fish after some have been killed by power plant entrainment.

The logical structure of figure 5.1 has also helped us understand that most stock-assessment models are, in fact, just generalizations of a very old and simple model known as a "depletion" or "removal" model. The idea in a depletion model is to remove a known number or biomass of animals from a population, while measuring how this removal affects the measures of relative abundance. As a very simple example of this idea, suppose 1000 tons of fish have been removed from an area, and this removal has reduced some index of abundance (catch per effort by fishers or by survey gear, acoustic target count, etc.) by 20%; we would then say that 1000 tons represents 20% of the initial stock, so the initial stock must have been 5000 tons. More generally, the removal estimator for multiple removals and samples from a closed population (no other gains or losses) is constructed by using the state and observation submodels:

$$N_t = N_o - K_t \tag{5.1}$$

where N_t = numbers or biomass at time t, K_t = total removals before t,

$$Y_t = qN_t \tag{5.2}$$

for the index of abundance at the start of the removal period t. Substituting the state equation for N_t into the observation equation for Y_t results in a simple linear regression model, or "Leslie method" (Leslie and Davis 1939) for estimating q and N_o. Note that this simple model can go wrong for two kinds of reasons: *biological*—there is gain or loss of animals over time not accounted for by the state equation; and/or *observational*—the index Y_t is not proportional to N_t either because it measures the abundance of a subset of relatively more vulnerable animals that are removed at early times (the index shows "hyperdepletion") or because it stays high as N declines (the index shows "hyperstability"). Much of the machinery of stock assessment has been developed by basically ignoring problems of the second kind and concentrating instead on representing components of biological change. Simple "surplus production" models are constructed, e.g., by replacing the closed-population state equation 5.1 with one that includes a function $f(N_t)$ for net production (recruitment + growth − natural mortality) as a function of population size:

$$N_{t+1} = N_t + f(N_t) - C_t \quad (C_t = \text{catch taken during time step } t) \tag{5.3}$$

which can be solved recursively over time for any starting N_o and parameters of $f(N)$ to generate a set of predicted states N_t and predicted observations Y_t (using equation 5.2). Note that as we construct more complex models like equation 5.3, we are still using the catches C_t to provide population scale information, but recognizing that we might be able to obtain information about $f(N)$ by observing whether changes in Y_t were less than predicted assuming no production, as Y_t declined following the initial C_t removals. An obvious next step has been to recognize that removals are likely to affect the population composition (relative numbers of animals of different ages and/or sizes) as well as the relative abundance, through the effects on mortality rates. To model these compositional changes (to predict observed changes in composition data), we replace equation 5.3 with a more complex accounting model for changes in size/age composition and extend the predicted observation submodel accordingly. But no matter how complicated we get, we remain reliant for most analyses (for which the historical abundance change has not been measured directly) on using the models to predict the relative impacts of changing the total removals C_t (or removals by size/age), i.e., we remain reliant on the basic idea of a removal estimator.

While biologists have tended to concentrate their efforts on developing better models for state dynamics, the main thing that appears to be going wrong in assessment with the general state dynamics/observation dynamics structure today is not with the population-dynamics models but, rather, with the observation models. Two main problems have caused dangerously misleading overestimates of abundance, recruitment, and net production during population declines and the onset of overfishing: (1) the use of commercial catch per unit effort (CPUE) or other relative-abundance indices that are not, in fact, proportional to abundance (Harley et al. 2001), and (2) changes over time in size/age selectivities that confuse the interpretation of population composition data. In particular, fishers often have the habit of targeting smaller fish as the abundance of larger ones declines; this can help to prevent the CPUE from declining and can shift the composition data so as to make it appear to a scientific observer that recruitment must still be healthy because plenty of small fish are still being caught. To a naïve reader, the recent stock-assessment literature appears to represent much progress in model development and the statistical representation of uncertainty (modern, equation-filled papers certainly look impressive), and few stock-assessment scientists have been forthright about pointing out that it is not better estimates that we are getting but, rather, better evidence that the data are actually telling us much less than we used to think, i.e., the estimates are often nonsense or even actively misleading. Unfortunately, this "defense of method" often makes it look as if we can get by in the future with the same kinds of data (abundance trend, composition) and, hence, directs attention away from the critical need to start gathering new kinds of data, especially direct estimates of exploitation patterns from methods like tagging experiments.

Still more unfortunately, many fisheries scientists are easy to fool: a lot of biologists and inexperienced assessment people seem to think that assessment methods will work if most of the assumptions about the data are correct, so long as there is plenty of data. That is, they seem to think that assumptions and observation-model components have an additive effect on assessment performance, similar to adding additional explanatory variables in a multiple-regression analysis. Unfortunately, this thinking is very wrong: all it takes to cause a misleading assessment and interpretation of stock changes is for even one key assumption to be wrong, particularly about stock trends as evidenced by the relative abundance indices. To put it vividly, a few bad assumptions, and sometimes even just a few bad observations, can poison a whole stock assessment. Indeed, it is a bit worrisome that the 2003 Lowell Wakefield symposium (21st Lowell Wakefield Fisheries Symposium Anchorage, Alaska, USA, October 2003; The Wakefield Symposia are a series that has been wonderful about bringing assessment scientists together to share methods and experience) pretends to be specifically about "data limited" assessment situations, as though these were special cases. In fact, when we examine the really critical data and observation-model assumptions that drive assessments, and how frequently the most basic data cannot be trusted, we really should be thinking about all fisheries as severely data limited situations.

Nowadays there is a lot of quibbling about the computer implementation tactics for doing general-state/observation-dynamic analysis. This has been driven largely by numerical problems with nonlinear estimation procedures when we get ambitious about the number of "nuisance parameters" that we try to include in any descriptions of observation processes and size/age-vulnerability patterns of fish-to-fishing and survey sampling. One school of thought is that we should continue to try to use spreadsheet implementations as much as possible, because they help to keep assumptions transparent, to present the results to wider audiences, and to move from historical reconstruction to future prediction (just add spreadsheet rows or columns to represent dynamic changes into the future). The other school of thought is that we should use the best available statistical and nonlinear search technology, best exemplified by David Fournier's AD Model Builder (AD = automatic differentiation, Otter Research 1994). Thankfully, software is gradually appearing that integrates the best features of both of these approaches.

5.3 ESTIMATION CRITERIA AND MEASURING UNCERTAINTY

Given a set of state-dynamics/observation-dynamics assumptions and their associated unknown parameter values, there are basically two ways to proceed with an estimation of the parameter values. First, we can seek the parameter values that maximize or minimize some intuitive measure of how well the model fits the data, such as weighted sums of squared deviations or absolute deviations between predicted and observed values. Second, we can adopt the maximum-likelihood approach of finding the parameter values for which the observations are most probable (from which the observations are most likely to have come). We now use the second approach most often, even when we cannot be confident about having chosen the right form of probability distributions for variability in the data. The likelihood approach offers several key advantages over ad hoc estimation criteria. These include

- formal methods for measuring possible overparameterization;
- a ready linkage to Bayesian methods for incorporating prior (e.g., other case) information about parameter values into the calculations and for providing outcome-probability statements for management-decision analysis; and
- the "automatic" and objective weighting of heterogeneous data types that may be combined in the likelihood function.

Unfortunately, it is no simple matter to construct a full likelihood function for data gathered from a nonlinear dynamical system that has been subject to nonlinear observation processes and both "process" and "observation" errors. While numerical methods for full, nonlinear "state-space" estimation methods are becoming available, most practical analyses are based on computationally convenient likelihood function approximations. Three main approximation approaches are now commonly used:

1. **Mean trajectory or observation error approach:** solve the state-dynamics equations omitting all process errors, as an approximation of the mean trajectory of the stochastic state dynamics, and treat the combined process + observation deviations from the predicted observations obtained this way as independent "observation errors."

2. **Nuisance parameter or errors-in-variables approach:** treat the process errors (δ_t in fig. 5.1) as unknown, arbitrary past disturbances, and try to estimate them as "nuisance parameters" along with the parameters representing the mean dynamic response. This generally involves making some a priori assumption about the relative magnitudes of observation and process errors, but at least avoids the often silly pretense that the process errors were, in fact, drawn from some simple statistical distributions (Mother Nature is not that generous).

3. **Full state-space approach:** calculate the total likelihood of each observation integrated over the possible state-variable values that may have resulted from process errors. This generally involves a complex recursive calculation of the likelihood function over time, with either an extended Kalman filter or numerical integration schemes for the state dynamics (Pella 1993; Reed and Simons 1996; Schnute 1994).

Recent simulation studies by Schnute and Kronlund (2002) compare approaches (2) and (3) and show that these approaches give similar results and estimation performance at least for stock-recruitment state dynamics. This hints that the errors-in-variables approach may generally be the better one to take, especially considering the computational difficulties with the full state-space approach and with its requirements to make dubious assumptions about the statistical properties of the process disturbances δ_t. Further, an informative byproduct of the errors-in-variables approach is the time-series estimates of the process errors, which can be critical in detective work about the causes of the "errors," forecasts of future variation, and the design of harvest strategies for coping with variation.

Whichever of the above approaches to likelihood formulation is taken, we generally have to deal with the likelihood components for two main types of observations: quantitative indices of abundance and/or total outputs (catches) and proportional contributions to observed composition data. Including observation errors, we most often assume observation models for quantitative variables to be of log-normal form:

$$Y_t = qX_t e^{v_t} \tag{5.4}$$

where X is any state variable or quantitative output, Y is an observation that we hope is proportional to X, and v_t is normally distributed with mean zero and some unknown variance σ_v^2. We use the log-normal form because most quantitative observations in fish dynamics arise as a product of component and proportional observation/collection processes, and the sum of the logs of such proportions is likely to be normally distributed because of Central Limit Theorem effects. For any collection n of such independent observations included in an analysis, the log-likelihood function component for the collection, evaluated at the conditional (on all other parameters) maximum likelihood estimates of q and σ_v^2 is just a constant that depends only on the data (and hence can be ignored for all parameter search and comparison calculations), plus the following term (see box 5.1):

$$logL = -(n/2)\ln(SS) \tag{5.5}$$

Here, SS is just the sum of squared deviations

$$SS = \sum_{i=1}^{n}(z_i - \bar{z})^2 \tag{5.6}$$

where $z_i = \ln(Y_i/X_i)$ and $\bar{z}$ (the arithmetic mean of the z_i) is the conditional maximum likelihood estimate of $\ln(q)$ (also SS/n is the conditional maximum likelihood estimate of σ_v^2). For predictions of absolute quantities ($q = 1$), omit $\bar{z}$ from the calculation of SS.

For observations of sample proportions (e.g., the proportions of catch that were of different ages), where n_i creatures of type i were observed and the state/observation model predicts that these should be a proportion p_i of the observations ($\sum p_i = 1$), the log-likelihood component needed for search and comparisons of the parameters that predict p_i from a state change is just

$$logL = \sum_{i} n_i \ln(p_i) \tag{5.7}$$

This equation must be used with caution because the effective sample size for composition sampling in fisheries is most often much less than the number of fish ($\sum n$) actually measured (Fournier et al. 1990; Fournier et al. 1991; Zheng et al. 1995; Quinn et al. 1998). Together, the simple equations 5.5 and 5.7 cover most log-likelihood components that are likely to be added up (over data types) to form the circled statistical fitting criterion in figure 5.1.

BOX 5.1

EFFICIENT CALCULATION OF LIKELIHOOD AND BAYES POSTERIOR
PROBABILITY FUNCTIONS

The likelihood function for a set of independent observations $Y = \{Y_1 \ldots Y_n\}$ given a set of parameters $P = \{P_1 \ldots P_m\}$ is just the product of the probability distributions of the individual observations given the parameter values, and the log-likelihood function is just the sum of the logs of these probability distributions. For example, if each observation Yi is assumed to have arisen from a normal distribution with mean $\mu_i(P)$ (i.e., μ_i is a function of the parameters P), so the probability of Y_i is

$$p(Y_i|P) = \frac{1}{\sigma\sqrt{2\pi}} \exp - \left[\frac{(Y_i - \mu_i)^2}{2\sigma^2} \right]$$

then the log-likelihood for the entire set Y is

$$\ln L = -n \left[\ln(\sigma) + \frac{1}{2}\ln(2\pi) \right] - \sum_{i=1}^{n} \frac{(Y_i - \mu_i)^2}{2\sigma^2}$$

When we want to compare alternative parameter estimates, either by likelihood ratios (log-likelihood differences) or in search procedures for the maximum $\ln L$, the calculations can be simplified in at least two ways. First, we can drop any multiplicative terms in $p(Y|P)$, or additive terms in $\ln L$, that depend only on the data, since these do not change during parameter searches and cancel in likelihood ratios. Second, we can evaluate "nuisance" parameters like σ^2 at their conditional maximum likelihood estimates given the other P, provided we can obtain analytical formulae for these parameters. For example, if we differentiate $\ln L$ above with respect to σ^2 and set the resulting derivative to zero, the maximum likelihood estimate of σ^2 is easily shown to be $\frac{SS}{n} = \frac{\Sigma(Y_i - \mu_i)^2}{n}$. Substituting this back into $\ln L$, then dropping any constant, additive terms that will not affect $\ln L$ comparisons across P values, we get the likelihood "kernel" $\ln \acute{L} = -\frac{n}{2}\ln(SS)$ since the second term in $\ln L$ above just becomes the constant $-\frac{n}{2}$, and the first term becomes $-(\frac{n}{2})\ln(\frac{2\pi SS}{n})$, which is equal to the constant $-(\frac{n}{2})\ln(\frac{2\pi}{n}) - (\frac{n}{2})\ln(SS)$.

The Bayes posterior probability for a particular parameter value P is defined as $p(P|Y) = L(Y|P)p_o(P)/k$, where $p_o(P)$ is a prior probability for P and k is a constant (the total probability of the data Y integrated over all P) that can be ignored in comparing probabilities for alternative values of P since it is the same for every P value. The log of this posterior probability is just $\ln L + \ln(p_o(P)) - \ln(k)$. In comparing posterior probabilities for alternative P estimates, we can again ignore both $\ln(k)$ and any additive terms in $\ln p_o$ that do not depend on P. So, e.g., if the prior probability for P_j is taken

(*Continued*)

(BOX 5.1 continued)

to be normal with mean $P_j^{(0)}$ and variance σ^2, the only term of this normal distribution that we need retain in comparisons of log posterior probabilities for alternative values of P is $-(P - P_j^{(0)})^2/2\sigma^2$.

Likelihoods are almost always very tiny values that cannot be meaningfully interpreted by themselves or computed accurately. That is why we always do calculations using logarithms of them, and why we use only ratios of them in comparisons of how well alternative parameter values describe the data. This also means we can subtract any convenient constant, e.g., lnL_{max}, the log-likelihood evaluated at the maximum likelihood estimates of P, from the log-likelihood or log-posterior probability when we want to summarize relative probabilities versus P rather than relative log probabilities, e.g., in plots of posterior probability distributions for key parameter values. ∎

With a bit of practice, one can learn very quickly to set up state/observation models in the format of figure 5.1 and to perform the nonlinear statistical searches needed to find the maximum likelihood estimates of those state/observation/disturbance parameters admitted to be uncertain in the problem formulation. But this is where the work really begins, because the next step in the analysis is to provide some honest assessment of how good the parameter estimates, particularly those with policy implications, really are. Two things can go wrong assessing of uncertainty in the parameter estimates:

Numerical uncertainty A very wide (or even infinite) range of parameter combinations may equally well explain the data (the same or nearly the same likelihood for a wide range of combinations); in particular, the data may be equally well explained by assuming they came from a large, unproductive population or from a small, productive one.

Structural uncertainty The presumption of knowledge implied by the choice of model structure (and whatever parameters are treated as fixed knowns) may lead to apparently very precise but actually biased parameter estimates.

To identify and approach the first problem, we somehow need to map how the likelihood function (or likelihood times prior for Bayesian analyses) varies with changes in the parameter estimates. To approach the second problem we need to construct a set of models of increasing complexity and determine whether "freeing up" potentially weak assumptions (allowing more parameters to vary) leads to different results.

There are basically four approaches to studying how the likelihood function varies with parameter values, so as to estimate parameter uncertainty: (1) *Brute force*—map the likelihood function over a grid of parameter combinations. (2) *Information matrix*—use measures of the curvature of the likelihood function near the maximum likelihood estimates to construct confidence regions. (3) *Profile*—vary one important ("leading") parameter across a grid of values, and plot the likelihood maximized over all other parameters for each value of this leading parameter. (4) *Sampling*—randomly sample parameter combinations using an importance sampling (SIR) or Metropolis-Hastings (MCMC) algorithm (Punt and Hilborn 1997; McAllister and Ianelli 1997), and display the frequency distributions of the important parameter values from this sample. Of these, the brute force approach is best for small problems (few parameters) but is impractical for most of the models used in fisheries assessment. The information matrix approach is the one commonly used to generate approximate confidence limits for parameters, but it can be misleading for fisheries problems because nonlinear model structures do not lead to confidence regions that are even approximately elliptic. The profile and sampling approaches typically give very similar results, provided the likelihood has one unique maximum for each value of the leading parameter. The choice among these approaches is largely a matter of what the analyst finds most convenient, except in cases for which the likelihood maximization procedure is very tedious or likely to fail entirely. Note that when the likelihood does not have a unique maximum (some parameters being completely confounded in their effects), the sampling approach amounts (deliberately or inadvertently) to assigning a Bayes prior probability distribution for the parameters, via at least the range of parameter values that are allowed in the sampling (e.g., a uniform prior distribution).

In our experience, it is very easy to waste time in the analysis of parameter uncertainty just to end up producing a collection of pretty probability distributions or confidence region maps that are hopelessly overoptimistic due to the failure to consider structural uncertainty. There has been much concern in recent years about how to measure structural uncertainty in assessments and about how to use alternative quantitative criteria for comparing alternative models and assumptions (Burnham and Anderson 1998; Helu et al. 2000; Patterson et al. 2001). Suggested approaches have mainly involved using simple statistical measures of prediction-error variance for alternative model structures, the Akaike (1973) criterion (AIC) or a very similar Bayes information criterion (BIC) suggested by Schwarz (1978).

$$AIC = -2lnL_{max} + 2p \qquad (5.8a)$$

$$BIC = -2lnL_{max} + pln(n) \qquad (5.8b)$$

where p is the total number of parameters treated as variable in the search for the maximum log likelihood lnL_{max}, and n is the total number of observations. Unfortunately, as we will see in the next section, there are two things that can go wrong with using such simple and scientifically appealing criteria when the goal of assessment is to find the best model *for providing management advice*. First, they help identify the model structure that will likely result in a minimum short-term prediction variance, which is not the same as obtaining the best estimates of particular parameters that are important for policy. Second, the variance of a parameter estimate can be misleading when there are asymmetric costs of estimation errors (e.g., when it is much more costly to overestimate a key policy variable such as MSY than to underestimate it). Ultimately, the only way to fully evaluate a proposed assessment model compared to other models is to simulate its "closed loop" performance: simulate the full data gathering, estimation, and harvest regulation process. That is a very complex and tedious matter, and it can also give misleading results if the closed-loop simulation procedure inadvertently makes overly optimistic assumptions about the state dynamics and observation processes.

It is important to understand that uncertainty about population-dynamics parameter estimates does not necessarily imply corresponding uncertainty about the best harvest policy. For example, if the goals of management imply that a fixed harvest rate should be maintained (see chapters 3 and 4), the estimates of the best harvest rate depend only on those population-dynamics parameters that affect per-capita productivity (e.g., the population intrinsic rate of increase "r" in surplus-production models, or the slope of the mean stock-recruitment relationship in age-structured models), and not at all on the parameters that define absolute population scale (e.g., the carrying capacity K of the logistic production model, or the mean asymptotic recruitment at high stock size). If the goal of assessment is to provide a reference point for overall harvest (e.g., MSY) that is a function of several population parameters, this function may be more or less uncertain than the component parameters. For example, in surplus-production models with rate and scale parameters r and K, the calculated MSY is roughly (depending on the shape of the assumed production function) $rK/4$. Thus, MSY appears to be poorly estimated since it is the product of two uncertain parameter estimates. But in fact, the estimates of r and K are often highly negatively correlated (we can explain the data equally well as having come from a large, unproductive stock or a small, productive one), and in this case the product rK can end up being quite precisely estimated.

5.4   MODELING OPTIONS

How complex should the state-dynamics model in figure 5.1 be, given options ranging from simple surplus-production equations (e.g., 5.1) to detailed age-structure accounting, and given various possible forms for observation relationships (e.g., proportional, hyperstable, hyperdepleting)? As noted in previous sections, there are two very different ways to answer this question:

1. **Scientific-statistical**—use a model that is optimally complex with respect to the available data (use the most parsimonious model that will do the "job");
2. **Policy relevance**—use whatever model is most likely to provide the best policy advice and/or to stimulate a search for better policy options.

What question (1) is really about is how to maintain one's credibility with scientific colleagues by appearing to be a knowledgeable statistical analyst and avoiding such mistakes as overparameterization. What question (2) is about is how to open doors for people who make policy decisions to compare their choices and to identify new choices that may be better than any used in past management, whether or not such choices can be immediately justified through a statistical analysis of the historical data.

Were policy relevance the only criterion for deciding assessment-model complexity, we would immediately recommend that every assessment model contain at least age/size-structured accounting (to allow for the exploration of policies that reduce the risk of growth and recruitment overfishing by avoiding the capture of juvenile fish). Considering that spatial policy options are becoming more popular in general (e.g., marine protected areas), we would even be tempted to recommend spatial models of the types discussed in chapter 8. But to move in these directions without careful thought about the aims of a particular assessment is as silly as to presume that model complexity should be dictated strictly by statistical concerns about overparameterization relative to the available data.

As noted in the previous section, one (difficult) way to compare alternative models for assessment is to simulate their "closed loop" performance when applied by a simulated fishery manager to a reference or "operating" model of realistic biological and observational complexity. Comparisons of this sort were first published by Hilborn (1979). He demonstrated the very surprising result that it can often be best from a management perspective to deliberately base assessments on an oversimplified model that is known to be wrong, if that model gives precise (low variance) estimates of policy parameters with a bias that is in a direction that helps avoid overfishing. Similar results have since been obtained using more elaborate likelihood estimation methods (Ludwig and Walters 1985), demonstrating that Hilborn's findings were not an artifact of the particular methods he used for simulating statistical analysis of the modeled data. Figure 5.2 shows an intuitive way to understand such results, involving a hypothetical comparison of three estimation methods: one that is biased in a costly direction, one that is less precise but is biased in a safe direction (as was the method studied by Hilborn 1979), and a third that is unbiased but realistic in terms of its representation of the sources of uncertainty (and hence prone to produce even more variable parameter estimates). In this example, the first method is clearly

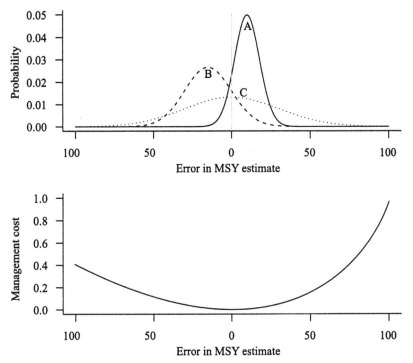

FIGURE 5.2: Variability and bias in the estimation of a key policy parameter (MSY) from three possible assessment models. Model A has very few parameters and so gives precise estimates, but biased upward in a way that would frequently result in a high management cost (due to overfishing) if the method were used in practice. Model B has more parameters and so gives a less precise estimate of MSY (admits more possible values of MSY given the data) that happens to be biased downward so as to give high-cost overestimates only rarely. Model C is a complex, fully "honest" representation of variability in the data; it is unbiased but even more imprecise than model B. Note that like Model A, Model C results in a relatively high probability of overestimating MSY and, hence, leads to a high management cost.

irresponsible: it results in a high probability of overfishing. The distinction between the other two methods is more subtle and becomes apparent only when the methods are applied many times (when MSY is estimated many times, as in multiple simulations to evaluate expected performance). The unbiased model with its realistic admission of uncertainty ends up with the lower average performance (and the higher average cost of applying it) because it more often results in high-cost (overestimation of MSY) outcomes.

An example of the precision-accuracy trade-off shown in figure 5.2 occurs in interpreting catch and effort data from the "one way trip" (Hilborn and Walters 1992) fishery developments shown in the simulated example

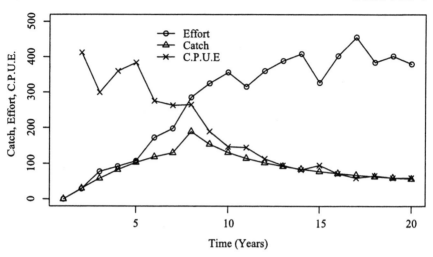

FIGURE 5.3: A simulated one-way-trip fishery development in which effort builds up enough to cause a severe decline in catch per effort, and apparent overfishing, over a 20-year period. Note that units are arbitrary in this example.

(fig. 5.3). Such data sets have typically been analyzed using simple surplus-production models and observation assumptions, e.g., the logistic biomass (B_t) model:

$$B_{t+1} = B_t + rB_t(1 - B_t/K) - qE_tB_t^\beta \qquad (5.9)$$

$$Y_t = qE_tB_t^\beta \quad \text{(catch)} \quad Y_t = qB_t^\beta \quad \text{(catch per effort)} \qquad (5.10)$$

Here the parameter β represents the possible changes in catchability with stock size ($\beta < 1$ implies hyperstability in the catch per effort, $\beta = 1$ implies proportionality, and $\beta > 1$ implies hyperdepletion). It is easy to demonstrate with simulated data that this model has far too many parameters (B_1, r, K, q, β) to be estimated uniquely from one-way-trip data. To reduce the number of parameters, we have typically assumed $B_1 = K$ (the stock initially near the unfished equilibrium) and $\beta = 1$. The "Gulland method" (Gulland 1961), which has been applied very widely in the developing world, invokes making an additional, even more restrictive assumption that development has been slow enough so that biomass has remained near equilbrium with respect to catch, i.e., $B_{t+1} = B_t$. In that method, the state equation 5.10 is replaced with the assumption $rB_t(1 - B_t/K) = qE_tB_t$, and a simple two-parameter linear regression relationship (CPUE $= a - bE_t$, $a = K/q$, $b = q^2K/r$) is used to estimate the effort E_{MSY} that would produce the maximum sustained yield on average (as the effort that would result in CPUE equal to $\frac{1}{2}$ the regression intercept; efforts are typically averaged so as to supposedly reduce the effect of the erroneous $B_{t+1} = B_t$ assumption).

Now, suppose we generate 100 20-year data sets as in figure 5.3 that exactly satisfy the state equation 5.10, i.e., no process errors or effects of population structure, and provide three alternative assessment procedures with simulated CPUE data where $Y_t = qB_t^\beta e^{v_t}$, $v_t =$ normally distributed with standard deviation 0.2 (a typical amount of unexplained variation in the catch rate). Let "Procedure A" be the Gulland method, let "Procedure B" be a general procedure based on figure 5.1 with the correct state equation (5.10) but assuming $B_1 = K$ and $\beta = 1$, and let "Procedure C" be the same as Procedure B except being "scientifically honest" by admitting uncertainty about β (in Procedure C, the conditional maximum likelihood estimates corresponding to equation 5.5 of $\ln(q)$ and β are the slope and intercept, respectively, of the linear regression of $\ln(Y_t)$ on $\ln(B_t)$; see Walters and Ludwig 1994). Since β cannot generally be estimated uniquely under Procedure C, suppose we provide the likelihood function with an additional term $-(\beta - 1)^2/0.3$ representing a normal prior-probability distribution for β with a mean of 1.0 and a variance of 0.15 (the metanalysis results of Harley et al. 2001 suggest we should assume this variance, but a less optimistic mean of 0.8).

Suppose we then judge the performance of the three procedures not by how well they fit the data (an award that most often goes to the Gulland method), nor by AIC or BIC criteria (which most often indicate Procedure B to be best), but by two management criteria: (1) how well they advise management about the best long-term fishing effort (for some objective like MSY) and the effort in year 20 relative to this optimum, and (2) how robust they are at continuing to provide good advice under the dangerous hyperstability case $\beta \ll 1.0$. As shown in figure 5.4, it is obvious that the precise and simple Procedure A throws the baby out with the bath water: it results in a dangerous upward bias in the estimates of optimum and current effort/optimum for all β, which just get worse and worse for more hyperstable CPUE cases. Surprisingly, Procedures B and C do not consistently result in overestimates of the optimum effort for hyperstable CPUE cases as might be expected, given that the simulated CPUE decline in these cases does not reflect the severity of actual stock decline; however, both procedures grossly underestimate the current (year 20) exploitation rate relative to its true value in the hyperstable cases, where catchability increases greatly at low stock sizes. The surprise about Procedures B and C is due in part to the way we generated the simulated data and is a warning about how the details of a simulation protocol for evaluating management procedures can affect the results: we used the same fixed, known exploitation-rate history for each simulation run, then calculated different effort sequences (depending on β) needed to have caused this fixed history; this protocol results in lower mean efforts for simulations with lower β and, hence, lower efforts needed to explain a CPUE decline. Despite a favorable protocol, Procedure C results in a higher variance in the optimum effort estimates among the 100 trials, meaning it has a somewhat

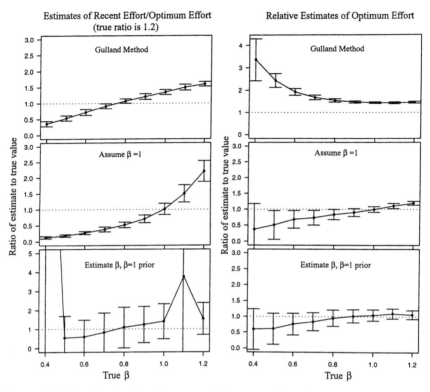

FIGURE 5.4: The simulated performance of three procedures for estimating optimum fishing effort and current exploitation rate for simulated data sets similar to figure 5.3. For each true value of the hyperstability parameter β, the results show the mean and range of estimates for 100 simulated development histories. Note that low values of the ratio of estimated to actual exploitation rates mean that the procedures underestimate the current impact of fishing.

higher probability of resulting in overestimates of the optimum effort (i.e., of leading to a recommendation that will result in overfishing). That is, it is similar to Model C in figure 5.2. If we change the prior distribution for β in Procedure C to have a mean of 0.8, to be more faithful to the empirical results of Harley et al. (2001), the procedure actually ends up doing worse in the sense that its bias is reduced for intermediate $\beta < 1$ cases, but it is much more likely to result in an overestimate of the optimum fishing effort.

In the above example, we have fixed the data available for assessment, then applied progressively more realistic models to these data. It is comforting in such examples that statistical measures such as AIC and BIC tend to agree with management-performance simulations in predicting that the optimum model should be one of intermediate complexity. However, this does not by any means complete the discussion about simple versus complex models: what happens to conclusions about optimum model complexity when we can not only compare simple versus complex models but also include additional "auxiliary" data to help estimate parameters for the more complex model options? In particular, even in poorly monitored fisheries there is often some information on changes in the average size of fish caught, and average size is often a good indicator of changes in mortality and/or recruitment rates. The simplest population-dynamics model that can predict such changes in the average size is the "delay-difference" model of Deriso (1980) and Schnute (1985, 1987), which exactly predicts changes in the vulnerable biomass (B_t) and vulnerable numbers (N_t) for age-structured populations, in which fish have an age-independent total survival rate S_t (the product of natural survival rate times 1-exploitation rate), grow over ages a according to the Ford-Brody body-weight equation $w_{a+1} = \alpha + \rho w_a$, and are equally vulnerable to fishing after some age k (see Hilborn and Walters [1992] for derivations):

$$B_{t+1} = S_t(\alpha N_t + \rho B_t) + w_k R_{t+1} \qquad (5.11)$$

$$N_{t+1} = S_t N_t + R_{t+1} \qquad (5.12)$$

Predicted mean body weight at time t: $\bar{w}_t = B_t/N_t \qquad (5.13)$

To generate surplus production dynamics with a recruitment overfishing risk under this relatively simple age-structured model, we must assume some functional relationship between mean recruitment R_{t+1} and past biomass(es) B_t, B_{t-1}, etc., e.g., the Beverton-Holt form $R_{t+1} = k_1 B_{t-1}/(1 + k_2 B_{t-1})$ if fish recruit at age $k = 2$. Even if we can accurately estimate the natural survival component of S_t and the growth parameters α and ρ from independent analyses of size and composition data, we apparently still need to estimate or assume values for at least the following parameters: B_1, N_1, k_1, k_2, plus any observation-model parameters (q and β in this case for representing the observation submodel for relative abundance). In this model, the recruitment-function slope parameter k_1 (along with the body-growth parameters) plays the role of "r" in the surplus-production models: it determines the intrinsic rate of population increase when abundance is very low, and is a key determinant of the optimum exploitation rate. The recruitment parameter k_2 plays the same scaling role as the carrying-capacity parameter K, determining the absolute maximum recruitment rate and, hence, the overall population size in the absence of fishing. We can use these facts to eliminate B_1 and N_1 from the estimation set if we are willing to assume the age-structured analog of $B_1 = K$ (the stock was initially at unfished equilibrium), i.e., we can calculate B_1 and N_1 from the other parameters for the unfished equilbrium case. But even under all these restrictive assumptions, we are still left with at least 3 or 4 parameters to estimate from time-series data: k_1, k_2, q, and possibly β.

Suppose we repeat the performance simulations described above for the logistic-production model but replace the state-dynamics with the delay-difference system (eq. 5.11 and 5.12) and include noisy annual measurements (c.v. = 0.2) of average body weight $\bar{w}_t$ in the likelihood function. This results in a spectacular improvement in both the precision and the apparent

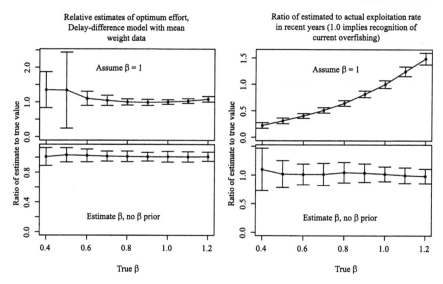

FIGURE 5.5: Performance simulations similar to those shown in figure 5.4 on page 106, but with an assessment procedure based on a delay-difference age-structured model that predicts (and is provided data on) changes in the average size of fish caught over time.

robustness of the assessment to hyperstability in the CPUE data (fig. 5.5). The procedure is able to estimate β quite well, and in fact, the simpler procedure with β assumed equal to 1.0 displays very obvious signals of failure when β is actually small, in the form of "contradictory data" (the model cannot predict both the correct CPUE trend and the mean-body-weight trend at the same time). For readers who worry about process versus observation errors, it is easily demonstrated that the procedure continues to perform quite well even if we include substantial process errors in the form of log-normal variations in the recruitment rates (c.v. 0.5), without even modeling these errors explicitly using one of the more elaborate likelihood formulations.

Simulation studies reported in NRC (1998) involved generating artificial data sets of increasing realism (violating simple model assumptions), then applying various age-structured stock-assessment methods including the delay-difference model (eq. 5.11–5.13) to these artificial data. Surprisingly, but as might be expected from the results in figure 5.5, the delay-difference model performed almost as well as the much more complex assessment methods discussed in the next section (at least at reconstructing stock-biomass histories), despite using only simple data on changes in average body size rather than full catch-at-age data.

Unfortunately, simulation tests as shown in figure 5.5 or reviews in NRC (1998) might well be hopelessly optimistic about an assessment performance with real data. Trends in the average size of fish measured from catches are actually influenced by four main factors: (1) total mortality rates (higher rates lead to smaller average size), (2) persistent recruitment trends (downward trends lead to larger average size), (3) changes in size selectivity due to fisher targeting practices (can change the average size in unpredictable directions), and (4) sampling procedures (it is difficult to collect a random sample of fish sizes from even one vessel, let alone a whole fleet). Factor (2) can easily mask the effects of fishing that should be evident via the effects of fishing on factor (1), if there have in fact been downward recruitment trends not correctly captured through assumed stock-recruitment relationships, e.g., due to persistent environmental "regime shifts." Factor (3) can also mask any mortality and recruitment effects, especially if fishers target smaller fish as larger ones become less abundant during stock declines (higher catches of smaller fish can easily be misinterpreted as evidence of recruitment increase). Factor (4) is not just a sample-size issue: when field personnel are allowed to grab fish "at random" for measurement from holds, buckets, or sorting trays, larger fish almost always end up being overrepresented in the sample in ways that can mask the impacts of changing mortality (small fish slither to the bottom of containers, larger fish come to hand first, etc.).

It is very easy to fall into the trap of worrying about simple versus complex models from only an ecological perspective (i.e., what state dynamics submodel to use), and indeed, we have done this in the paragraphs above by emphasizing comparisons of surplus-production versus age-structured assessment models. We have acknowledged problems with observation submodels (assumptions about the data) by examining the effects of hyperstability in the relative-abundance indices and by making a few points about the difficulties of interpreting average-size/size-composition data and about effective sample sizes in multinomial composition sampling. And the examples certainly support the usual assumptions that an "optimum" model complexity depends on what data are available for analysis and that adding additional types of data (and corresponding complication in models to predict these types) can, in principle, help correct for bias problems caused by incorrect assumptions. But it is hard to escape the intuitive feeling that all we are doing in such examples is chasing assumptions with more bad assumptions, grounded largely in wishful thinking about both population dynamics and available fisheries data. This is not just an issue in relation to the analysis of fishery-dependent data; we should probably be equally suspicious about fishery-independent survey data. Such data commonly show interannual variances far higher than would be expected on the basis of variability among observations within surveys and may be prone to hyperstability problems in catch rates due to incorrect definitions of sample survey areas (frames) relative to changing fish distributions.

In a variety of recent assessment exercises, we have found it productive in terms of both data analysis and policy development to abandon the simpler state models (e.g., eq. 5.10–5.12) in favor of general age-structured models initialized using the leading parameter structure discussed in box 3.1 on page 56. In that structure, we must provide a collection of age schedules (vulnerability, body size, fecundity), an assumed form of mean stock-recruitment relationship, and two key leading parameters representing population scale (unfished biomass B_o) and recruitment performance at low stock size (compensation κ parameter), which play the roles of K and r in simpler models. Assuming that the age schedules can be estimated through an analysis of even fragmentary historical data on the fecundity and size/age compositions of catches (such data are almost always available these days), the age-structure model can still be used to predict relatively simple observations including the catch per effort and the mean size of fish harvested. With fixed vulnerability and growth schedules, such models behave essentially the same (in terms of estimation performance) as delay-difference models. But they have the added advantages of (1) options to explore the impacts of past and future changes in age-selectivity patterns even if such changes were not precisely measured; (2) the representation of smoothing effects on the recruitment of partial vulnerability patterns; and (3) options to examine the impacts of possible past changes in growth patterns, e.g., possible genetic impacts that selective removal of larger fish may have had early in the fishery's development (Conover and Munch 2002).

5.5 USING COMPOSITION INFORMATION

Most major fish stock assessments today are based on the analysis of age- and size-composition data. Such data have been painstakingly collected since at least the late 1970s to early 1980s, when age-structured assessment models first became widely available and showed great promise for improving our ability to reconstruct the historical impacts of fisheries. The basic idea in such models is to combine information on trends in relative abundance from fishery or survey catch rates, with information on changes in age and/or size composition. While some models attempt to compare observed versus predicted size compositions directly, in most cases we use estimated age compositions obtained by combining the sampling of ages with larger size-composition samples (using length-age keys to infer ages from sizes). Table 2 shows a typical catch-at-age data set for the main cod stock off Newfoundland (2J3KL or "Northern cod" stock, Baird et al. 1992).

   Such catch-at-age tables generally show four basic kinds of obvious "treatment effects": (1) age selectivity, with young fish usually appearing to be less vulnerable than old ones; (2) yearly exploitation, with higher overall catches in some years than others; (3) cumulative mortality, with older fish ultimately being less numerous than younger ones; and (4) recruitment or cohort, with some year-classes or cohorts (down diagonals of the table) being stronger than others. The published equations used for representing these effects often look very complex, with exponential terms and such for dealing with the overlap in the time of fishing and natural mortality losses over each year. We generally obtain about the same results with a very obvious and simple

TABLE 2: Catch numbers at age (in thousands) used for stock assessments on the Northern cod stock (2J3KL) off Newfoundland (data from Baird et al. 1992).

Year	2	3	4	5	6	7	8	9	Age 10	11	12	13	14	15	16	17	18	19	20	Sum
1962	301	8666	26194	64337	58163	47314	27521	20142	18036	10444	9468	7778	5785	4669	3888	3955	2161	232	403	319457
1963	1446	5746	27577	60234	118112	58996	29349	15520	11612	8248	4204	3942	2933	2928	1737	1263	1352	328	182	355709
1964	2872	19338	27603	57757	60681	100147	50865	20892	12264	8698	6352	4989	4036	2703	1456	1918	1154	501	312	384538
1965	85	5177	28709	46800	66946	64360	68176	33819	14913	6945	3729	3948	3730	2722	1859	575	971	183	226	353873
1966	819	14057	65992	93687	62812	59312	30423	23844	8762	4528	2280	1825	1186	967	806	416	279	486	178	372659
1967	790	15262	77873	100339	96759	54996	38691	17146	16084	5949	3367	2108	1529	685	424	193	107	72	211	432585
1968	288	6142	94291	205805	150541	83808	39443	23171	10984	5591	5249	1939	1334	818	610	127	89	83	26	630339
1969	59	4330	39626	100858	163228	107509	52661	19651	12370	6389	4479	3004	1557	622	567	319	100	46	99	517474
1970	6819	18104	60102	82357	101249	85696	29218	10857	3825	2000	1200	507	224	214	244	124	32	10	34	402816
1971	33	12876	71557	95384	98111	57865	25055	11732	4470	2223	1287	1140	720	355	474	124	128	148	78	383760
1972	236	6737	79809	116562	76196	55984	29553	11750	6393	2987	1660	1388	725	748	606	452	136	195	36	392153
1973	0	3963	40785	94844	59503	35464	27331	14153	7566	3815	2153	1173	450	278	309	85	27	38	8	291965
1974	473	3231	13201	34927	74403	60539	35687	18854	10492	5818	2934	1078	652	249	338	162	113	45	20	263216
1975	420	3968	14101	25370	34426	39105	36485	13421	7514	2315	1179	808	372	165	82	5	8	22	1	179767
1976	15	13767	33727	28049	20898	16811	16022	10931	4637	1462	631	292	251	100	50	40	64	30	20	147797
1977	108	7128	65510	40462	12107	5397	3396	2730	1381	532	296	149	75	42	21	20	14	2	6	139376
1978	0	1323	17556	39206	20319	7711	3078	1530	1083	437	219	105	62	40	21	7	8	2	7	92714
1979	0	1152	12361	37493	29202	10982	3460	1300	757	560	183	116	51	43	38	7	7	4	9	97725
1980	92	2554	12025	28814	30016	18017	4830	1217	520	232	229	56	65	37	13	10	14	4	10	98755
1981	0	2185	7172	13191	24800	22014	11848	3175	779	309	195	125	48	14	28	20	5	5	5	85918
1982	0	1702	31286	19003	14397	25435	16930	11936	1923	338	156	90	153	40	12	13	4	0	0	123418
1983	18	2585	13616	42602	19028	12044	14701	8934	6341	1018	248	90	41	29	11	9	6	2	3	121326
1984	3	782	14871	31760	38624	12503	7246	8910	4227	2536	451	146	48	41	30	7	7	4	3	122199
1985	0	650	14824	36614	33922	28006	7050	3836	5162	2905	1681	254	107	39	20	17	1	3	5	135096
1986	1	831	15219	44168	45869	26025	14722	3104	2000	1977	1101	574	116	29	18	11	9	2	2	155778
1987	42	2329	9217	32340	49061	28469	19505	5818	1346	676	873	391	200	37	22	3	1	4	0	150334
1988	25	2779	14651	20184	47917	45725	18608	9026	4337	774	422	366	223	100	32	5	10	5	5	165194
1989	8	1696	17639	21150	25212	38708	28499	8696	3640	1695	572	244	180	94	43	4	9	0	1	148090
1990	58	7693	40557	36410	22695	16390	17940	9156	2865	1084	478	103	98	36	25	8	7	1	0	153604
1991	35	3111	31654	53805	29553	9064	6164	4745	1696	641	250	88	39	21	9	3	2	2	0	140882

equation for changes in the number of fish of each age (a) each year (t), based on pretending that the harvest is taken in a short season at the start of the year:

$$N_{a+1,t+1} = s_a(N_{a,t} - C_{a,t}). \qquad (5.14)$$

Here, s_a is a (possibly age-dependent) survival rate of the age-a fish that are not caught, and $C_{a,t}$ is the catch. In the model state equations, $C_{a,t}$ can be treated either as a known removal each year (model "conditioned on catch") or as a predicted output using an equation of the form

$$C_{a,t} = v_{a,t}U_t N_{a,t}, \qquad (5.15)$$

where $v_{a,t}$ is a relative vulnerability of age-a fish in year t to the overall exploitation rate U_t for that year. When $C_{a,t}$ is predicted from U_t, we may in turn either try to estimate each U_t as an unknown parameter, or try to predict it from some measure of fishing effort or area swept by gear in year t. The s_a are almost never treated as variable over time unless many tagging data are available and, in most cases, are treated as age independent ($s_a = s = e^{-M}$, where M is an instantaneous natural mortality rate). It is not possible to estimate the s_a or M from catch-at-age models unless the exploitation rates U_t can be well predicted (with strong contrast in the data) from the fishing effort or area swept.

Patterson et al. (2001) provide an excellent review of case studies and alternative approaches that have been taken to reduce the number of parameters for a practical estimation using equations similar to equation 5.14, and for measuring uncertainty in the resulting estimates. From a statistical and logical perspective, these equations admit far more unknowns than measurements. If the catch-at-age table has data for A ages over T years, then it appears that we need to estimate $A \cdot T$ vulnerabilities $v_{a,t}$, T annual exploitation rates U_t, and $A + T - 1$ recruitments and initial abundances $N_{a,1}$ and $N_{1,t}$. Even when we can add considerable additional information in the form of age-specific relative abundances $Y_{a,t} = q_a N_{a,t}$ from surveys that have stable age-selection patterns, the parameter set still needs to be severely restricted to obtain estimates with meaningful (less than infinite) uncertainty. One popular way to simplify the parameter set has been to assume a "separation" of v and U effects, i.e., assuming $v_{a,t} = v_a U_t$; we warn against this assumption, since about the only cases for which it appears to have worked well are a collection of severely overfished stocks in the North Atlantic. For most fisheries, the $v_a U_t$ assumption is likely to result in misleading stock reconstructions due to temporal changes in the way fishers have targeted fish of different sizes and ages. Other options for reducing the number of parameters to be estimated, particularly the $v_{a,t}$, depend on how the state-dynamics equations are solved over age and time.

There are two apparently quite distinctive ways to solve or propagate the state equation 5.14 (fig. 5.6). In "stock synthesis" or "statistical catch-at-age (SCA)" models, we solve equation 5.14 forward over times and ages

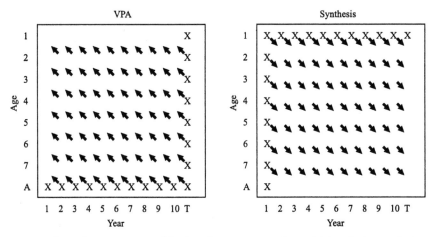

FIGURE 5.6: Assessment models for the reconstruction of historical population changes using catch-at-age data can be solved either backward in time (VPA case) or forward in time (Synthesis case). The flow of the calculations for each of these methods is shown by the arrows, and X's show unknown initial states that must be estimated somehow in the assessment procedure. Note that for the VPA case, we can often assume the X's representing the oldest fish every year to be zero.

just as equation 5.14 shows, either subtracting the observed catches as fixed removals or predicting them from the vulnerability and U_t parameters. In "virtual population analysis (VPA)" models, we propagate the equations backward in time by solving equation 5.14 for $N_{a,t}$, given $N_{a+1,t+1}$, as

$$N_{a,t} = N_{a+1,t+1}/s_a + C_{a,t}. \tag{5.16}$$

There are two main differences between these approaches: (1) error propagation, and (2) which $N_{a,t}$ need to be estimated as boundary-condition parameters (or perhaps predicted from a stock-recruitment relationship in the SCA case), as shown by X's in figure 5.6. Also, VPA cannot be used when there are many missing years in the catch-at-age data table. In the VPA case, errors in the estimation of $C_{a,t}$ are propagated with expanding effect backward in time (small errors at time t are successively divided by numbers less than 1.0, the survival rates, as the calculation proceeds backward over ages), whereas such errors decay over time under forward propagation. But balancing this possible pathology in error propagation, there are usually far fewer boundary $N_{a,t}$ values to be estimated with VPA, since numbers in the oldest age $N_{A,t}$ (where A is the oldest age) can usually be assumed to be zero. Further, except when the exploitation rate is very low (less than 0.1), the numbers in all ages in the last year ($N_{a,T}$) can often be modeled in the VPA setting as

$$N_{a,T} = C_{a,T}/(v_{a,T}U_T). \tag{5.17}$$

Especially if many of the $v_{a,T}$ can be assumed to equal 1.0 (older fish being equally vulnerable to the harvest rate U_T), the equation 5.17 "terminal harvest" method for setting $N_{a,T}$ can again substantially reduce the number of parameters to be estimated (to U_T and a subset of $v_{a,T}$). Note that when equation 5.17 is used in VPA, only the year-T terminal vulnerabilities $v_{a,T}$ need be specified as unknown parameters, since all other $v_{a,t}$ ($t < T$) are implied by the backcalculated exploitation rates $C_{a,t}/N_{a,t}$.

Though the origins of SCA methods can be traced to Doubleday (1976), by historical accident much of the recent development of these methods has been by scientists working on the Pacific rim of North America (e.g., Deriso et al. 1985; Fournier and Archibald 1982; Methot 1990; Schnute 1994). In contrast, much of the development of VPA has been by scientists working around the Atlantic, particularly the widely used software ADAPT (Gavaris 1988). It says much about the culture of stock-assessment science, and perhaps about whether scientists really follow the literature, that a recent NRC stock-assessment panel (NRC 1998) found that almost all assessments for Atlantic fisheries were being done with VPA, while almost all Pacific assessments were being done with SCA models. There was much consternation when test-simulated data sets were distributed widely to stock-assessment scientists on both coasts by the NRC panel, and ADAPT was found to sometimes perform poorly compared to SCA methods. This came on the heels of retrospective studies showing that precursors of ADAPT had helped contribute to the Newfoundland cod collapse through gross overestimates of stock size and recruitment (Walters and Maguire 1996). The net result of these findings has been something of an intellectual stampede from VPA- to SCA-based assessments and, in particular, to the use of the AD Model Builder software (Otter Research 1994) that provides extremely efficient and reliable nonlinear search procedures for fitting nonlinear statistical models with large numbers of nuisance parameters (e.g., vulnerabilities $v_{a,t}$).

Unfortunately, there has been a recent tendency to condemn VPA methods in general, on grounds ranging from error propagation to the particular bias problems of ADAPT, without regard to the obvious advantages of using a method for which relatively few parameters need be estimated. Indeed, some workers even appear to take pride in announcing that they are fitting an SCA with more than 100 parameters. They seem to believe either that admitting more unknowns and uncertainties is a virtue in itself (honesty) or that having more parameters is sure to result in "better" estimates of important leading parameters such as the mean recruitment rate and recruitment trend (by accounting for more sources of variation in the data). There has been much retrospective examination of cases in which the much simpler VPA approach has been found to give very misleading stock reconstructions and estimates of recent abundance, i.e., the Newfoundland cod stock. In these cases, the main problem has not been error propagation or some other intrinsic bias at all but, rather, the two specific difficulties mentioned at the start of this chapter about observation-submodel assumptions:

1. changes over time in the vulnerabilities at age, $v_{a,t}$, leading to poor estimates of the terminal vulnerabilities $v_{a,T}$ used in the terminal numbers calculation equation 5.16; and
2. fitting (or "tuning") the models using relative-abundance data that is not proportional to stock size, in particular the use of hyperstable fisheries CPUE data.

(Even VPA does not escape the need to use relative-abundance data in estimations; without information on recent abundance trends, there is no way to decide a best estimate for the terminal U_T in eq. 5.16). Exactly the same problems have plagued SCA in some cases, e.g., the Pacific halibut stock (Parma and Deriso 1990a) in which there has been a large change in the $v_{a,t}$ associated with changes in body growth patterns. Both SCA and VPA are prone to give misleading results in any case in which the $v_{a,t}$ are changing rapidly in the most recent few years of data, e.g., during stock declines when fishers often start to target smaller fish; in these cases, there is no way in principle of using only the catch-at-age data to decide whether catches of younger fish have been sustained because of relatively high recruitments or because of relatively high $v_{a,t}$.

Neither SCA nor VPA methods have lived up to our early expectations about how valuable catch-at-age data would be in providing useful information for management. Consider for a moment the three most important "products" that we hoped they would provide: (1) estimates of the historical exploitation rates $U_{a,t} = C_{a,t}/N_{a,t}$, to evaluate trends in fishery impact and overfishing risk; (2) estimates of trends in recruitment, both to help detect recruitment overfishing and to provide information for scientific studies of the causes of recruitment variation; and (3) estimates of current stock size and age composition, to use in implementing harvest strategy rules. Both approaches are fine for meeting objectives (1) and (2), provided we look only at the results for $t \ll T$ (and in SCA, do not assume away much of the $U_{a,t}$ variation by assuming a constant $v_{a,t}$ over time). Both approaches are dangerously misleading for objective (3) when the $v_{a,t}$ may be changing rapidly and when we cannot be certain that available age-specific relative-abundance indices $Y_{at} = q_{a,t}N_{a,t}^{\beta}$ are not in fact hyperstable (i.e., when we cannot be sure that $\beta = 1$ and $q_{a,t} = q_a$ independent of t). But these are just the cases in which good management advice is most important!

Even for the analysis of long-term trend patterns, SCA methods have not turned out to be such a wonderful generalization of early catch-curve techniques (plots of catch over age to obtain total mortality estimates) as we had hoped. They do not really lead to a clear separation of the four types of "treatment effects" mentioned at the start of this section. Changes in catches over age at any time still exhibit the confounding effects of changes in selectivity, past recruitment trend, and past effects of fishing on mortality rates. Changes in catches over time and age (along the dynamic catch curve for each cohort) still have the confounding of effects of changing selectivity (with both age and time) and of changes in the exploitation rate, despite the subjecting of these observations to the same recruitment "treatment." In some cases, declines in vulnerability with age of older fish (which we usually assume away by treating all older fish as equally vulnerable) have likely masked even the crude trend signals of overall changes in the mortality rate due to fishing. About the only thing that SCA methods sometimes do well is to recover information on strong versus weak recruitments in successive years, something that VPA sometimes misses due to the "smearing" of fish across cohorts caused by aging errors in the $C_{a,t}$ assessment.

So here is our general advice to assessment workers who are worried about whether to use VPA or SCA methods. Use VPA whenever there are reasonably large catch-at-age samples for every year and in places where historical exploitation rates have been relatively high (at least 0.1), so as to take advantage of its compact parameterization, but pay very careful attention to the estimation of the terminal vulnerabilities $v_{a,T}$ and the terminal exploitation rate U_T. Use SCA methods only when there are large gaps in the data, when the $C_{a,t}$ data are highly suspect due to inadequate sample sizes, and/or when recent exploitation rates are suspected to have been very low. When in doubt about whether to trust the $C_{a,t}$ data or the U_t, try both approaches, with particular attention to whether the VPA results suggest changes in vulnerabilities $\bar{v}_{a,t} = (C_{a,t}/N_{a,t})/\overline{U}_t$ (where $\overline{U}_t$ is the mean of $C_{a,t}/N_{a,t}$ over "fully vulnerable" ages a) that it has been necessary to assume away in order to obtain unique maximum likelihood estimates with SCA. For both methods, never pretend that the estimates for recent years ($t > T - A/2$) are accurate or reliable, despite what any statistical measures of goodness of fit or prediction-error variance (AIC, BIC) may suggest.

Some assessment scientists appear to believe that all we need do to make the SCA/VPA methods work well is to invest in collecting good survey data on age-specific relative abundances. This is a very naïve belief that has been used to justify some very expensive but still suspect survey programs (variances are higher than they should be, with possible hyperstability in the survey catch rates). Such advice needs very careful rethinking, to consider whether we should be going after the $v_{a,T}$, U_T problem with other more direct monitoring methods such as tagging programs.

A variety of models now attempt to deal directly with (to predict observations on) length-composition data, hence avoiding possible problems with the estimation of catch-at-age composition from the length data (Fournier et al. 1998; Fu and Quinn II 2000; Sullivan et al. 1990). New methods for analyzing growth from tagging experiments promise better estimates of growth curves (box 5.2). Unfortunately, there is a catastrophic flaw in these models. None of them explicitly represents the cumulative effect of fishing

BOX 5.2

ESTIMATION OF BODY GROWTH PATTERNS FROM AGE-SIZE AND TAGGING
DATA

Assessments of body growth are critical in many aspects of fisheries assessment, from the interpretation of length distribution data to evaluation of effects of size limits/size selection on yield (yield per recruit analysis) to the prediction of natural mortality rates when such rates cannot be measured directly. Typically, growth in body length $l(a)$ with age a is described by models of the form $l(a) = l_\infty f(a)$, where l_∞ is the asymptotic (maximum) body length and $f(a)$ is a monotonic increasing function of a, approaching 1.0 as a increases. The most common model is the Von Bertalanffy, for which $f(a) = 1 - e^{-K(a-a_o)}$, where the metabolic parameter K is a useful predictor of the natural mortality rate and a_o represents an apparent age at the time of hatching. Other suggested forms for $f(a)$ mainly involve changes in K with age for young fish (e.g., Porch et al. 2002). Typically, we expect to see considerable variation among individual fish in the asymptotic length parameter l_∞ but relatively little variation in the metabolic parameter K.

The average growth parameters l_∞ and K for fish populations are typically estimated by assuming that representative samples of fish of each of a range of ages has been collected, then either fitting the growth function to these samples directly or to data on changes in the length of fish between the times of tagging and later recapture. Unfortunately, there are two potentially serious problems with the basic assumption of representative sampling: (1) in fact, fish are first collected, then are aged or tagged for later recapture, so that the proportions of fish at age a are not set beforehand "by design"; (2) most fish collecting methods are highly size selective, so that most often the fish of younger ages that do appear in samples are the fastest-growing (highest individual l_∞) individuals. Problem (2) is serious: it causes an upward bias in estimates of K and a downward bias in estimates of the mean l_∞. Recent work, particularly by Laslett et al. (2002), has built on earlier developments by Wang et al. (1995), Wang (1998), and others to show how to deal with the age-composition-sampling effects along with the individual variation in l_∞ in maximum likelihood estimations. However, there has as yet been no general recommendation about how to cope with biases caused by size-selective sampling. Simulation tests with the likelihood function suggested by Laslett et al. show that it can provide nearly unbiased estimates of K from tag-recapture data even when the size at tagging for younger fish has been quite size-selective, provided K is not treated as variable with age.

One solution to the size-selectivity problem when only size-at-age data are available is to estimate a size-selection function, along with relative abundances at age, while estimating the growth parameters. Suppose the growth

(*Continued*)

(BOX 5.2 continued)

data consist of a matrix of observed numbers of fish by discrete length l and age $n_{l,a}$, and that these data represent a random sample from the vulnerable fish $V_{l,a}$ present in the population. Suppose further that $V_{l,a}$ is predicted to be $V_{l,a} = v(l)N_a W(l|a)$, where $v(l)$ is a simple (few parameters) length-vulnerability function, N_a is the relative total number of age-a animals in the population, and $W(l|a)$ is a normal probability density function (or integral over the width of the discrete length measurement interval) with mean $l(a)$ given by the Von Bertalanffy or other growth function (and typically standard deviation proportional to the mean length, implying only one parameter for describing the individual variation in growth). Under these assumptions, the likelihood for the multinomial sample of observed lengths can be written as

$$logL = \sum_l \sum_a n_{l,a} \ln[V_{l,a}/V_T]$$

where V_T is the total vulnerable population ($V_T = \sum_l \sum_a V_{l,a}$). Parameters are then estimated by maximizing $logL$ over the parameters of $v(l)$ (vulnerability), $l(a)$ and $W(l|a)$ (growth), and relative numbers at age N_a. We have done extensive simulation tests of this approach and find it surprisingly good at finding the underlying growth parameters even when $v(l)$ implies the sampling of only larger fish. However, here is an important caveat to these findings: the estimation can be fooled if an incorrect shape is assumed for the $v(l)$ function, e.g., it is assumed asymptotic when it is, in fact, dome-shaped.

There are at least three ways do deal with the abundance-at-age "nuisance" parameters N_a. The first is to assume a simple functional form for these, e.g., $N_a = e^{-Za}$ and include parameters of this form (e.g., Z) in the estimation. The second is to include the $logL$ likelihood terms for each $n_{l,a}$ sample in the overall likelihood function for some stock-assessment model that includes the estimation of numbers at age (a stock synthesis or VPA model), so that the growth and the size-vulnerability parameters are estimated along with recruitments, etc. The third is to treat each N_a as an arbitrary nuisance parameter and eliminate these nuisance parameters by evaluating $logL$ at their conditional maximum likelihood estimates whenever estimating the vulnerability and growth parameters. It is easily seen that by differentiating $logL$ with respect to N_a and setting the derivative to zero, the maximum likelihood estimates of N_a must satisfy $N_a = n_a V_T/(v_a n)$, where n_a is the total number of age-a fish in the sample, n is the total number of fish aged, and $v_a = \sum_l v(l) W(l|a)$ is the mean vulnerability of age-a fish. Substituting this condition into $logL$ (and dropping the terms involving sample sizes that do not affect the likelihood comparisons) results in the "reduced" likelihood function

(Continued)

(BOX 5.2 *continued*)

$$logL' = \sum_l \sum_a n_{l,a} \, ln[v(l) \, W(l|a)/v_a]$$

which depends only on the vulnerability and growth parameters. Simulation tests show that this reduced likelihood gives considerably more variable parameter estimates than can be obtained when it can safely be assumed that the N_a vary in some structured way with a. ∎

on the size distributions of animals of different ages. The sampled length frequency at age is assumed to be distributed around some mean (possibly density-dependent) growth curve, but the growth curve is not represented as dependent on the fishing mortality rate. In reality, individual fish tend to be persistently either fast or slow growing, and the faster-growing individuals usually become vulnerable to fishing at earlier ages (i.e., each age cohort consists of growth type "subcohorts" with different growth curves and age-vulnerability patterns). The cumulative removal of these individuals results in a decrease in the mean size and growth rate of the remaining individuals. Thus the apparent growth curve, and length-frequency distributions at age, are likely to be strongly dependent on the fishing mortality rate and size-selection pattern (Sinclair et al. 2002a, 2002b), whether or not there is genetic selection over long periods for slow-growth types as reported by Conover and Munch (2002). Fishery-induced changes in the growth curves can, of course, be monitored on a year-to-year basis to avoid a bias in predicting length-frequency patterns (or an alternative approach can be used to predict length-composition data, see box 5.3 for details), but with the sampling effort required to do this one might as well use age-structured methods in the first place.

CHAPTER 5

BOX 5.3
MODELING CUMULATIVE EFFECTS OF INDIVIDUAL VARIATION IN GROWTH

When we are trying to interpret historical size-distribution changes or predict impacts of size-selective harvesting, it can be misleading to assume that the mean and variance of size at age are independent of mortality rate. Individual differences in growth rates tend to be persistent, so, e.g., if growth is represented by the Von Bertalanffy equation $l_a = l_\infty(1 - e^{-Ka})$, we generally expect individuals to exhibit different maximum body lengths l_∞ but a similar metabolic parameter K (Wang et al. 1995; Laslett et al. 2002). This implies that if larger individuals are subject to higher mortality rates, the mean length at age of surviving fish will shift downward as the fishing mortality rate increases, and the distribution of lengths at age may not remain normal.

To simulate such cumulative effects, we somehow have to keep track of changes in both the size and growth patterns of surviving animals, over age and time. Moving to a pure length-transition probability model does not solve the problem, and moving to a full individual-based model (IBM) is computationally very costly. A simple accounting tactic that avoids most of the cost of an IBM approach is to divide each cohort or year class into a set of growth-type groups or subcohorts g, where the subcohort index g varies from, say, $g = -10, -9, \ldots, 0, \ldots, 9, 10$. Then for each of these subcohorts, construct a cohort-specific length-at-age table $l_{g,a}$, from a model like $l_{g,a} = (l_\infty + \Delta g)(1 - e^{-Ka})$, where Δ represents a length bin or interval width for the individual variation in l_∞. Use a length-vulnerability function $v(l, t)$ to calculate a subcohort-and-age table of vulnerabilities to harvest, so that in model year t the fish of subcohort g and age a are assigned the exploitation rate $u_{g,a,t} = v_{g,a,t}u_t$, where u_t is an overall ("fully vulnerable") rate for the year. Also construct a table of weights and a length-weight relationship, and use these to predict a fecundity table $f_{g,a,t}$ so that the total egg production ϵ_t can be predicted each year by summing $f_{g,a,t}$ times the subcohort abundances $N_{g,a,t}$.

This approach simplifies the accounting of an individually based model by representing the stock as a collection of different populations that differ only in their asymptotic length L_∞. Another way to think of it is that each of the growth-type groups comes from a family, or distribution, of asymptotic sizes, and size-selective fishing mortality is explicitly represented for each growth-type group g. Then the assessment model estimates the distribution of asymptotic lengths in the population (assuming that variation in length increases with age), rather than estimating standard deviations in length for each age class.

(Continued)

(BOX 5.3 continued)

Time simulations then proceed just as in age-structured models, but with each annual recruitment $R(\epsilon_t)$ distributed initially over the subcohorts and all survival calculations carried out over all subcohorts g as well as ages a:

$$N_{g,1,t+1} = R(\epsilon_t)p(g) \quad \text{all } g$$

$$N_{g,a+1,t+1} = s_{g,a,t}(1 - v_{g,a,t}u_t)N_{g,a,t} \quad \text{all } g,a$$

Here, $p(g)$ is a vector of proportions of the total recruitment assigned to subcohorts, with values most likely just proportional to a normal probability density. An important issue is whether to treat the annual natural survival rates $s_{g,a}$ as dependent on subcohort and age and, in particular, as lower for fish with higher growth rates ("Lee's phenomenon"); not including such variation is equivalent to treating growth rate as having zero heritability, since if the $s_{g,a}$ are independent of g, the contributions of fast growing and early maturing fish to ϵ_t will be differentially high.

Note that the useful equilibrium calculations for dynamic-model initialization, yield, etc. involving Botsford's incidence functions over age (box 3.1) apply equally well when there is a subcohort structure in the accounting. Simply include sums over the g index in all incidence function (ϕ) calculations. ∎

5.6 DEALING WITH PARAMETERS THAT AREN'T

The things we call "parameters" in population-dynamics models are not, in fact, fixed physiological or physical properties of organisms but, rather, are the complex resultants of myriad fine-scale processes. For example, the slope of a mean stock-recruitment relationship (chapter 7) is a product of the short-time survival rates of juvenile fish through various early life-history stages or stanzas. The interactions that result in each survival rate involve habitat/predation/competition factors that may have a "random" variation but are also likely to exhibit persistent change (nonstationarity, Walters 1987) as well. Likewise, even scientific observation parameters including survey q and v_a are the results of a complex interaction between survey fishing gear and fish, subject to persistent changes (even if the gear is held constant) because of changes in size, behavior, and distribution of fish relative to the survey area.

In stock assessment we have tended to assume away persistent changes in order to get any answer at all (or have failed even to recognize the possibility of them). We have, instead, spent much time agonizing about how to measure and cope with the essentially trivial variation due to the random,

time-independent statistical effects caused by sampling and by singular ecological events such as the occasional production of strong year classes. This
is not because it is wise or even necessary to be concerned about simple statistical variation but, rather, because it is what we know how to do. Further,
assessment scientists have wasted much time in trying to counter mindless
criticisms of the form "things are so complicated and there is so much variation out there that your models cannot possibly work," coming mainly from
people who barely passed introductory biometrics courses and hence have
no clue about how to deal with variation of any kind. Of course, pure statistical noise can overwhelm any assessment technique, but that is not generally
what happens. If you read over the warnings earlier in this chapter about
things that can go wrong with assessment methods, you will see that most
of these things involve persistent changes, i.e., parameters that aren't.

There are at least two assessment tactics for trying to find out whether the
available data are consistent with the hypothesis that particular parameters
are not (i.e., that these parameters have changed over time in ways that would
result in misleading assessments). The simplest tactic is to assume some
functional form for the time dependence, then try to estimate the parameters
of this form. A more complex approach is to treat the parameter(s) as
state variables with incremental, random change-per-time steps in state-space
likelihood formulations, then try to track these changes using state-space
filtering and smoothing operations (Gelb 1974).

Plausible functional forms to assume in estimating parameter changes over
time include (1) piece-wise constant, (2) polynomial (including spline functions), and (3) Fourier (sum of sine curves). The easiest of these to implement
in the general estimation framework of figure 5.1 is the piece-wise constant
option: divide the historical data period into reasonable blocks, and generate
the state/observation predictions with different q's, r's, or whatever for each
block. When change is likely to have been smooth over time, e.g., over a
period of gradually changing gear or habitat alteration, an easily interpreted
way to represent the main components of change via a polynomial function
is to use the form $P_t = P_o + a_1(t - t_o) + a_2(t - t_o)^2 + a_3(t - t_o)^3 + \ldots$, where
P_o is the parameter value at some base reference time t_o (typically $T/2$),
and the parameters a_1, a_2, etc. represent linear, quadratic, and higher-order
components of temporal variation. Similar patterns interpretable as "low
frequency" versus "high frequency" variation can be achieved with Fourier
series. Since all of these functions are purely descriptive (with no particular
biological or physical meaning or derivation), it is really a matter of convenience for the analyst about which is easiest to interpret and explain. It is
not uncommon to see various functional approaches used in statistical catch-
at-age (SCA) models, particularly to represent known temporal changes in
the vulnerability patterns ($v_{a,t}$) due to gear changes (both survey and fishery)
and regulations (e.g., Hampton and Gunn 1998).

Peterman et al. (2000) demonstrate using simulated data that state-space
filtering methods may be a very useful way to track (recover information

about past) climate-induced changes in recruitment parameters. Several workers have suggested using state-space methods to track possible parameter changes in statistical catch-at-age models (Sullivan 1992; Schnute 1994; Millar and Meyer 2000).

For both the functional form and state-space approaches, it is generally necessary to make some a priori assumption about relatively how much of the variance in observations is due to possible parameter changes as opposed to observation errors and independent (uncorrelated) process errors. This means setting more or less restrictive prior-probability distributions on parameters of functional forms, and variances for random walk processes in state-space models. We cannot, of course, know ahead of time how restrictive to be. The best practical strategy may be to follow a "relaxation" approach, starting out with very restrictive assumptions and relaxing these to move more and more of the explained variation into an estimated change over time in the parameter values. This strategy will in effect "sketch" out a set of alternative hypotheses about how to interpret the historical data, and these hypotheses are quite likely to have very different policy consequences. In particular, admission of the possibility of nonstationarity in the stock-recruitment parameters means admission that any observed stock declines might have been due to two very different mechanisms: environmental changes/regime shifts or recruitment overfishing. So scientists who attempt to be honest about alternative ways to explain the available data should be aware that their results will contribute to the "Thompson-Burkenroad" (see chapter 7) or "environment vs. fishing" debates that surround almost every major fishery decline. Perhaps some honest modeling will help show that such debates should be fundamentally changed, from arguments about what happened to arguments about how to manage the risks and costs implied by not being able to decide which hypothesis is correct on the basis of the historical data. It is pointless to argue about what the future will bring in such situations, and the wise thing to argue about is what policy to follow given that either outcome is equally likely (equally consistent with available historical data).

CHAPTER 6

Foraging Arena Theory (I)

IMPLICATIONS FOR RECRUITMENT PATTERNS

RESEARCH IN THE FIELD of behavioral ecology over the past three decades has focused on how natural selection has shaped animal behaviors to deal with two general problems: getting enough food to grow and reproduce, and avoiding predation while doing so (Krebs and Davies 1981). For most fish species, only a tiny proportion of individuals actually manages to solve these problems; most die (mainly by predation) when they are quite small.

There is a horrific linkage between getting food and being food, and this creates a severe trade-off relationship. In order to feed, most fishes need to engage in foraging behaviors that involve movement to areas of higher prey density. But these behaviors expose individuals to predation risk. Sitting quietly under a rock, or in water too shallow for predators, or in the middle of a school, or in dark abyssal waters (figure 6.1) are great ways to avoid being eaten, but they are not effective ways to get enough to eat. So most fish, especially juveniles, have to partition their most precious resource, which is time, between resting/hiding and dispersing/feeding. The predation risk per time spent feeding is often very high, so small individuals dare spend only a few minutes or hours feeding each day. Because resting/hiding places are not randomly distributed over the aquatic environment, a restricted feeding time also means that foraging activities tend to be concentrated in space. Most feeding occurs in relatively restricted "foraging arenas" like the ones that the senior author's son William was easily able to identify when he was only nine years old (figure 6.1).

To a field biologist, the spatial and temporal organization of behavior shown in figure 6.1 has some immediate and largely annoying consequences. It is a key driving force in making fish distributions highly patchy, adding much to the difficulties and costs of finding fish for biological sampling and census work. It concentrates fish in ways that fishers can exploit to make fisheries more efficient and dangerous. It causes much confusion for biologists who expect fish distributions to be shaped largely by simple physical "habitat factors" such as current velocity that are predictably related to physiological characteristics like swimming speed. It causes small-scale patchiness in prey distributions, with prey concentrations depressed in foraging arena sites, making it hard to study "food availability" just by measuring the densities of food organisms at spatial scales and locations that are convenient for biologists to sample. It means that much of the time, many fish (the ones that are hiding) haven't eaten anything in a while and hence have

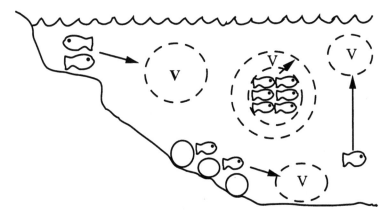

FIGURE 6.1: William Walters, age 9, drew this figure to show his father some of the ways that juvenile fish might moderate predation risk by hiding and moving (redrawn from Walters and Juanes 1993). Foraging arena theory has been developed largely by pursuing the implications for trophic interactions of the spatial organization shown in this figure.

empty stomachs, making it hard for us to sample their stomach contents for diet composition and estimate feeding rates. And when we do put in the extra effort needed to characterize diets and food populations, the spatial organization of the fish's interactions makes it very difficult for us to link these data to make useful predictions of how many fish the food supply can support. That is, the food supply that the fish actually see can be very different from the apparent supply that we can see with our sampling devices.

The concentration of foraging behavior and predation risk in foraging-arena arrangements also has severe consequences for those of us who insist upon trying to construct models for predicting how fish populations should respond to changes in trophic conditions. Until very recently, most of our models for predicting "trophic linkages" (feeding and predation-mortality rates) were built upon simple assumptions borrowed from chemistry and physics about spatial distributions and encounter patterns among organisms. In particular, we borrowed the "mass-action" assumption of well-stirred reactions in chemistry, because encounter rates (reaction rates) between chemical "species" are predictably related to (just proportional to) the mean densities of the interacting species. This leads to predictions such as "doubling the number of predators should lead to a doubling in the predation component of the natural mortality rate" (M being proportional to predator density), or "doubling the food supply should result in a doubling of the feeding rates, unless those rates become limited by handling time or satiation effects." We have never been able to find much empirical justification for such predictions (except under very carefully controlled laboratory conditions allowing us to stir things up like chemists do), and they lead to models that cannot explain most of the complexity, diversity, and apparent limits to production that we see in the field.

Imagine taking an introductory physics course at a school where the lab-oratories have been organized by a character who enjoys torturing students, and going to the lab on Boyle's law. This is where you would ordinarily heat up a closed chamber with some gas molecules in it, to show that molecules hit the chamber walls proportionally more often (exert more pressure) as the temperature increases. But suppose the character makes you use long, thin metal chambers stuffed with sand and other debris, makes you heat up only one end of each chamber, places the pressure-sensing device at the other end, then makes you measure the transients in pressure over time as the heating proceeds. Then the character demands that you compare not your final (hopefully equilibrium) results with those obtained by other stu-dents but, rather, the time-transient patterns. To make matters worse, the character asks you to make inferences about what kind and how much junk got stuffed in the chambers before you arrived at the lab. For this exer-cise it obviously will not do you a whole lot of good to know that Boyle's Law has still applied at fine (interstitial) spatial scales within each of the chambers over time. You are going to have to deal with a whole mess of other complicating factors (like diffusion rates) in order to explain how your macro-system has behaved over time. This is just the sort of problem that we face when we try to understand data from field fisheries situations, especially data on the dynamics of juvenile survival that lead to the net number of recruits that enter harvestable populations over time. We can try to take the recruitment "chambers" apart and try to characterize all of the junk and processes that have gone on in them, but with limited time and human resources we must also try to find simpler ways to make useful predictions about the net effects of all that fine-scale complexity. Finding useful simplifications is the aim of the foraging-arena theory introduced in this chapter.

The imagery of figure 6.1 invites us to partition spatial patterns into three main components: hiding places, foraging arenas where feeding and preda-tion rates are concentrated, and larger "refuge" areas where food organisms are protected from depletion by the fish because of the high predation risk. This is obviously a gross oversimplification; there are small and large hiding places, foraging and dispersal activities are not restricted to specific, well-defined volumes or areas, and factors like body size have a big effect on how individuals can exploit hiding and feeding opportunities. But we can still use the imagery to build useful predictions about how real interaction patterns are likely to differ from predictions based on even more simplis-tic assumptions about mass-action, random encounters between organisms, just as chemists have used the imagery of the Bohr atom to build useful predictions about chemical reactions.

Foraging arena imagery can be used to help understand two types of larger-scale patterns that "emerge" from the fine-scale dynamics of feeding and risk taking:

1. the shapes of stock-recruitment relationships, more specifically why most fish populations show recruitment relationships that are flat across a wide range of the parental stock sizes (i.e., recruitment is independent of the parental stock size);
2. the forms of relationships between predator and prey abundances, predator feeding rates, and predation-mortality rates on the prey, which most often appear to involve less variation in the prey mortality rate than would be expected on the basis of simple "top-down" control (predation-mortality rates seldom appear to be just proportional to predator abundances).

This chapter discusses the first of these patterns, and chapter 10 examines the second one in the context of building more useful models for food-web interactions and multispecies dynamics.

The discussion begins by reviewing a discovery by Beverton and Holt (1957) of a very simple functional model that often appears to describe nicely the mean relationship between recruitment and the number of parental spawners or eggs contributing to that recruitment. Next we examine an alternative model that explains recruitment limits as arising from the availability of suitable rearing sites (habitat limits rather than trophic limits). Then we show that the imagery of figure 6.1, when combined with some simple ideas about how most fish must grow a lot in order eventually to reproduce themselves, leads directly to the functional model proposed by Beverton and Holt. That is, we show how their model arises almost inexorably from the fine-scale dynamics of feeding and being fed upon. Along the way, we discuss some generalizations of the idea, in particular to situations in which (1) foraging arenas are fine-scale entities (each arena being only large enough to provide a feeding opportunity for one or a few individual fish) so that predation risk is incurred mainly in dispersing among such arenas; (2) the predation risk during foraging is variable over time and is caused mainly by conspecifics (cannibalism, leading to another form of recruitment curve proposed by Ricker); and (3) habitat-component sizes (refuges, arenas) can change dramatically over time due to "physical forcing" changes. For more details about the material reviewed in this chapter, see Walters and Juanes (1993) and Walters and Korman (1999).

A key reason for delving into the material presented in this chapter is to clear the air about whether we should expect to see a predictable relationship when we examine data on spawning-stock abundance and subsequent recruitment. Whether such relationships exist is obviously an absolutely central, overriding management issue. If there is assumed to be no such relationship, so that recruitment is truly independent of parental abundance, then there would be no real reason for advocating conservative harvest strategies to protect spawning stocks. That is, there would be no real reason to worry about recruitment overfishing. Advocates of that peculiar position must resort to arguments about how fish have millions of eggs, so parental abundance could in principle be nearly irrelevant to predictions of recruitment. These advocates seem to forget that recruitment is determined by a product, of fecundity times survival rate, and that recruitment can be independent of egg deposition only if survival rates improve dramatically when egg deposition is low. That is, it is the change in survival rate that matters when fishing reduces adult abundance, not adult fecundity per se. What this chapter argues is that there are important evolutionary reasons for expecting not just some relationship between abundance and survival rate but, in fact, quite particular forms of relationships, even when there is much variation in recruitment due to physical and biological "habitat factors."

6.1 Beverton-Holt Model for Stock-Recruitment

Beverton and Holt (1957) examined early data on the relationship between parental spawning stock and subsequent recruitment, and data on the early mortality rates of juvenile fish. They noticed that the spawning stock is generally a very poor predictor of recruitment (recruitment being independent of parental abundance) except at relatively low parental stock sizes, and they correctly inferred that this must mean that juvenile mortality rates must in general be strongly density dependent. This is a key point that still seems to confuse a lot of fisheries students and even experienced scientists. In order that total recruitment not increase when the spawning stock is larger and more potential recruits (eggs) are being produced, the average survival rate from egg to recruitment has to decrease with increasing spawning stock size. (If there were no change in the average proportion surviving, the increased egg numbers would necessarily, when multiplied by this proportion, give more surviving recruits.) Most ecologists define density dependence as a change in per-capita birth and survival rates with changes in abundance, so we would say that the recruitment process is strongly density dependent (there is a decrease in juvenile survival with increasing abundance) despite the fact that the net resultant recruitment is independent of parental abundance.

Beverton and Holt then developed a very clever argument and model, as follows. First, they supposed that we follow the number N_t of juvenile fish over time (or age) for a single cohort or year-class, from their initial number of eggs or larvae N_o to an eventual number of recruits $N_T = R$. That is, R is the final number reaching some *arbitrary* age T that we define as the "age at recruitment" for an assessment measurement or the age when the fish become vulnerable to fishing. We can write the die-off process for these fish by using a simple rate equation,

$$dN_t/dt = -M_t N_t \qquad (6.1)$$

where M_t is the instantaneous death rate at any moment t during the development of the cohort.

Second, based on bits of empirical data and general arguments about the likely effects of competition and predation, they argued that M_t might be linearly related to juvenile density at any moment. That is, they asked what the implications would be if $M_t = M_o + M_1 N_t$, where M_o and M_1 are parameters for the linear relationship (M_o = base mortality rate, M_1 = the density effect on the rate). Putting this argument into the basic die-off relationship results in

$$dN_t/d_t = -M_o N_t - M_1 N_t^2 \qquad (6.2)$$

Note that there is no implication here that the density-dependence parameter M_1 (or the product $M_1 N_t$) is "large" compared to M_o, i.e., that the density-dependent effects are in any simple or additive sense more "important" than the density-independent ones represented by M_o. For many fish, M_o for at the least the first year of life must be quite large, on the order of 5–10, implying a very low survival rate even when there is little competition. Further, even small changes in M_o can cause big changes in the overall survival rate and recruitment. For example, if M_o is around 5, then even a "small" change to $M_o = 6$ in some year would result in the total survival dropping by about 63% (year-class being only a third of the average in abundance at age 1).

To see the implications of equation 6.2, Beverton and Holt then applied a bit of basic calculus to integrate the rate equation so as to predict net numbers $N_T = R$ of recruits to age T. You can do this by separating variables or by using symbolic math programs such as ©Maple or ©Mathematica. The result, after rearranging the solution into a convenient format for prediction, is

$$R = N_T = N_o e^{-M_o T} / \left[1 + (M_1/M_o)(1 - e^{-M_o T}) N_o \right] \qquad (6.3)$$

We can write this model in a more readable form as just

$$R = s_{max} N_o / (1 + k N_o) \qquad (6.4)$$

by defining the short-hand parameters

$$s_{max} = e^{-M_o T}, \text{ the maximum average survival rate}$$
$$\text{absent density effects, and}$$

$$k = (M_1/M_o)(1 - e^{-M_o T}), \text{ a "scaling" parameter}$$
$$\text{representing density effects.}$$

Equation 6.3 or 6.4 says just that we expect the average recruitment to be a maximum survival rate times the initial juvenile or egg number ($s_{max} N_o$), with the survival rate modified downward when N_o is large by the denominator factor $1 + k N_o$.

Now the Beverton-Holt proposal in equation 6.2 gets very interesting. If you plot the form of the predicted relationship between recruitment and the

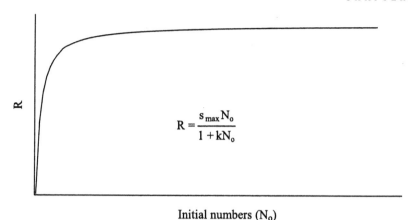

$$R = \frac{s_{max} N_o}{1 + k N_o}$$

Initial numbers (N_o)

FIGURE 6.2: The relationship between recruitment (R) and initial numbers (N_o) using equation 6.4.

initial juvenile numbers using equation 6.4, you might expect the negative quadratic rate effect from equation 6.2 to cause a dome-shaped prediction. Instead, the prediction is always (for all positive N_o, M_o, M_1 values) asymptotic (fig. 6.2) and looks very much like the typical pattern that we can now examine for many, many fish populations (fig. 6.3) thanks to the efforts of Ransom Myers and his colleagues to collate data from around the world (Myers et al. 1999).[1]

The story does not quite end at this point. There is at least one thing wrong with the basic rate equation (eq. 6.2) used to generate equations 6.3 and 6.4, namely, that most juvenile fish undergo a complex life history ontogeny, utilizing a set of different habitats and being subject to quite different mortality agents as they grow. We can think of this as a series of stages or stanzas (Fuiman and Werner 2002). If there are $i = 1 \ldots n$ such stages, each of duration T_i and with the mortality rate parameters $M_o^{(i)}$ and $M_1^{(i)}$, what might happen to the form of the net (overall) relationship between N_o and N_T when these occur (are applied) in sequence to finally result in recruitment? Beverton and Holt provided an answer to this question: the overall relationship is still expected to be exactly of the form predicted by equations 6.3 and 6.4, but the overall parameters s_{max} and k have a more complex interpretation. The slope parameter s_{max} is easy enough to calculate, as just the product of the maximum survival rates through the n stages:

$$s_{max} = exp[-M_o^{(1)} T_1 - M_o^{(2)} T_2 - \ldots - M_o^{(n)} T_n] \qquad (6.5)$$

But k becomes a complex function of the M_1's and M_o's, representing such possibilities as that just one stage might have high density dependence and

---

[1]Data are available at: *www.mscs.dal.ca/~myers/welcome.html*.

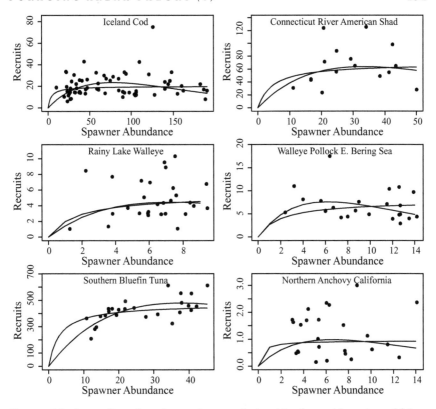

FIGURE 6.3: A sampling of stock-recruitment relationships for a wide variety of fishes, from the Ransom Myers database at *www.mscs.dal.ca/~myers/welcome.html*.

override the effects of weaker density dependence at other stages. That complex function is not worth even trying to write down, because in practice we have to determine the overall k value empirically in any case. The key point is that complex life histories are not fatal for the Beverton-Holt formulation but, rather, just force us to think about the model parameters as complex functions of fine-scale time dynamics during juvenile development.

Beverton and Holt were vague about why and how the density-dependent component M_1 might arise through the fine-scale dynamics of individual fish trying to make a living. Biologists have been largely unsuccessful at demonstrating that M_1 represents something simple like an increased probability of starvation in the presence of an increased number of competitors (starvation is apparently very rare), and it is not obvious that increases in N per se should attract higher predation risks (e.g., by predator switching behaviors). Further, we generally see very poor correlations between both egg and larval abundance and subsequent recruitment, suggesting that the density dependence occurs after these life stages and, in particular, is something that first appears when juvenile fish start to show active behaviors such as schooling and resting/feeding site selection. This observation about the involvement of behavior in creating density dependence motivates much of the analysis presented below.

6.2   ALTERNATIVE MODELS BASED ON JUVENILE CARRYING CAPACITY

When biologists see data such as those shown in figure 6.3, where the same average recruitment seems to result from a wide range of spawning stocks, they often interpret the pattern as implying the existence of some sort of juvenile habitat "limit" or "carrying capacity," perhaps representing a maximum number of physical sites that can sustain and protect juveniles. The Beverton-Holt derivation tells us that this appearance can be misleading: a similar pattern results just from linear density dependence in the mortality rate, without any implication that there is a particular maximum number of juveniles that the system or habitat can support. Further, when pelagic fish spawning is distributed over a wide area (e.g., the tuna example in fig. 6.3), it is hard to see what one might mean in the field by there being any particular number of juveniles that can find rearing sites or be supported.

But there certainly are cases in which small juveniles do appear to flood the rearing environment in large initial numbers, and in which there are indeed only a limited number of small microsites where these juveniles can find refuge from physical dispersal forces and predators. Often each microsite can hold (and feed) only one or a few juveniles. Examples would be streams where anadramous salmonids spawn, rocky bottom areas where juveniles of many demersal fishes seek refuge, and coral reefs where the juveniles of many tropical fishes, including planktonic feeders, are concentrated.

It turns out that in such cases we may see a Beverton-Holt relationship shape. But we may be much more likely to see a pattern that Barrowman and Myers (2000) have called the "hockey stick," in which recruitment increases almost linearly with the initial abundance, then suddenly plateaus at a carrying capacity. To see how such a pattern might arise from assumptions similar to the Beverton-Holt but including the idea of a limited number of safe rearing sites, let us construct a model with the following assumptions:

1. an initial number of juveniles N_o are widely distributed by spawning or larval drift over a potential rearing habitat in which there are only k ($<< N_o$) suitable rearing sites, each capable of rearing only one juvenile safely;
2. all juveniles start out in a "dispersive" behavioral state, then move into a "resting" behavioral state if/when they find a suitable rearing site;
3. the instantaneous mortality rate, say M_d, is very high for juveniles while in the dispersive state, and drops to a low base rate M_o for juveniles that have found a rearing site;
4. juveniles search randomly for rearing sites and encounter sites at a "mass action" rate proportional to the number of juveniles still searching for a site and to the number of unoccupied sites still remaining.

These assumptions lead to a pair of differential equations, one for the number $N_t^{(d)}$ of juveniles still in the dispersive state and one for the number $N_t^{(r)}$ of juveniles that have found resting sites:

$$\frac{dN_t^{(d)}}{dt} = -M_d N_t^{(d)} - a N_t^{(d)}(k - N_t^{(r)}) \tag{6.6a}$$

$$\frac{dN_t^{(r)}}{dt} = -M_o N_t^{(r)} + a N_t^{(d)}(k - N_t^{(r)}) \tag{6.6b}$$

Here, the parameter a represents the search rate by dispersive juveniles for rearing sites, and the term $a N_t^{(d)}(k - N_t^{(r)})$ represents the flux of juveniles from the dispersive into the resting state.

Integrating equations 6.6a and 6.6b over time from 0 to T, with $N^{(d)}$ set to N_o initially and $N^{(r)}$ set to zero initially, we can predict the net recruitment to age T as $N_T = N_T^{(d)} + N_T^{(r)}$. If M_d is large, N_T will consist almost entirely of settled or resting juveniles. Unfortunately, there does not seem to be a convenient analytical form for the integral (or at least the authors have been unable to find one using ©Maple). But it is quite easy to integrate the equations numerically using a spreadsheet (box 6.1).

Some numerical integration results for equations 6.6a and 6.6b are shown in figure 6.4. The bottom line of these results is a predicted stock-recruitment relationship that looks very similar to the Beverton-Holt model (steep at a low spawning stock, then flattening), but with more of an abrupt transition between increasing and flat recruitment than predicted by the Beverton-Holt equation. This is just the sort of "anomaly" pattern reported for some data sets by Barrowman and Myers (2000). Notably, their test cases (and those reported in a companion paper by Bradford et al. 2000) were for coho salmon (*Oncorhynchus kisutch*), which very obviously disperse and take up rearing sites/territories much as assumed above.

More abrupt transitions (a better fit to the hockey stick than the Beverton-Holt model) are predicted from equations 6.6a and 6.6b when M_d is large, and when the site search rate a is also large. Higher M_o values (a high mortality risk even in resting sites) make the relationship look more similar to Beverton-Holt even for high values of M_d and a. So there is no obvious rule for deciding when one or another of the formulations might be better to use for predictions such as the number of spawners needed to "fully seed" a rearing environment. To make matters worse, in practice there are seldom enough $N_o - N_T$ stock-recruitment observations concentrated near the stock sizes at which the models make different predictions to allow strong statistical tests of which works best.

A possibly more realistic alternative to random search for settlement sites as assumed in equations 6.6a and 6.6b would be a model based on assuming that juveniles are initially concentrated in one or a few source (spawning, larval settlement) sites, then spread out over time in search of settlement sites. For example, we might assume that the juveniles that have not yet

---

BOX 6.1
SOLVING THE ORDINARY DIFFERENTIAL EQUATIONS THAT COMMONLY ARISE
IN ECOLOGICAL SYSTEMS USING SIMPLE SPREADSHEET METHODS

In ecological modeling we often need to numerically solve nonlinear ordinary differential equation systems of the form $dx/dt = f(x)$. It is a good idea to do this even when an analytical solution appears to have been found, to check the solution equation. Solution methods generally involve dividing the total solution time T into short steps Δt, then using the rates evaluated before and during each interval Δt to provide a discrete approximation of the solution over time. A very nice discussion of various methods is in Patten (1971). The simplest (and very bad) version of this is the "Euler method," in which we take x at time $t + \Delta t$ to be just x at time t plus $f(x)$ evaluated at time t times Δt:

$$x(t + \Delta t) = x(t) + f(x(t))\Delta t$$

At an opposite extreme, the most accurate method that we use a lot is the "4th order Runge-Kutta," which divides each interval Δt in half and involves evaluating $f(x)$ at the start, middle, and end of the period so as to provide a polynomial approximation of how $f(x)$ is changing over the interval. But in our experience with equations that appear in ecological dynamics, a relatively simple method called the "Adams-Bashforth" almost always performs just about as well as the Runge-Kutta (Shampine 1994). In this method, we use the derivative at time $t - \Delta t$ along with the derivative at time t to predict rate changes over the interval Δt and the net result of these changes, $x(t + \Delta t)$. The Adams-Bashforth recursive equation is just

$$x(t + \Delta t) = x(t) + \frac{\Delta t}{2}\left[3f(x(t)) - f(x(t - \Delta t))\right]$$

That is, the change in state from t to $t + \Delta t$ is approximated by Δt times a weighted average of the derivative evaluated at time t, $f(x(t))$, and the derivative evaluated at time $t - \Delta t$, $f(x(t - \Delta t))$, with the weights being $+3/2$ on the current derivative and $-1/2$ on the previous derivative.

To set up the Adams-Bashforth method in a spreadsheet, just set up a column for t (increasing by Δt from each row to the next), a column for each x (rows representing x over t), and a column for each $f(x)$ evaluated at the time t and $x(t)$ for that row. Take the second row x (for $t = \Delta t$) to be the x for the first row, plus Δt times the $f(x)$ from the first row (the Euler method for the first time step). Then for each following row, make x be the x from the previous row, plus $\Delta t/2$ times the difference of 3 times the derivative from the previous row minus the derivative from two rows above. ∎

---

settled spread so as to test (and occupy) S new settlement sites per unit time, until all dispersing juveniles have either died or settled or until the total number k of possible rearing sites has been occupied (imagine juvenile salmon dispersing downstream from a spawning area, with dispersers

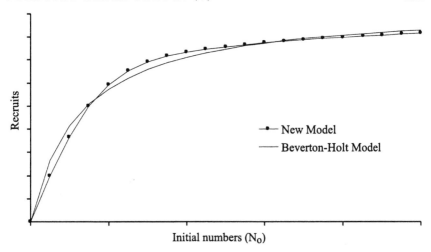

FIGURE 6.4: Comparing recruitment predictions using a numerical integration of equations 6.6a and 6.6b versus using the Beverton-Holt model.

moving downstream each day enough to "see" S unoccupied rearing sites). We can model this assumption just by replacing the mass-action settlement-rate term $aN_t^{(d)}(k - N_t^{(r)})$ by the constant rate S in equations 6.6a and 6.6b, then integrating the rate equations only up to that time T_d when $N_t^{(d)}$ goes to zero (all dispersers have settled). Solving the simpler differential equations up to this time limit, we find

$$T_d = \frac{-\ln\left[S/(N_oM_d + S)\right]}{M_d} \tag{6.7}$$

and the number of "successful" (settled) juveniles $N_{Td}^{(r)}$ at this time is given by

$$N_{Td}^{(r)} = (S/M_s)\left\{1 - [S/(N_oM_d + S)]^{M_s/M_d}\right\} \tag{6.8}$$

This model looks similar to the solution to equations 6.6a and 6.6b and also tends to produce more of a hockey-stick pattern than the Beverton-Holt model; it predicts a sharp break in the recruitment relationship and a flat top at k juveniles if we assume that the settlement rate drops suddenly from S to zero when $N_t^{(r)}$ reaches k (and if we ignore any density-independent mortality at rate M_s after all sites have been initially occupied).

But our aim at this point is not to try to build a better equation for a stock-recruitment prediction. Rather, it is to introduce methods of thinking more carefully about the role of juvenile behavior, particularly movement and mortality risk related to movement, in shaping juvenile survival rates and recruitment relationships. The main message to take away from this derivation is that we should be able to do better than to think simplistically in terms of just "bottom-up" habitat factor/capacity effects; we need to think carefully about how these factors interact with behavior to result in juvenile mortality. Equations 6.6a and 6.6b, and the spreading model equation 6.8, imply a strong density dependence in the mortality rate, but the expression of this effect is in terms of how fast juveniles can find safe places rather than directly in terms of mortality parameters like M_1 of the Beverton-Holt model.

6.3   USING FORAGING ARENA ARGUMENTS TO DERIVE THE BEVERTON-HOLT MODEL

Let us return now to the foraging arena imagery of fig. 6.1 and see how the intense localized food competition and predation risk implied by this figure can result in the mortality rate and recruitment equations proposed by Beverton and Holt. The notion here is not to disparage models based on the dispersal of animals to fill safe habitat sites as discussed in the previous section but, rather, to show that behavioral arguments still predict the same general relationship even when there is no habitat structure that juveniles might "fill," e.g., species in which juveniles form schools that move widely over the environment.

The basic argument proceeds in three stages. First, we develop a simple model to predict food concentrations in the restricted foraging arenas of figure 6.1, as a function of the abundance of juveniles using these arenas. Next, we argue that juveniles must respond to changes in these food concentrations in order to maintain growth rates, or else accept slower growth and spend more time at small body sizes at which they are more vulnerable to predation. Finally, putting the first two arguments together along with the assumption that most predation occurs while animals are foraging, we show that attempting to maintain a constant growth rate is likely to result in a linear increase in the juvenile mortality rate with increasing juvenile density. We show further that the alternative of just accepting a longer time at small body sizes has the same net effect.

Consider how the concentration of food, say C_t, varies over short time scales within a relatively small foraging arena being used by a cohort of juveniles of abundance N_t. We would expect the concentration dynamics to be dominated by physical (mixing) and biological (dispersal, emergence) exchange processes, along with exploitation by the juveniles, rather than by food production dynamics within the arena. Also, we would expect juvenile fishes to search more or less at random while foraging in the arena. Using these assumptions, we can write a dynamic equation for changes in C_t over very short time scales, as

$$dC/dt = vC_o - \acute{v}C_t - (a/A)C_tN_t \qquad (6.9)$$

Here, v represents the movement rate of food organisms into the arena at rates proportional to a "global" food concentration C_o that is not accessible to the juvenile fish (due to the spatially restricted movement in response to predation risk), $\hat{v}$ represents the movement rate of vulnerable food organisms out of the arena, a represents the volume (or area) searched for food per time by each of the N_t individual fishes using the arena, and A is the total volume (or area) of the arena. For small arenas (A small), the rate constants $v, \hat{v}$, and a/A are likely to be large compared to any food-production-rate processes. Note here that we are also ignoring any handling time/satiation effects on juvenile behavior, i.e., assuming that C_t is generally not high enough for juveniles to regularly gorge themselves or be limited in food attack rates by the time needed to pursue and handle each prey item.

Assuming a fast equilibration of C_t (on time scales of minutes to hours while juveniles are feeding) relative to how fast N_t changes, we can predict the average equilibrium food concentration by a "variable speed splitting" (see box 6.2) assumption that dC/dt stays near zero most of the time. The two time-dependent variables in equation 6.9 are operating at different time scales. The variable-speed-splitting argument is used to approximate the fast variable (C_t) at an equilibrium value as a function of the slow variable (N_t). To obtain this prediction, we set $dC/dt = 0$ in equation 6.9 and solve for the equilibrium food concentration:

$$C_t = \frac{vC_o}{\hat{v} + (a/A)N_t} \qquad (6.10)$$

Equation 6.10 states that the average food concentration in foraging arenas should be positively proportional to the mixing/immigration rate v and the overall food concentration C_o, and negatively proportional to the food mixing/emigration rate and to the number of juveniles that use the arena. It predicts the food density that juvenile fish should actually see in the foraging arenas that they use (as opposed to the food supply C_o that we might sample on larger scales), and as such defines "food supply" more precisely than biologists often bother doing. It involves an element of "supply" as a rate (vC_o) and also as an instantaneous density of available food organisms (C_t). The form of equation 6.10 is central to the arguments that follow, because this form predicts potentially strong impacts of N_t on C_t, even and especially when N_t is small. An example of this form of relationship is shown in figure 6.5, using data from experiments in which juvenile rainbow trout were stocked into British Columbia lakes (Post et al. 1999) and using the juvenile growth rate as an "assay" for the apparent food concentration seen by the juveniles in the absence of high predation risk. Note that equation 6.10 does not imply that C_t as a "local equilibrium" value is constant on the longer time scales over which we would expect N_t (and perhaps C_o) to change. On longer time scales, equation 6.10 predicts how C_t should on average "track" changes in N_t (and C_o).

Equation 6.10, along with simple bioenergetic arguments about food-conversion efficiency, predicts that the daily growth rate g_t (the body weight gain per day) achieved by a juvenile fish should be proportional to how

BOX 6.2
VARIABLE SPEED SPLITTING

For almost every large system in nature, we must somehow try to model the interactions of variables that can change at very different speeds. This becomes especially important as we try to understand systems that are spatially complex, such that some things can change rapidly in local spatial areas (like foraging arenas), and also have important consequences for dynamics at larger scales. Every ecologist who has played with anything but the simplest predator-prey models has already used formulations that involve speed-splitting assumptions, most likely without even realizing it.

The basic idea in most of our applications is as follows. Suppose we have to deal with a system of equations of the form

$$\frac{dx}{dt} = f(x, y)$$

$$\frac{dy}{dt} = g(x, y)$$

where the rate constants in $g(x, y)$ imply that the variables y can change much more rapidly than the variables in x (dy/dt much larger than dx/dt). Suppose further that the rate equations $g(x, y)$ imply stable equilibria for y, given fixed or slowly changing values of x. Then the overall system (x, y) will "closely track" the approximation obtained by solving dx/dt explicitly, while setting the y variables to the algebraic or numerical values implied by solving $g(x, y) = 0$ for y, i.e., by having the dy/dt rates stay near zero over time while using the resulting equilibrium values of y in solving dx/dt. The mathematical analysis of such approximations is a topic in singular perturbation theory; for further reading on the more general conditions and methods of that theory, see O'Malley (1974) and Kevorkian and Cole (1996).

One very widely used ecological example is the predator-prey models with type II functional responses of the predators, e.g., the Michaelis-Menten equation for enzyme rate kinetics (often also used to predict the feeding rates of filter-feeders, predators) or the Holling (1959) "disc equation" that includes the effects of handling time/satiation on predator feeding rates. The type II response equations are derived by assuming that fast chemical-physiological variables (e.g., the proportion of enzyme molecules in a reactive vs. bound state, the gut contents of predators) quickly reach equilibrium in relation to local substrate or prey concentrations. Other examples are "ideal free distribution" equations for predicting the spatial distributions and densities of animals, assuming that the animals quickly shuffle about so as to end up distributed so that no individual can improve its circumstances by further movement. ∎

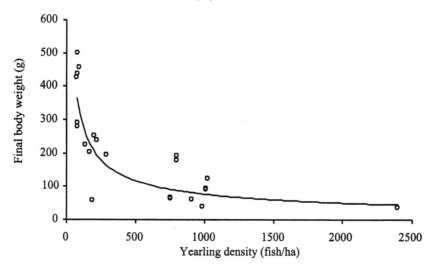

FIGURE 6.5: Final body weights after one year of growth versus the initial stocking density for several small British Columbia rainbow trout lakes. Higher densities result in the rapid depletion of food resources, leading to reduced growth rates.

much volume it searches per day, the food density in that volume, and the food-conversion efficiency. If the fish searches volume a per day, spends a proportion p_t of its time actually searching, encounters a food concentration C_t per unit volume searched, and converts its intake to growth with an efficiency e_g, we would predict its growth rate to vary as

$$g_t = e_g a p_t C_t \qquad (6.11)$$

Here, $a p_t C_t$ is the daily ration.

Equations 6.10 and 6.11 warn us that high densities of juvenile fish, as in years of higher larval settlement N_o or while a cohort is still numerous after settlement, could result in a severe depression of the food densities C_t actually seen by juveniles foraging in restricted arenas, and hence in the growth rates g_t. We seldom observe the high variation in body-growth rates that such depression might cause, if fish were purely passive and did not adjust their feeding behaviors, particularly the time spent feeding p_t. Rather, fish most often seem to compensate in some way for changes in the food availability. In an evolutionary sense, there has likely been harsh natural selection to maintain at least some minimal growth rate, in order to complete ontogenetic changes such as movement into adult feeding areas, migration, and egg production. Evolutionary optimization models suggest that fitness should be maximized by trading off growth and survival, with the optimum trade-off point being at a growth rate just high enough to successfully reproduce (Werner and Hall 1988; Walters and Juanes 1993; Anholt and Werner 1998). Let us borrow from these ideas and assume that juvenile fish that must forage in dangerous foraging arenas will spend at least enough time foraging to maintain some base or target growth rate g_o (units of weight gain per time). Substituting this minimal growth rate "requirement" into equation 6.11, we see that the fish needs to adjust its behavior so as to maintain the relationship

$$g_o = e_g a p_t C_t \qquad (6.12)$$

Notice that we are not being very careful here about how g_o, the search rate a, and the growth efficiency e_g likely change over time as the fish grows. Such changes can be incorporated into the derivation later, by dividing the juvenile life into a set of stanzas or stages, and using the notion discovered by Beverton and Holt that linking such stages in sequence does not change the basic form of the recruitment relationship.

If juvenile fish do, in fact, alter their behavior so as to try and meet a growth "constraint" (eq. 6.12), and if the only factor in the equation that they can alter is their feeding-time proportion p_t, then equation 6.12 can be rewritten as

$$p_t = \frac{g_o}{e_g a C_t} \qquad (6.13)$$

That is, the proportion of time p_t that fish "need" to spend feeding in order to meet a basic growth rate "goal" g_o is positively proportional to that goal and inversely proportional to the growth efficiency, search rate, and food concentration. This brings us to the crux of the argument about density dependence. Suppose we substitute the arena-scale relationship equation 6.10 for food density C_t in relation to juvenile abundance N_t into the required-feeding-time relationship equation 6.13:

$$p_t = \frac{g_o}{e_g a v C_o/[\acute{v} + (a/A)N_t]} = \frac{g_o}{(e_g a v C_o)} [\acute{v} + (a/A)N_t] \qquad (6.14)$$

While this relationship looks messy because of the various parameters that we have included in the component calculations, it can be rewritten as a linear relationship between p_t and N_t, i.e.,

$$p_t = k_1 + k_2 N_t \qquad (6.15)$$

Where

$k_1 = g_o \acute{v}/(e_g a v C_o)$ represents the base proportion of time spent feeding at low N, and

$k_2 = g_o/(e_g A v C_o)$ represents the effect of increasing competition on feeding time.

What equation 6.15 tells us is that if the food concentration is inversely proportional to the competitor abundance, and if competitors react to this effect by trying to maintain a constant growth rate, then the average time spent feeding should be positively proportional to the competitor abundance.

If you have followed the derivation of equation 6.15, it should be pretty obvious what the next step in the argument will be: suppose that the juvenile mortality rate M_t is proportional to the proportion of time that animals spend feeding in relatively dangerous foraging arenas, i.e.,

$$M_t = Pp_t \qquad (6.16)$$

where P is the instantaneous predation mortality risk per time spent feeding. This is similar to the M_d mortality rate suffered by dispersing animals in the derivations of the previous section, but expressed in terms of a risk-per-time parameter P and a behavioral variable p_t representing the proportion of time spent in a dispersing/active/vulnerable/feeding behavioral state. Then including (substituting) the "evolutionary imperative" for p_t in order to maintain growth (eq. 6.15) into equation 6.16, we finally obtain

$$M_t = Pk_1 + Pk_2 N_t \qquad (6.17)$$

This is exactly the original rate assumption of the Beverton-Holt derivation, but with the parameters M_o and M_1 now "articulated" more precisely in terms of parameters representing the food supply, selection to maintain growth rate, and predation risk. Specifically,

$$M_o = \frac{Pg_o\acute{v}}{e_g avC_o} \qquad (6.18a)$$

$$M_1 = \frac{Pg_o}{e_g AvC_o} \qquad (6.18b)$$

These relationships tell us some pretty obvious things, such as that base mortality rates (M_o) ought to be higher for juveniles that face higher risks per time feeding (P), try to eat more or need more to grow faster (g_o/e_g), use foraging arenas with high food-loss rates ($\acute{v}$), or live in environments with lower food-supply rates (vC_o). And density-dependent effects (M_1) ought to be larger for juveniles that face more risks, try to grow faster, have lower food-supply rates, and have less total habitat volume/area (A) over which to distribute their feeding activity.

Walters and Korman (1999) have shown that the basic Beverton-Holt recruitment equation form is still obtained even when fine-scale (daily) foraging time p_t is fixed (so that g_t varies with N_t due to changes in the food density C_t), but what then changes is the total time T needed for juveniles to reach any particular, arbitrary size at which we would call them recruits. That is, we get the same answer whether we view the density-dependent component of juvenile mortality as arising from fine-scale time partitioning into feeding/dangerous versus resting/safe times, or look at just the total of these times eventually needed to recruit to the harvestable or adult population.

Perhaps the main weakness in the arguments of this section is about the assumption that behavior can be partitioned into two simple components, feeding/risky versus resting/safe, that results in the $M_t = Pp_t$ step in the derivation. It is easy enough (though algebraically more messy) to allow for some risk even for resting individuals, and to assume that at least some food intake can be obtained while in resting/hiding places. But the important missing piece in terms of a mortality prediction is about the relatively rare but potentially very risky behaviors that are involved in the development of mesoscale habitat-use patterns. To see this issue, compare the arena-scale derivation with the derivation based on assuming that individuals undertake risky dispersal behaviors in order to find suitable (less risky, energetically more favorable) rearing sites. The field reality for most species is probably a complex mixture of the effects represented in these models: there is a fine-scale risk-management (time-management) behavior within locales where most diurnal behavior is exhibited, and there are also less frequent larger movements to find better (less crowded) rearing sites. The less frequent, likely riskier large-scale movements may, in fact, be driven by changes in the required local feeding times p_t, so that the total mortality risk remains proportional to p_t. But we have not yet found a simple, analytical way to model the dampening of arena effects (an increase in the arena area A used by juveniles and the corresponding decreases in p_t) expected from this combination of behaviors.

Another potential weakness arises from among-cohort effects when more than one juvenile cohort of a species uses essentially the same foraging arena. In this case, older juveniles (and adults) may not only impact the food resources C_t directly but also cause a significant component of the predation risk P. Most fish seem to cannibalize small juvenile conspecifics whenever there is opportunity. Traditionally, we have modeled such effects by using the Ricker stock-recruitment equation $R = s_{max}N_o e^{-bN_o}$, where the bN_o term represents the mortality rate due to cannibalism and N_o is used as an index or predictor of the number of older fish (e.g., spawners that produced N_o) present. But the "risk ratio" component P/C_o of equation 6.18 argues for the possibility of more exaggerated effects, if older fish contribute to changes in C_o as well as P. If it is older juveniles that cause a significant part of the risk-ratio change, we expect to see a periodic recruitment variation with short periods, e.g., 1–3 years; such patterns are, in fact, common in the Myers data base (see, e.g., his summary of recruitment time series for walleye [*Stizostedion vitreum*] stocks, as some but certainly not all of the stocks show 2-year cyclic variation). One extreme example of this is the barramundi (*Lates calcifer*) studied by Roland Griffin in the Northern Territory, Australia. Griffin has found a violent 2-year cycle in abundance of age-0 and age-1 juvenile barramundi in the Mary River, Northern Territory (fig. 6.6, also see Ley et al. [2002] for another example of recruitment suppression in barramundi). In this region, the species is a catadromous, protandric hermaphrodite: spawning occurs in estuaries, and juveniles migrate upstream into freshwater rearing areas where they spend 2–3 years before maturing as males and moving back downstream

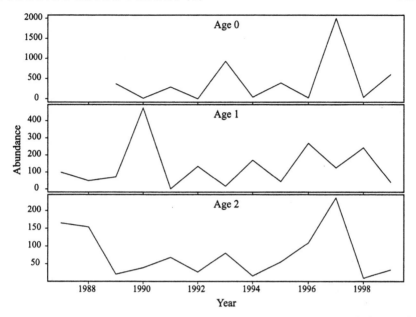

FIGURE 6.6: The abundance of age-0 to age-2 barramundi (*Lates calcifer*) in the Mary River, Northern Territory. (Data courtesy of Roland Griffin.)

(where they spend another 2–3 years as males before changing to females). This life history means that the freshwater rearing areas are heavily seeded every year with age-0 juveniles (body lengths 5–30 cm). But when these age-0 fish encounter a large cohort of age-1 juveniles (body lengths 35–50 cm), their survival is extremely low, due likely to a combination of cannibalism and food competition effects. So the next year this weak cohort "allows" the next year's age-0 fish to survive well, and this dominant cohort then suppresses the next year's cohort. Similar cyclic behaviors have been reported for vendace (*Coregonus albula*) (Helminen and Sarvala 1994, 1997). There are strong hints of cyclic juvenile production in other species as well, e.g., the Pacific hake (*Merluccius productus*) uses juvenile rearing areas off California for roughly its first 3 years of life and seems to produce only one strong cohort about once every 3–4 years. Three-year cycles attributed to cannibalism have been reported for a British ocean bass (*Dicentrarchus labax*) (Henderson and Corps 1997), for brown trout (*Salmo trutta*) (Borgstroem et al. 1993; Nordwall et al. 2001), and for blue crab (*Callinectes sapidus*) (Lipcius and van Engel 1990). These sorts of patterns obviously cannot be reproduced just by using the basic Beverton-Holt model, unless we explicitly couple the risk rate P to abundances of particular older age classes.

Combining equations 6.3 and 6.18 allows us to develop a recruitment model that allows the explicit examination of how time-varying cannibalism (or other predation) risk and habitat capacity (represented by the arena area A) ought to influence recruitment over time:

$$R_t = \frac{N_{ot}e^{-a_1 P_t}}{1 + N_{ot}a_2(1 - e^{-a_1 P_t})/A_t} \qquad (6.19)$$

Here the "empirical" parameters a_1 and a_2 represent the combined effects of behavioral ecology and population measurement scale effects, N_{ot} represents the initial recruitment of eggs or larvae to the foraging arena interaction in year t, and the variables P_t and A_t represent indices of relative predation risk and habitat size in year t (the absolute values of these factors are "absorbed" into the units of the parameters a_1 and a_2). Using this formulation in a population accounting framework (e.g., a simple spreadsheet), we can easily represent hypotheses such as strong cannibalism effects of age-1 fish on age-0 fish just by making P_t depend on past recruitments. For example, setting $P_t = P_o + R_{t-1}$ represents the proposition that predation risk has two components, a base rate P_o plus a rate dependent on age-1 abundance indexed by R_{t-1}. In this example, we are tacitly indexing the noncannibalism predation risk P_o in units of age-1 cannibalism risk (i.e., P_o is the noncannibalism predation risk per unit of predation risk caused by an age-1 recruit from last year). In this formulation, a general rule is that increasing the "background" risk P_o has a stabilizing effect on recruitment, while decreasing it results in an increased likelihood of cyclic behaviors such as those documented for barramundi. A simple model to demonstrate these effects is developed in box 6.3.

BOX 6.3

DEMONSTRATING THE POSSIBLE EFFECTS OF INTER-COHORT JUVENILE
CANNIBALISM ON RECRUITMENT VARIATION

We can use a simple population model to demonstrate how cannibalism of age-1 juveniles on age-0 juveniles may produce complex cyclic patterns despite the sort of "risk management" adjustments in foraging time by the age-0 juveniles used in this chapter to derive the Beverton-Holt recruitment relationship. To do this, first set up a spreadsheet balance model for age-1 and older total population size N_t, of the form

$$N_t = SN_{t-1} + R_t$$

where the annual survival rate S is set to a moderate value like 0.7; R_t represents age-1 recruits to N_t. For each year, predict the initial age-0 recruits N_{ot} from a simple mean-fecundity relationship $N_{ot} = fN_t$, where $f = 100$ represents the mean effective fecundity per age-1 + fish. For each year $t > 1$, predict R_{t+1} from equation 6.19,

$$R_{t+1} = \frac{N_{ot}e^{-a_1 P_t}}{1 + N_{ot}a_2(1 - e^{-a_1 P_t})/At}$$

where predation risk P_t is set to a background risk parameter P_o plus R_t: $P_t = P_o + R_t$. Including a spreadsheet column for P_t makes this easier and allows for the exploration of broader hypotheses about the impacts of changing P_o over time as well as the impacts of age-1 recruits. Also include a column for the relative habitat area A_t so that the model can be used to explore the effects of changing the habitat size, and initially set $A_t = 1$ for every year.

To ensure that the model has reasonable parameter values given our general experience about the steepness of stock-recruitment relationships (s_{max} in eq. 6.4), set up two leading parameters:

1. an equilibrium "mean" recruitment rate R_e, set to, say, 1000 (population scale);
2. a maximum reproductive rate κ in the range of 3 to 20.

From these, you can calculate the following:

3. the equilibrium mean population size $N_e = R_e/(1 - S)$ and juvenile production $N_{oe} = fN_e$;
4. the recruitment curve slope $s_{max} = \kappa R_e/N_{oe}$;
5. the predation risk $a_1 = -\ln(s_{max})/(P_o + R_e)$;
6. the habitat capacity $a_2 = (\kappa - 1)A_1/[N_{oe}(1 - s_{max})]$ (where $A_1 =$ the habitat area in year 1).

Notice how the equations for a_1 and a_2 contain or "remove" the units of measurement of P, R, and A from the recruitment prediction; i.e., the predictions of R_t depend only on the relative values of $P_o + R_{t-1}$ and A_t over time.

(Continued)

(BOX 6.3 continued)

Notice further that these parameter settings will result in a population that is exactly balanced (does not change over time), provided the first year recruitment R_1 is initialized to exactly R_e. To explore possible cyclic dynamics, the initial R_1 should be set to a nonequilibrium value, e.g., $R_1 = 0.9R_e$; if the parameter values predict cyclic dynamics, these should then develop over time and be quite clear within 20–40 simulation years.

Hypotheses about the relative importance of compensatory foraging time adjustments and cannibalism can then be explored by changing the κ and P_o parameters. High κ values imply strong reductions in the predation risk and first-year mortality with decreases in age-0 juvenile abundance, and should have a stabilizing effect on recruitment despite any cannibalism effects. Increases in P_o relative to R_e imply a decreasing relative importance of cannibalism, and should also be stabilizing. To obtain violent cycles of the sort observed for barramundi, κ should be set relatively low (< 10), and P_o should be 3000 or less (i.e., about 3 or fewer other piscivores attacking age-0 fish for every age-1 fish that attacks the juveniles, if the age-1 average abundance $R_e=1000$).

This simple model predicts an interesting and perhaps unrealistic effect of increasing the foraging arena size A_t, e.g., by providing artificial reefs or other protected areas from which juveniles can move out to feed (and be predated). If κ and P_o are set high enough to barely predict stable recruitment over time, an increase in A_t results in the appearance of cyclic recruitment, i.e., "overcompensation" such that the increase in A_t permits a stronger recruitment that then causes a higher mortality among the next cohort. This is caused by a "hidden assumption" in the basic formulation, namely, that increases in A_t do not "dilute" the impact of each age-1 fish on the predation risk, i.e., increasing the arena size does not reduce the effective density of age-1 cannibals. This effect can be avoided by assuming area-dependent dilution of age-1 fish, modeled by setting $P_t = P_o + R_{t-1}/A_t$, i.e., by diluting the age-1 recruits over the larger habitat area.

The model can be used to see how cannibalism affects the performance of marine enhancement programs. Modern aquaculture techniques are permitting the "growout" of larger, even age-1 juvenile marine fish for stocking to provide fishing opportunities as well as the supposedly faster rebuilding of wild stocks (see chapter 13). Usually, these larger stocked fish survive much better than stocked fry or small fingerlings and so are often favored by proponents of marine enhancement. This can be represented in the spreadsheet model by adding a column representing hatchery additions to the age-1 recruitment, so $R_t =$ (prediction from eq. 6.19) $+ R_{hatchery}$. In the presence of significant cannibalism, the model will then predict increases in the adult abundance N_t, but possibly severe declines in the contribution of wild recruits to R_t. ■

6.4   Implications for Recruitment Research and Prediction

So what? Even if the many assumptions leading to equations 6.18a and 6.18b were precisely correct, no one is ever likely to be able to go out and actually measure all of the constants contributing to M_o and M_1 with any accuracy. We can repair some of the weaknesses by using a multistage calculation (eq. 6.5), more detailed food production-exchange calculations, more detailed bioenergetics models, more elaborate optimization models for predicting foraging/dispersal times p_t, and more detailed predictions of the predation risk P. Doing these things would not change the basic argument about how localized competition can drive down food availability, force changes in feeding times, and hence cause density dependence in mortality rates. But the trophic factors P and C_o and habitat factors representing A are unlikely to be stable over long enough periods to be treated as useful constants for numerical prediction. Is there anything more we can learn from the modeling exercises presented above, apart from the obvious point that relatively simple overall recruitment relationships may arise from a variety of very complex behavioral ecologies?

One comforting implication is that we can now reasonably claim to see how recruitment relationships are "embedded" in larger trophic circum-stances (A, P, C_o). So if we want to build multispecies models that predict the components of recruitment variation caused by changes in predator abun-dances and food supplies, we can now see how to "insert" these effects into the Beverton-Holt equations (make A, P, and C_o variable in equations 6.18a and 6.18b), at the same time accounting for how juvenile fish respond at very fine space-time scales to changes in risks and opportunities. That is, we have a way of seeing recruitment dynamics across multiple space-time scales, ranging from the meters-hours scales of foraging arenas to decadal scales of variation in R due to changes in A, P, and C_o (fig. 6.7). Perhaps this will allow us to better prioritize research on the bits and pieces of biology and oceanography that contribute to the overall cross-scale picture.

Another comforting implication is that we can now see why average re-cruitment rates are seldom related in any obvious way (i.e., are not pre-dictable from) to gross measures of potential food supplies and predator abundances. The asymptotic average recruitment (s_{max}/k in eq. 6.4) is not just a messy function of M_o and M_1, but also M_o and M_1 are related in messy ways to behavioral and trophic parameters (eqs. 6.18a and 6.18b). And the mess gets much worse if we try to look at the recruitment process in terms of multiple stages (eq. 6.5).

As a practical aside, foraging arena arguments might have helped us pre-vent a gross blunder in the 1970s when we were asked to make predictions about whether the Pacific Ocean would be capable of supporting increases in Pacific salmon populations through artificial enhancement (hatcheries). Using crude food-abundance and production data, we confidently predicted

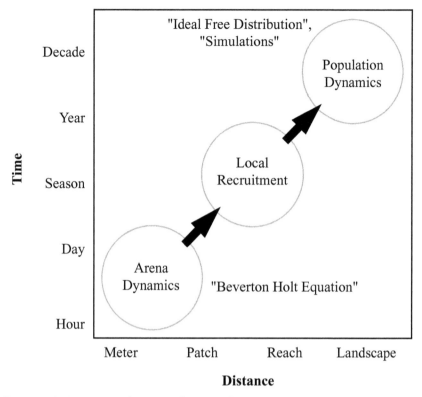

FIGURE 6.7: Moving predictions to larger scales.

(Walters et al. 1978) that the ocean should quite handily support 5 times the then-current abundance of chinook and coho salmon. We were proven wrong: enhancement was used to at least double the smolt (ocean migrant) production from Oregon through British Columbia, yet there was little or no corresponding increase in the adult abundances (see review in Walters 1995). Had we recognized that competition between hatchery and wild fish might be highly concentrated in foraging arenas (surface waters, near benthic refuges), we might have anticipated this apparent density-dependence in marine survival rates.

One very large implication of the foraging arena derivation is that we are unlikely to ever be able to "solve" the recruitment-prediction problem just by using more detailed, individual-based models (IBMs). The arena derivation claims that density-dependent effects arise through habitat relationships and behaviors at very, very fine space-time scales (meters, minutes), much finer than any scales for which we have yet attempted to model individual distributions and behaviors using IBMs. There is no good reason to believe that a simple integration of these effects across scales, by using differential equations for the total numbers as we have in this chapter, is going to produce a much less accurate result than the massive computations required for tracking many individuals through their early life histories. Indeed, we have a bad feeling that IBMs have lulled some fisheries scientists into a feeling of having captured "all" of the important variation, while in fact these have not yet even touched the most important scales for determining mortality rates.

There is also a very large warning for scientists who would hope to explain or predict recruitment variation from simple correlations between recruitment rates and physical factors such as oceanic regime shifts. This kind of correlative research is a central activity in the research field that is now widely called "fisheries oceanography," and the main thing that practitioners of it have learned is that you need to be quick about publishing your results. Most correlations break down soon after discovery (Drinkwater and Myers 1987; Walters and Collie 1988) and are quite likely to have been spurious in the first place (there are lots of time series to compare, and few statistical degrees of freedom due to autocorrelations in the series). But more important, equation 6.18 warns us to look for variation caused by the ratios of predation and food-production effects, in particular, the "risk ratio" P/C_o, with juvenile survival varying as e^{-cP/C_o} and the constant c being quite large (survival always being quite low). For most fish, the average value of the survival exponent cP/C_o is on the order of 5–10. This means that even a small proportional change in P and/or C_o can easily cause a change of (say) 20% or 1.0 in the exponent. But a change of 1.0 in the total instantaneous mortality rate during the juvenile stage translates into a change of almost 300% upward (if M drops by 1.0) or 70% downward (if M increases by 1.0) in the number of surviving recruits. It may well be that the risk-ratio factors P and C_o are indeed determined broadly by changes in oceanographic and limnological conditions, e.g., when an incursion of warm water masses into a region allows a new suite of predators to invade that region. But in such circumstances, both factors are likely to change! If the changes are in opposite directions, survival variation would be amplified. But if they are in the same direction (more predators, more food), the effect may be dampened or even reversed depending on the quantitative details of which factor (P or C_o) changes the most. To make matters worse, there can be very substantial time lags in the response, particularly of predation risk P, to physical changes.

When we think about how variable the risk ratio factors P and C_o are likely to be from year to year, and also the usable area A of the foraging arenas, it becomes obvious that the remarkable thing about fish recruitment is not how variable it is but, rather, how stable it is (Ursin 1982; Beyer 1989; Samb and Pauly 2000). We bemoan recruitment variance because it makes statistical analysis and model fitting difficult and because it causes an apparently unpredictable variation in the availability of fish to harvesters. We worry particularly about how low recruitments can trigger bouts of depensatory overfishing and long population recovery times. We put much research effort into trying to explain the "anomalies" and predict them from all sorts of environmental and ecological factors. We certainly must design monitoring and harvest-regulation programs that detect unexpected variation as it arises, and respond so as to at least prevent depensatory pathologies. But from a research perspective, perhaps we should be devoting much more effort to finding out why recruitment varies so little: is stability caused or permitted by a positive covariation in P and C_o? Is the spreading of fish over microhabitat choices a better model than the foraging arena imagery? Are there compensatory behaviors that stabilize the mortality risk more than expected on the basis of simple trade-off (you eat or you die) behavior?

Problems in the Assessment of Recruitment Relationships

IT IS COMFORTING TO IMAGINE that foraging arena theory and related ideas in behavioral ecology and population biology now appear to be giving us a better understanding of the processes that drive density-dependent changes in juvenile survival rates and often make recruitment appear to be independent of spawner abundance until or unless the spawning stock has been very severely reduced. But this understanding is not helpful when we need to address specific, quantitative policy questions about the risks or impacts of recruitment overfishing, i.e., when we need to advise about how large spawning stocks should be to prevent overfishing. The theory tells us that there should be a compensatory response, but it does not tell us how strong that response should be. Indeed, the theory warns us that the response in any particular case involves a very complex interaction of behavioral, habitat, and trophic factors; we must discover the net quantitative effect of these interactions, i.e., the strength of the response, through direct experience. That experience might be gained by actually driving spawning stock down to low enough levels to see the recruitment response, or else by relying upon the growing body of metaanalysis results from populations that have been driven to low levels, i.e., have already been recruitment overfished (Myers and Barrowman 1995; Myers and Barrowman 1996; Myers et al. 1999).

When we examine empirical stock-recruitment data, we most often see a frightening "shotgun" scatter of points, with little indication of an effect of spawning abundance except for a few observations at the lowest spawning-stock sizes. So in most cases, statistics texts would tell us to conclude that "there is no significant effect of spawning-stock size on recruitment, so spawning stock size can safely be ignored as a predictor variable." It is not uncommon to hear biologists agree with this statement but then turn right around and agree with equal vigor with statements about the need to protect spawning stocks and to define spawning-stock goals. These statements seem contradictory: how can it be important to protect spawning stock if spawning stock can safely be ignored? There are two answers to this question: (1) it may be that spawning stock can indeed be ignored provided there is no intent to allow lower spawning stocks than observed historically, i.e., there is no need to extrapolate beyond the range of past data; (2) there are two very distinct uses of stock-recruitment relationships, to make short-term recruitment forecasts versus to estimate long-term mean performance under alternative spawning-abundance targets or reference points.

It is a mistake to confuse forecasting and mean-performance assessment goals by claiming that high recruitment variability (which usually makes spawning stock pretty useless as a forecasting variable) also implies that the mean effect of spawning stock can be ignored in developing long-term harvest strategies and goals. A variation on this mistake is to make the more precise claim that because some stock has been observed to produce a strong year class or two when the spawning stock was low, the spawning stock should not be a key policy concern. In the following discussion, we concentrate mainly on assessment problems related to long-term management performance, for which mean effects are critical. Short-term recruitment forecasting is mainly a policy problem for semelparous and short-lived species of which most of the harvest each year is of new recruits (e.g., Pacific salmon); that policy problem can best be addressed by using alternative forecasting variables besides spawning stock and by deploying "in-season" adaptive management systems that allow the achievement of longer-term spawning-abundance goals by quickly learning about the current stock size as each fishing season proceeds (see chapter 4).

This chapter examines a few key problems that have been encountered in the "practical" analysis of recruitment relationships for fisheries policy design. It begins by pointing out that we need to think of the stock-recruitment relationship not as a single curve or equation but, rather, as a family of probability distributions that represent both the mean effects of spawning stock and the variation due to factors other than parental abundance. Next it discusses two of the main prediction problems encountered in an analysis aimed specifically at policy development, namely, reproductive performance at low stock sizes and the capacity to rebuild in stocks that have been historically overfished. It then reviews two of the most critical statistical problems that have arisen in stock-recruitment data analysis, the errors-in-variables regression problem and the time-series-bias problem. It closes with words of caution regarding claims that are becoming common today about our ability to explain and predict recruitment variations using environmental information.

7.1 Which Parameters Matter?

When trying to develop long-term abundance goals and strategies for dealing with unpredictable variations due to environmental factors, it is crucial to think of the stock-recruitment relationship not as a deterministic curve or equation but, rather, as a collection of probability distributions as shown in figure 7.1. This view avoids the pretense that spawning stock is a good short-term recruitment predictor but helps avoid the mistake of assuming that the means of the distributions can be ignored.

There are three main properties that we might hope to find in such a collection of probability distributions, whether the distributions have a high or low variance:

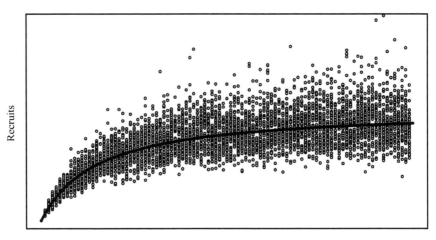

Spawners

FIGURE 7.1: The "stock-recruitment relationship" is best defined as a collection of probability distributions (represented here as a distribution of points for each spawning-stock level), where the means of the distributions hopefully vary in a relatively simple way with the spawning stock.

1. *Continuity*—If we could obtain repeated observations at two "nearby" spawning-stock levels, the means of the two distributions should not differ greatly (ensuring that a relatively simple, smooth function can be used to connect the means and to interpolate the mean performance when we have only a scatter of observations from many of the distributions). It is an apparent paradox that simplicity in the mean relationship is most likely to result from high complexity in the recruitment biology. We would expect a violation of the continuity assumption only in very simple ecological settings, in which very similar juveniles utilize a homogeneous habitat. For example, consider a hypothetical situation in which juveniles need to reach some critical size or energy storage by the end of their first growing season, in order to survive a stressful winter. Suppose there is a fixed total food supply each summer. Then at lower spawning stocks, each juvenile will get enough of this supply to reach the necessary fall size, and survival will be good. But if there are just a few too many juveniles (due to a small increase in the spawning stock), and if the juveniles were very similar to one another, then none would reach the critical size and all would die over the winter. The stock-recruitment relationship would then show a discontinuity, with recruitment suddenly "failing" at the spawning-stock size large enough to "overexploit" the food supply. We do not, in general, see such discontinuities, in part because of natural complexity: there is no single fixed food supply; individuals vary a lot in feeding rates and growth; etc. These complexities at the individual response level imply statistical smoothing and simplicity at the population response level.

2. *Repeatability*—If we could hold spawning stock at some level S1 for some time and observe the distribution of recruitments at that level, then do the same "experiment" at another quite different level S2, then return to make further observations at the original level S1, the second set of recruitment observations at S1 should have the same distribution as the first set. In simpler terms, the historical ordering of observations should not matter to the analysis. This assumption might be violated, e.g., when moving to high spawning abundance results in the persistent depletion of food organisms, thus resulting in lower recruitment performance after the spawning-stock reduction. Another example would occur when moving to a low spawning stock results in the loss of some spatial substock structure associated with fish homing to local areas to spawn, without these areas being immediately recolonized when the total spawning stock is allowed to recover.

3. *Stationarity*—This is basically the issue of whether historical data are useful predictors of the means of future recruitment distributions. We say that a recruitment relationship is nonstationary (Walters 1987) if the means of the distributions change persistently over time, e.g., due to irreversible habitat loss or persistent change in the physical environmental conditions (e.g., a global climate change).

In the short data series that are typically available, especially those collected during the progressive stock declines associated with fishery development, we obviously cannot distinguish between nonrepeatability and persistent nonstationarity.

An interesting point about the distributions in figure 7.1 is that ecologists have interpreted the observed recruitment variances in two opposite ways. One interpretation is to say that the recruitment variance is high, and to bemoan the difficulty that this creates in trying to see the underlying mean relationship. The opposite interpretation is to say that the recruitment variance is remarkably low, considering how variable it could be given the factors that determine it. In the second interpretation, we recognize that recruitment is a product of egg production times the survival rate, i.e., $R_t = \epsilon_t s_t$, where ϵ_t = total egg deposition and s_t = survival from egg to recruitment. Typically, ϵ_t is very large and s_t is very small; so for fixed ϵ_t (fixed spawning stock), even small "random" variations in the survival rate s_t could (and sometimes do) produce huge changes in the numbers of recruits. Indeed, the strongest evidence that compensatory mechanisms, such as those discussed in the last chapter, are very important is that we see much less variation in the survival rates s_t than might be predicted from the observed variations in some factors that affect survival, such as the physical patterns of larval transport that affect potential juvenile settlement into rearing areas.

For a practical recruitment analysis we generally assume a very simple functional form for the relationship between mean R and S, with the functional form chosen to force the mean relationship through the "known" point 0,0 (no spawners, no recruits). The most common forms are the two-parameter (s_o, b) models:

$$\text{Ricker: } R_t = S_t e^{s_o - bS_t + w_t}$$

$$\text{Beverton-Holt: } R_t = S_t e^{s_o + w_t}/(1 + bS_t)$$

In these models, e^{s_o} represents the maximum reproductive performance per spawner at low spawner abundance, i.e., the slope of the stock-recruitment curve near the origin. The b parameter represents the density-dependent effects and determines the population size scaling (how large R can be in absolute terms). The w_t term represents a time-varying, mean zero stochastic or environmental effect (i.e., the distribution around the mean relationship defined by s_o, b). In some cases, we add a third parameter to represent recruitment failure at very low stock size (Allee effects or depensatory predation effects), by just raising the leading S in the above equations to a power >1.0 (e.g., depensation can be represented in the Ricker model by $S^c e^{s_o - bS + w}$). Another three-parameter family of recruitment curves can be generated by raising the denominator of the Beverton-Holt form to a power, $Se^{s_o + w}(1 + bS)^{1/d}$; the joint variation of b and d in this "Deriso" or "Shepherd" model (Deriso 1980; Shepherd 1982) generates a family of curves ranging from Beverton-Holt ($d = -1$) to Ricker ($d = 0$) to quadratic ($d = 1$). Inclusion of a c to represent low-density recruitment failure does improve fits to roughly 10% of the high-contrast data sets in the Myers database (Liermann and Hilborn 2001). We know of no example in which inclusion of a d term for Ricker to Beverton-Holt shaping has resulted in a noticeably better fit than one or another of the two-parameter models. Significantly better fits can sometimes be obtained with a "hockey stick" two-parameter model (Barrowman and Myers 2000), which has a more rectilinear shape (linear from the origin to a plateau, then constant) than the Beverton-Holt. When we do not need to make predictions of performance at a very low S, another form that often fits just about as well as the two above is the power model $R = \acute{a}S^b e^w$, in which $\acute{a}$ represents scaling and b represents density dependence (with strong compensatory effects for $b << 1.0$), but a problem with this model is that it predicts "infinite compensation": R/S approaches infinity as S decreases to zero, if $b < 1.0$.

Most often we assume that variation around the mean $R - S$ relationship is log normal, i.e., that the recruitment "anomalies" w_t are normally distributed. There is good empirical and theoretical support for this assumption, except in cases in which there is obvious nonstationarity or high autocorrelation in the anomalies (Peterman 1981). The theoretical basis for assuming log normal variation is quite simple. Recruitment can be viewed in temporal detail as a product of egg production (dependent on S), and a product of a series of stage-i or time-i or stanza-i specific survival rates s_i: $R = Es_1 s_2 \ldots s_n$. The logarithm of recruitment is then $\ln(R) =$

$\ln(E) + \ln(s_1) + \ln(s_2) + \ldots \ln(s_n)$. We expect to see an "environmental" variation in each of the $\ln(s_i)$ in this sum. If such variations are at least partly independent of one another ($\ln(s_i)$ can be good at stage i even if it has been bad at stage j), we expect the sum to be approximately normally distributed (Central Limit Theorem of statistics) even if the individual $\ln(s_i)$ have peculiar (e.g., only low or high values, or highly skewed) distributions. This theoretical argument can, of course, fail if there is some single, dominant survival stage i for which $\ln(s_i)$ is a major component of the sum (s_i very small), and if the distribution of this s_i is very different from normal. But the reader can easily see how uncommon such situations apparently are, just by scanning through the Myers stock-recruitment data base and looking for complex anomaly patterns (e.g., clumped or highly skewed after log-transformation).

For some assessments it is important to be careful about the choice of the spawning-stock measure S. Most published assessments take S to be either the number of spawners (typical for semelparous salmonid studies) or the spawning-stock biomass (typical for iteroparous species). In general, we want a measure of S_t that is proportional to egg production ϵ_t, so that we do not confuse the effects of changes in fecundity with the effects of changes in the survival rate in the relationship $R_t = \epsilon_t s_t$. Even for iteroparous species, ϵ_t is not necessarily proportional to the spawning-stock biomass (Blanchard, et al. 2003). We often do see a linear increase in fecundity with body weight, but with fecundity falling to zero at a nonzero fish weight (i.e., larger fish are disproportionately more fecund that small ones). In such cases, increases in the fishing mortality rate can cause severe decreases in the mean fecundity of the spawning fish; if a decrease in fecundity then results in decreased recruitment despite a high spawning biomass (of smaller, low-fecund individuals), we could easily misinterpret this recruitment decline as having been caused by a survival effect rather than by recruitment overfishing due to inadequate egg production. To avoid this risk, stock assessments for iteroparous species should use a spawning index S_t that is a sum of products of the numbers of spawners at age times the mean relative fecundity at age, rather than weight at age. This may greatly complicate the use of metanalytical results or comparative data from other stocks/species in the assessment.

As noted in chapter 3, much has been learned about the harvest-policy implications of treating recruitment relationships as collections of probability distributions. Using mainly the methods of stochastic optimal control (stochastic dynamic programming), we have made some surprising discoveries about how to manage populations that are subject to unpredictable recruitment variation, so as to maximize particular objectives such as the maximum average annual yield (MAY, the stochastic equivalent of MSY) or risk-averse measures such as the sum of ln(catch) over time. Recall from chapter 3 that there are three main findings from the various optimization studies to date about how to deal with various types of uncertainties about the recruitment relationship:

1. When the mean recruitment relationship and variances are well known, and when the stock size is "known" at the time of each harvest decision, the optimum harvest to take each year is a fixed and simple feedback function of the exploitable stock size B_t that year. In particular, if the management objective is MAY, the feedback function is the so-called "fixed escapement" or "fixed base stock" rule: If the stock size B_t is greater than an optimum base value B_o, take catch $= B_t - B_o$, otherwise do not harvest (B_o can be computed by dynamic programming methods but is usually well approximated by calculating the B_t that will give MSY in the corresponding deterministic model). If the management objective is the risk-averse measure sum of ln(catch) (a strong negative penalty for catching nothing, diminishing returns from high catches), then the feedback function is even simpler, namely catch $= u_o B_t$, where u_o is a fixed optimum harvest rate that again can be found by dynamic programming or well approximated by equilibrium analysis. Somewhat more complex feedback functions or decision rules may be optimum when there is a depensatory recruitment failure at low stock sizes (Spencer 1997), or when the harvest of small, fast-growing individuals cannot be avoided (pulse harvesting can be best in that case; see Walters 1969). But even for these cases, near-optimum performance is generally predicted by using either the simple fixed-escapement or fixed-harvest rate rule.

2. When the mean recruitment relationship is poorly known due to a lack of historical data or a poor contrast in the spawning-stock levels in the data, each harvest decision has a "dual effect of control": it produces immediate value and can also drive S_t to an informative level that may help future management achieve a better value. In this case, the optimum decision rule may be "actively adaptive" (Walters 1986; Parma and Deriso 1990b) and may involve deliberate probing experiments to vary S_t so as to gain more information about the recruitment responses outside (higher or lower S_t) the range of historical experience.

3. When the mean recruitment relationship appears to be changing in a persistent way (with nonstationarity or long-period regime shifts represented by a very high autocorrelation in the recruitment anomalies w_t), the best policy may still be a fixed-escapement or fixed-exploitation rate rule, but with the parameters of the rule (B_o or u_o) adjusted each year by essentially assuming that the most recent anomalies (or parameter values) will persist into the future (Walters and Parma 1996; Spencer 1997).

Much more work needs to be done on optimum harvest strategies, particularly on (3) above because persistent changes in recruitment relationships appear to be common today. But we do firmly understand at least this much: optimum-harvest decision rules are likely to be quite simple functions of the current stock size, of recruitment parameter values derived from the analysis of historical data, and of how uncertain we are about those parameter values.

Note that if the management objectives favor using a fixed exploitation-rate (or fixed fishing-rate F) decision rule, then the central concern in the stock-recruitment data analysis should be the estimation of the stock-recruitment curve's slope parameter s_o. That is, the optimum exploitation rate does not in general depend at all on the Ricker or Beverton-Holt density-dependence parameter b that sets the stock-size scale. To see why this must be so, consider changing the units of stock-size measurement; a units change will certainly affect the calculation of b but must not affect the units-independent calculation of s_o. Otherwise, we would be saying that the estimate of the optimum exploitation rate, which as a ratio with units stock/stock (per time) cannot depend on the units of stock-size measurement, could be affected by the units we happen to choose for stock measurement—we cannot affect nature that way. More specifically, if we examine equations for the optimum exploitation rate and optimum stock size, e.g., table 7.2 in Hilborn and Walters (1992), we will see that u_o (called u_{MSY} in Hilborn and Walters) is given approximately for the Ricker model (for semelparous species) by just $u_o = 0.5s_o - 0.07s_o^2$, and exactly for the Beverton-Holt model by $u_o = 1 - e^{-s_o/2}$. See also Schnute and Kronlund (1996), who show how to replace the (s_o, b) parameterization entirely by equations expressed directly in terms of u_o and S_o (the optimum exploitation rate and the optimum spawning escapement).

7.2 PREDICTING REPRODUCTIVE PERFORMANCE AT LOW STOCK SIZES

It is often assumed that the recruitment curve's slope parameter (s_o) can be estimated only by actually overfishing, reducing the spawning-stock size enough so that low recruitment observations are obtained at low spawning-stock sizes. Indeed, when you look only at a plot of the stock-recruitment observations, and those observations do not include low enough spawning-stock sizes for any recruitment effect to be immediately and visually apparent, it is easy to see that only lowering the stock size would give the needed observations.

However, suppose we examine not the raw stock-recruitment plot but, instead, a plot of (R/S) versus S. What we are then examining is how an index of per-spawner juvenile survival rate (R/S) is literally the survival from egg to recruitment if S is an estimate of the total egg production rather than a spawner count or spawning-biomass estimate) varies with the spawning stock. In such plots, we can generally see apparent density dependence in the survival rate (but beware of errors-in-variables and time-series bias effects, as discussed later in this chapter), as either a straight-line (Ricker model) pattern or a weakly curved (Beverton-Holt) pattern (fig. 7.2).

Note in figure 7.2 that the $\ln(R/S)$ regression intercepts for the Ricker and Beverton-Holt models are estimates of the s_o parameter (and hence the optimum exploitation rate). But the constant-mean-recruitment model predicts an unbounded increase in $\ln(R/S)$ as S decreases (no intercept, no apparent need for a limit to the exploitation rate). As shown in the example, fitting

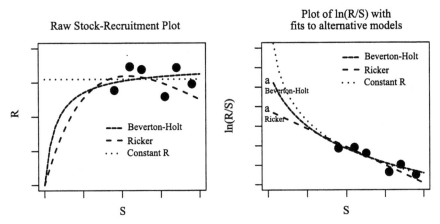

FIGURE 7.2: The lack of an apparent spawning-stock size effect in stock-recruitment plots does not mean that the survival index $\ln(R/S)$ shows no effect. The plot on the right shows three extrapolations beyond the range of $\ln(R/S)$ data, using the Ricker model, the Beverton-Holt model, and a constant mean recruitment assumption (R truly independent of S).

the Beverton-Holt model more commonly results in higher estimates of s_o, and hence of the optimum exploitation rate, than fitting the Ricker model; indeed, the Ricker model has been advocated for use even when the data show no dome-shaped S-R pattern, just because it does tend to produce more conservative policy estimates due to the linear extrapolation to $S = 0$.

Because they use $\ln(R/S)$ regression information on compensatory changes in survival rates even at high stock sizes along with assumptions about the form of the compensatory relationship, stock-recruitment fitting procedures often tell us that we have quite precise estimates of the s_o parameter despite the lack of direct information on R at very low S. Should we ignore these "reasonable" extrapolations of compensatory response at low stock size? Experience with fitting data when observations are indeed available at very low S (there are about 100 convincing cases in the Myers data base) tells us that we should not. There are very few empirical cases in which the $\ln(R/S)$ response at very low S shows much stronger compensation than that predicted by the Beverton-Holt and Ricker models. So the onus is very much on those who would advocate throwing out the simple stock-recruitment extrapolations of s_o (and the limits on the exploitation rate) to demonstrate in any particular case why that case should involve much stronger compensation than has been seen in the vast majority of cases. In short, it is just not true that we must always overexploit in order to establish the steepness of stock-recruitment relationships; very often there is plenty of information about the compensatory responses that define and limit the steepness, in data sets for which observations are available only at relatively high stock sizes.

It is wise to be cautious about using $\ln(R/S)$ information in the analysis of compensatory change because an apparent compensatory pattern will appear in the R/S ratios even if the R and S data are garbage, i.e., if they are purely independent random-number sequences. But if both R and S have been measured accurately, i.e., are not a collection of random numbers that bear no relationship to actual historical recruitments and spawning abundances, then the independence of R and S does indeed imply a strong compensatory change.

7.3 PREDICTING CAPACITY TO RECOVER FROM HISTORICAL OVERFISHING

The opposite situation from that of fig. 7.2 is that for which there are data at a very low stock size following severe historical overfishing and/or other disasters, so R appears to be just proportional to S, but no data to say how large R might be if S were allowed to increase. A case of this type involving Fraser River sockeye salmon was the prototypical example used by Walters and Hilborn (1976) to develop their basic concepts of adaptive management: we argued that there was no way either to analyze the available R and S data so as to guarantee a correct extrapolation of the stock rebuilding effects, or to predict these effects by studying the underlying biology of the stocks. Hence, management would have to become actively involved in providing data over a wider range of spawning stocks, by a "probing experiment" toward higher spawning-stock size. We supported our contention that a Fraser sockeye experiment was a good management gamble with arguments about possible bias problems in any analysis of the historical data, and with evidence from historical catches that recruitments must have been much higher in the past. Later limnological studies of juvenile sockeye carrying capacities in Fraser rearing lakes tended to support our arguments (Williams et al. 1996), as did a multistock comparative stock-recruitment analysis (Collie et al. 1990). Adaptive optimization calculations suggested that the best experiment in such situations is an "all-or-nothing" change in S: shut down the fishery and get strongly informative data as quickly as possible, so as to minimize the duration of the experimental testing period. The staff of the International Pacific Salmon Commission argued that they had already been doing a gradual, less economically painful, and less risky "titration experiment" by regulating harvest rates so as to allow a slow rebuilding of the spawning runs while continuing to provide fishery benefits.

While debate continued about whether to subject British Columbia's most valuable salmon fishery (Fraser sockeye) to an adaptive escapement rebuilding experiment, an opportunity arose to test the concept on sockeye salmon in Rivers Inlet, a smaller system where fishery closures to increase spawning would have much less economic impact (Walters et al. 1993). Again, the stock-recruitment analysis and historical catch data had indicated a considerable potential to rebuild the stock, and in this case it was decided to close the fishery entirely for at least five years (one complete life cycle) and decide on the basis of initial recruitment responses whether to continue the closure for even longer. From both a scientific and ecological perspective, this experiment was basically a failure: in designing it, we failed to think carefully about the improvements in R and S monitoring that would have been needed in order to detect responses quickly, and the crude monitoring data that were collected showed no evidence of any positive recruitment response to the closure. Worse, a few years after increased recruitments might have started to appear, there was a catastrophic decline in recruitment apparently caused by changes in marine survival rates, which also affected the nearby Smith Inlet sockeye stock that we had been viewing as something of an experimental "control."

Just as the evidence of response failure at Rivers Inlet was starting to appear, the Canadian Department of Fisheries and Oceans decided to proceed with experimental escapement increases for Fraser River stocks. But by some peculiar twist of bureaucratic planning, increased escapement goals were set not for the "off cycle" low stocks for which we had originally advocated adaptive experiments but, rather, for the "dominant" cycle lines for which stock-recruitment analysis had clearly indicated that further recruitment increases are very unlikely. This experiment has caused considerable pain for the fishing industry and, predictably, has not resulted in increased salmon abundance.

These sockeye salmon examples emphasize the possible importance of "nonrepeatability" in interpreting historical data and designing experiments involving stock rebuilding. Large fish stock collapses probably always involve at least some loss in the spatial stock structure, with a loss of genetic substocks and/or local migration/homing groups of fish. Many apparently suitable spawning areas for sockeye have not yet been recolonized following very low stock levels early in the twentieth century. The Newfoundland cod collapse was associated with the virtual disappearance of large, offshore migratory spawning shoals, so the remaining spawning is apparently occurring mainly in isolated bays that may always have had distinct, nonmigratory substocks. A severe range contraction has accompanied the major clupeoid stock collapses (MacCall 1990). In such cases, the stock-rebuilding process may require a two-stage dynamic: (1) an increase in density (with protection from fishing) may initially cause overcrowding in the remaining spawning and rearing areas, and the appearance of recruitment limits far below the historical maxima; (2) crowding and localized competition may then cause increased dispersal rates, resulting in a slow dynamics of recolonization of the original mosaic of habitat use. The slow (and unpredictable) dispersal/colonization dynamics may then cause the overall recruitment relationship to move upward over time, perhaps in fits and spurts as particular sites and movement patterns become reestablished.

It is a key and completely open policy question whether the dispersal and recolonization rates of stocks with an eroded spatial structure are, in fact, strongly density-dependent and might, therefore, be hastened by the complete protection of remaining spawning aggregations. In the Fraser sockeye case, what fishers see today is the apparent "overcrowding" of spawning areas and the apparent "waste" of fish that could have been harvested. In Newfoundland, fishers see harvest opportunities in the remaining local cod stocks. It will be difficult to implement long-term rebuilding policies in the face of this myopia, especially since we have so little empirical data on large-scale recolonization processes.

7.4   THE ERRORS-IN-VARIABLES BIAS PROBLEM

Generally, we try to pose problems in the estimation of mean recruitment relationships as statistical regression problems, so that we can use the tools of regression theory in calculating parameter estimates. So, e.g., the Ricker stock-recruitment model $R = Se^{s_o - bS + w}$ can be written as a linear regression by log-transforming the data to $\ln(R/S) = s_o - bS + w$. Here, $\ln(R/S)$ becomes the "dependent variable," and S becomes the "independent variable" for ordinary linear regression. It is easy to verify with simple simulation experiments with fake data that when standard regression assumptions are met, regression procedures are surprisingly good at finding the underlying mean recruitment curves even in data that are so noisy due to high recruitment variations that no visual pattern is apparent when the data are plotted (statistical methods can come close to finding the right curve of mean values even when we cannot come close to "eyeballing" the right curve).

While most biologists have agonized over the apparently unpredictable variation in the dependent variable recruitment (or $\ln(R/S)$), one of the most serious stock-assessment problems has, in fact, arisen from errors in the measurement of the independent variable spawning-stock size. Situations in which regression relationships are measured with random error in the independent variable measurement are called "errors-in-variables" problems. Elementary statistics texts generally do not talk about such problems, except to list "independent variable measured without error" as one of the statistical assumptions used in deriving standard regression estimators.

To see why random errors in spawning-stock measurement can have misleading effects from a fisheries policy perspective, consider the simple, extreme example shown in figure 7.3. In this example, suppose that the "true" R and S data have been collected from a stock that has actually had stable spawning abundance, perhaps because of regulatory measures or dynamic fishing-effort responses involving fishers "giving up" when the stock has been reduced to about the same spawning level each fishing season. If we had accurate spawning-stock measurements, our basic conclusion would be that there is no information in the data to help decide whether a higher spawning stock is needed or whether it would be safe to fish the stock down to a lower spawning level. But when we plot the observed data, the data points are

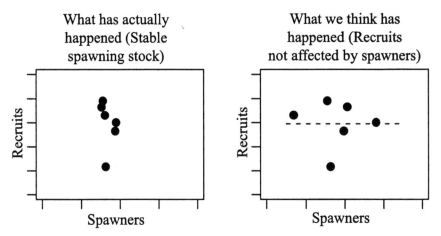

FIGURE 7.3: A pathological errors-in-variables situation, in which we should conclude that there is no information about the effect of spawning stock on recruitment, but would instead conclude that recruitment is stable (on average) over a wide range of spawning-stock sizes, i.e., fishing can reduce the spawning stock without causing recruitment overfishing.

spread out horizontally due to measurement errors, and it appears to us that recruitment does not (on average) depend on spawning stock. Taking the data at face value, we would then draw the stronger conclusions that increasing the spawning stock would not result in enhanced recruitment (no need for stock rebuilding) and that we could fish harder (reduce the spawning stock) without causing recruitment overfishing. These are clearly pathological conclusions from a policy perspective!

What random measurement errors in spawning abundance do is to create the false appearance of informative contrast in the spawning-stock data and, in particular, the appearance that recruitment does not depend on the spawning-stock size. Obviously, the problem is not severe when the data cover a very wide range of actual spawning stocks, e.g., when the spawning stock has been greatly reduced over time by fishing. It is mainly an issue when we need to make inferences from relatively short time series of data about which, for one reason or another, we cannot be confident that there has been as much variation in the spawning abundance as is indicated by the "raw" spawning-stock data.

Early attempts to deal with the errors-in-variables problem in stock-recruitment contexts (Ludwig and Walters 1981; Walters and Ludwig 1981) involved using assumptions about the measurement error variance (or the ratio of recruitment "process" error variance to the measurement error variance) to construct likelihood functions for the unobserved actual spawning stocks as well as the mean relationship parameters (e.g., Ricker s_o, b parameters). That is, instead of just finding the maximum likelihood estimates for s_o, b given the data, we tried to estimate s_o, b, and the time series of "true"

$S_1 \ldots S_T$. This approach did not reliably result in unbiased estimates and just amounted to "squeezing" the spawning-stock data toward the observed time mean; the approach helped to show that measurement error was creating a false contrast and that the s_o, b estimates should be assigned a higher uncertainty.

Much better statistical models for the problem have been developed recently using state-space modeling methods (Meyer and Millar 2000; Millar and Meyer 2000; Rivot et al. 2001; Schnute and Kronlund 2002). Interestingly, the simplest explanation of the probability reasoning behind these methods as applied to a stock-recruitment analysis has appeared outside the fisheries literature, in an *Ecological Monographs* article by de Valpine and Hastings (2002). Basically, the idea behind the state-space methods is to construct a marginal likelihood function for the data given the mean relationship parameters (s_o, b, etc.), integrated over the possible values of the "true" spawning stocks over time. That is, the likelihood of each recruitment observation is calculated as an integral of the form

$$P(R|data, s_o, b) = \int_{S_{true}} P(S_{true}|S_{observed})P(R|S_{true}, s_o, b)dS_{true}$$

In time-series settings, the exact calculation of the first probability following the integral involves complex recursive calculations (and numerical integrations) using all past observations to define the uncertainty about the observation at each time. Rivot et al. (2001) develop a simpler approach, using the idea that the above integral is also the expected value of $P(R|S_{true}, s_o, b)$ over $P(S_{true}|S_{observed})$ and that this expected value can be approximated just by taking a sample of $P(R|S_{true}, s_o, b)$ values from the assumed $P(S_{true}|S_{observed})$ distribution over all times.

The state-space methods are all computationally intensive, so they have not yet been subject to enough testing with simulated data (known parameters) to determine whether they reliably give unbiased estimates, especially for the more dangerous situations as shown in figure 7.3. Until these tests are carried out, we recommend that policy parameter findings from the methods, especially conclusions about whether or not spawning-stock objectives should be altered substantially, be treated with considerable suspicion. For most stocks that have been exploited historically, the main suspicion should be whether the assessment fails to recognize the opportunities for stock rebuilding (fig. 7.4).

Whenever possible, long-term data should be examined for indications of whether historical stock sizes were much higher or lower than those used in the assessment and, hence, would imply considerably different mean recruitment rates than those predicted to be possible from the estimation procedures. It may also be helpful to examine the habitat and trophic indices of potential recruitment, again to help decide whether the assessment may be giving an incorrect range for potential stock sizes under altered management.

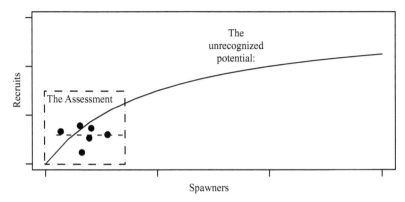

FIGURE 7.4: Stock-recruitment assessments where there may be errors-in-variables effects in spawning-stock measurement, and where the stock has been exploited heavily in the past, should be treated with suspicion.

7.5 THE TIME-SERIES BIAS PROBLEM

In the early 1980s we discovered, quite by accident while developing simulations of "closed loop" management performance (simulations that included the interplay of assessment and regulatory response to assessment advice), that much of the bias that we (Ludwig and Walters 1981; Walters and Ludwig 1981) had thought was due to errors-in-variables effects had, in fact, been caused by a much more pathological effect: nonrepresentative sampling of the recruitment pattern. We would start each closed-loop simulation by generating a short (5–15 year) time series of stock and recruitment "data" under a constant, "historical" harvest policy. Then we would simulate an initial stock-recruitment analysis of these data, choose a harvest policy based on the results, and proceed forward in time with annual simulated reassessments and policy choices. We discovered, to our surprise, that the first assessments were systematically biased toward overestimating productivity at low stock size, and this caused the simulated management system to initially overharvest in most simulation trials.

The time-series bias problem is not just a hypothetical issue that arises in models. It is common to encounter situations in which inferences about stock-recruitment relationships are based on short (5–20 year) time series of data collected under relatively stable historical harvest rates. The most common situations occur with:

1. *no data*: data were not collected early in the fishery's development;
2. *rejection of data*: older data are considered unreliable, so there is a tendency to use only the most recent "reliable" data (an obvious instance of the "shifting baseline" problem in fisheries assessment);

3. *semelparous species*: populations with fixed generation times are divided into distinct "cycle lines" that might each have a different productivity (e.g., sockeye salmon with a 4-year lifespan have 4 cycle lines, implying only 10 observations for each cycle line if there are 40 years of data for the stock);
4. *nonstationarity*: there is a known or suspected persistent change in the production "regime" (habitat or ocean productivity change) that makes older data "irrelevant" to the estimation of current productivity.

Nonstationarity is the most worrisome of these situations. As fisheries scientists have obtained more long-term data series on fishes, we have found that persistent productivity changes are not only common but, in fact, may be nearly universal. This can make older productivity data irrelevant to management (as a manager would say, "Thanks a lot for telling me what I should have done 20 years ago; now tell me what I should do today!").

Probably the easiest way to understand the time-series bias problem is to think of Nature as acting as a "mad scientist," generating experimental variations in spawning stocks by a set of "design rules" that practically guarantee that the experimental recruitment response will not be sampled in a representative way. The rules are as follows: whenever the current experimental treatment is a low spawning-stock size, wait for the first high recruitment, then change to a higher spawning-stock treatment. Whenever the current treatment is a high spawning stock, wait for a poor recruitment, then move to a lower spawning-stock treatment. An experiment carried out with these rules could obviously produce a highly nonrepresentative data set, with high recruitment observations at low spawning-stock sizes and low recruitment observations at high spawning-stock sizes both being overrepresented in the results (fig. 7.5).

Why does Nature act this way when conducting "natural experiments" to provide supposedly informative variations in spawning-stock sizes? The answer is quite simple: the spawning stock in any year arises from previous recruitment(s). So if a previous spawning stock produced a recruitment higher than average, and if a constant proportion of these recruits are harvested, then the spawning stock produced by the recruits will, in turn, be higher than average, i.e., the spawning-stock "treatment" will move upward in response to the high recruitment. In more technical terms, the so-called "independent variable" of the stock-recruitment statistical regression is not independent of past recruitment observations and deviations. The data are linked over time through the dependence of spawning stocks on past recruitment deviations.

To see how severe the time-series bias can be, consider a very simple example in which the spawning-stock size of each generation is measured exactly, and the mean recruitment relationship can be well described by the power model $R_{t+1} = S_t^b e^{\hat{a}+w_t}$ (models of this form are often written as $R = aS^b$), the w_t are independent normally distributed environmental effects, and the fishing-mortality rate F is constant so that $S_t = R_t e^{-F}$. This

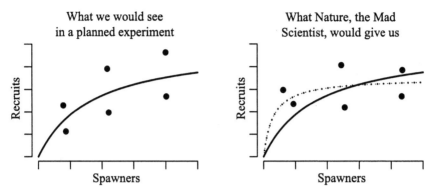

FIGURE 7.5: A simple way to visualize how time-series bias arises. When "Nature" is left to choose spawning stocks over time, high recruitments at low spawning stocks (and low recruitments at high spawning stocks) will be overrepresented in the data set, since the spawning stock will move up (down) after each such observation. The solid lines in both graphs are the "true" relationship between mean recruitment and spawning stock; the dotted line on the right is the relationship that we would likely fit to the sample.

model has two convenient features as a test case: (1) the fishing-mortality rate that would maximize the average annual yield is easily calculated, as $F_{MSY} = -\ln(b)$; and (2) the state dynamics are linear when analyzing log-transformed spawner data, avoiding many difficulties in statistical analysis. Performing a log transformation of the model and taking the data to be the log spawning stocks $s_t = \ln(S_t)$, we obtain the linear state-dynamics relationship:

$$s_{t+1} = á - F + bs_t + w_t \qquad (7.1)$$

Two approaches have been used for the estimation of $á$ and b in this re-lationship: (1) a simple least-squares regression of s_{t+1} on s_t ($á =$ intercept, $b =$ slope; in time-series analysis terms, the least-squares estimates are called the "Yule-Walker" estimates), and (2) a maximum likelihood estimation based on state-space derivations. The log-likelihood function for obser-vations $s_1 \ldots s_T$ for this problem can be written exactly by following the recursive arguments used to derive likelihood functions for state-space mod-els, with the extra simplification that s_t is observed without error. That is, we first write $p(s_1 | á, b)$, then find $p(s_1, s_2 | á, b)$ as $p(s_1 | á, b)p(s_2 | s_1, á, b)$ and proceed forward in time to obtain the full likelihood $p(s_1, s_2, \ldots, s_T)$ by re-cursion (see de Valpine and Hastings 2002; Millar and Meyer 2000; Rivot et al. 2001; Reed and Simons 1996). Given the linear structure of equation 7.1 and assuming that the system is at stochastic equilibrium with respect to F, $p(s_1 | á, b)$ has a normal distribution with a mean of $(á - F)/(1 - b)$ and a variance of $\sigma^2/(1 - b^2)$, where σ^2 is the variance of w. Each $p(s_t | s_{t-1}, á, b)$

is also normal, with a mean of $\acute{a} - F + bs_t$ and a variance of σ^2. Hence, the overall log-likelihood function for the spawning-stock time-series data is

$$
l\left(s_s \ldots s_T | \acute{a}, b\right) = -\frac{1}{2}\ln\left(\frac{2\pi\sigma^2}{1-b^2}\right) - \frac{1-b^2}{2\sigma^2}\left(\frac{\acute{a}-F}{1-b} - s_1\right)^2
$$
$$
-\frac{T-1}{2}\ln\left(2\pi\sigma^2\right) - \frac{1}{2\sigma^2}\sum_{t=2}^{T}\left(\acute{a} - F + bs_{t-1} - s_t\right)^2
$$

(7.2)

The first two terms of this likelihood function represent $\ln(p(s_1|\acute{a}, b))$, and the remaining terms represent the log probabilities for observations $s_2 \ldots s_T$. Treating s_1 as fixed and known (omitting the first two terms of equation 7.2) would result in the ordinary regression estimators for $\acute{a}$ and b.

Suppose that we now generate a collection of fake data sets that each exactly meet the statistical assumptions leading to the likelihood function equation 7.2. That is, we generate s_1 from a normal distribution with a mean of $(\acute{a} - F)/(1 - b)$ and a variance of $\sigma^2/(1 - b^2)$, and each successive s_t from a normal distribution with a mean of $\acute{a} - F + bs_t$ and a variance of σ^2. Suppose we then fit these data sets by ordinary regression analysis and by maximizing equation 7.2 (this exercise can easily be done with Excel, using Solver for the nonlinear maximization of l). A typical set of mean results over $n = 100$ simulation trials with a constant with 10 years and 30 years of time-series data is shown in table 3.

TABLE 3: Mean parameter estimates from 100 simulated data sets where parameters were estimated using ordinary regression and maximum likelihood methods. True parameter values used to generate the data are: $\acute{a} = 2, b = 0.5, \sigma^2 = 0.3$, and a constant fishing mortality rate $F = 0.8$ (note the true $F_{MSY} = -\ln(0.5) = 0.69$).

	Ordinary regression	Maximum likelihood
T=10		
$\acute{a}$	1.76	2.45
b	0.26	0.3
F_{MSY}	1.92	1.96
T=30		
$\acute{a}$	1.4	2.18
b	0.41	0.42
F_{MSY}	1.02	1.01

Consider the pathologies in these results: (1) F_{MSY} is grossly overestimated (telling us in the $T = 10$ case that we should harvest about 85% of the recruits each generation, when in fact we should take about 50% for MSY); (2) F_{MSY} is biased even for the long ($T = 30$) data series even though we know that maximum likelihood estimators should be "asymptotically unbiased" (the estimates of b are biased enough to cause a severe overestimation of F_{MSY} in the long-data-series case even though the bias in b does not look that bad); (3) the population scale parameter $á$ is underestimated when ordinary regression is used, meaning we would underestimate how much the stock has been reduced by fishing (or how much it would recover if fishing were reduced). The reader can easily confirm by repeating the simulation experiment that the pathologies are not reduced by assuming lower "noise" (environmental effect variance) in the data; in fact, the biases are even worse for $\sigma^2 = 0.1$, due to "Nature's" providing less informative contrast in the s_t series.

We might hope that the time-series bias is not so severe if we are observing stocks that are either initially unfished or historically overfished. Table 4 shows the means of the maximum likelihood estimates in the example case above for $T = 10$ when we vary F (where $F = 0$ corresponds to collecting a decade of "natural reference" data before allowing the fishery development):

TABLE 4: Maximum likelihood parameter estimates from $n = 100$ simulated data sets for unfished stocks ($F = 0$), stocks fished at F_{MSY}, and over-fished stocks ($2F_{MSY}$).

Mean Fishing Rate	Mean á (true á = 2.0)	Mean b (true b = 0.5)	Mean F_{MSY} (true F_{MSY} = 0.69)
$F = 0$ (no fishing)	2.79	0.3	1.96
$F = 0.69$ (F_{MSY})	2.65	0.25	2.12
$F = 1.38(2F_{MSY})$	2.24	0.28	1.96

The variation in mean estimates with F here is mainly due to chance variations in the estimates over 100 trials; there is no systematic improvement in the estimates for any historical case. In particular, it does not help to see the slope of the stock-recruitment relationship at low spawning stocks to have only data from the heavily fished case ($F = 2F_{MSY}$), because the potential improvement in information about the slope is tangled with the bias in the apparent maximum recruitment rate (the stock-recruitment plot still looks "flat" around the mean recruitment, despite this mean being far down on the left descending limb of the real long-term mean relationship).

An important point about this simple example is that biases appear despite the very favorable statistical setting: a very simple and single-variable population dynamics, independent and normally distributed environmental effects, no measurement errors of any sort, and complete knowledge of the historical impact of fishing as measured by F. It is not correct, as might be inferred by reading between the lines of recent papers advocating elaborate state-space modeling methods, that we are on the verge of "solving" the estimation problem by using better statistical tools. The problem is already there when we hardly need such tools, and it can only get worse in more realistic situations, e.g., when there are severe measurement errors and/or difficulties in using auxiliary information such as population composition to help estimate rate parameters like F.

These poor results are not specific to the power model; similar biases are seen in simulation experiments aimed at finding bias-correction factors for the Ricker and Beverton-Holt models (Walters 1985; Korman et al. 1985; Myers and Barrowman 1995). We have not had much luck with these bias corrections, basically because the process of nonrepresentative sampling is destructive, i.e., is very good at "hiding" the underlying average relationship from us. The problem goes away very quickly when we include informative contrast in the data due to variations in the fishing impact F over time, or when we begin observations at extreme (unfished or very depressed) abundance levels, then systematically observe recruitments produced by a wide range of spawning stocks. It is also not expected to be particularly bad for long-lived (iteroparous) stocks that have not been severely overfished (Myers and Barrowman 1995). But these caveats are of little comfort to scientists charged with saying something about the stock status and optimum fishing for the many cases in which the only trustworthy data span a short period during which fishing mortality rates have been near constant (because of an environmental change or the lack of past measurements).

Unfortunately, statements continue to appear in the fisheries literature that would lead unwary users of statistical methods to ignore the risk of time-series bias. For instance, Quinn and Deriso (1999) state that least-squares regression provides unbiased parameter estimates when the spawning stock is measured without error, and Rivot et al. (2001) recommend that spawning-stock sizes be treated as state variables that have not been observed directly (due to measurement errors) but have taken arbitrary values over time unrelated to past recruitments. When there is doubt about this risk, a quick way to check whether to be concerned is to plot both the stock-recruitment data ("SR plot") and the relationship between recruitment and the spawning stock in the same year ("RS" plot), as in figure 7.6.

Whenever the RS relationship is strong, i.e., whenever past variations in recruitment have led to immediate changes in spawning abundances, careful simulation tests should be carried out to estimate the likely magnitude of the upward bias in the estimated productivity at low stock size, and the downward bias in the estimated carrying capacity or unfished abundance. That is, use both an SR relationship with known parameters and an RS relationship similar to the data to generate a collection of fake data sets; apply SR estimation methods to these sets and compare the mean and distribution of these estimates to the known parameter values. Note that this amounts to simulating both the recruitment biology (SR) and past harvesting and management (RS) impact.

There is one possible approach to dealing with the time-series bias problem that does not require the calculation of questionable bias-correction factors but, instead, involves "filtering" the data so as to "see" smaller recruitment anomalies in the likelihood function. In likelihood functions like

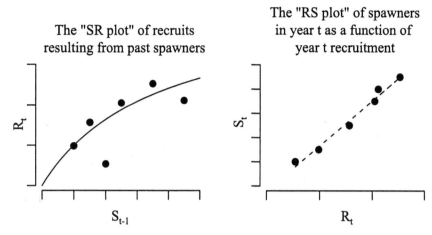

FIGURE 7.6: Plots of stock-recruitment information should include not only the usual "biological" stock to recruitment (SR) relationship but also the "management" relationship (RS) between the recruitment and spawning stocks "allowed" the same year. Time-series bias should be suspected whenever the RS relationship is strong.

equation 7.2, conditional (on $á$ and b) recruitment anomalies appear as the deviations $w_t = á - F + bs_{t-1} - s_t$. Suppose we could partition each w_t into two parts, a "known" component w_t^* and an unknown component $\hat{w}_t$, i.e., $w_t = w_t^* + \hat{w}_t$. Then in the likelihood function we would replace w_t with the smaller deviations $\hat{w}_t = á - F + bs_{t-1} - s_t - w_t^*$. Table 5 shows what would happen to the bias in the example above for $T = 10$ and $\sigma_w = 0.3$, when we generate fake data with two random deviation draws for each time, $w_t = w_t^* + \hat{w}_t$, and then give the likelihood function the known w_t^* values (basically, just vary $á$ and b to minimize the sum of squares of the unexplained $\hat{w}_t$ values).

TABLE 5: Mean estimates of $á, b$, and F_{MSY} from $n = 100$ simulated data sets where recruitment anomalies are partitioned into two parts, known w_t^* and unknown $\hat{w}_t$. Considerable reduction in the estimate of F_{MSY} occurs when independent estimates of w_t^* explain $> 50\%$ of the recruitment variation.

σ_w^*/σ_w	Mean $á$ estimate (true $á = 2$)	Mean b estimate (true $b = 0.5$)	Mean F_{MSY} est. (true $F_{MSY} = 0.69$)
0	2.61	0.25	1.93
0.5	2.33	0.36	1.4
0.9	1.98	0.5	0.68

These results are not particularly encouraging, because they indicate that we would have to independently explain well over 50% of the deviations $(\sigma_w^*/\sigma_w >> 0.5)$ in order to effectively eliminate the bias in estimated b and

F_{MSY}. But this is a very uninformative case, with only 10 observations and no contrast at all due to variations in F over time, and even for the 50% case there is considerable reduction in the overestimate of F_{MSY}.

How might we obtain such independent estimates of the recruitment anomaly components w_t^*? There are at least two possibilities:

1. use environmental covariate(s) to explain at least part of w_t (being careful not to use the biased estimates of w_t obtained from the maximum likelihood estimates of $\acute{a}$ and b while searching for such covariates in the first place) as suggested by Caputi (1988);
2. simultaneously analyze data from several stocks that are thought to share environmental effects.

As an example of the environmental-factor approach, suppose we are able to find a standardized environmental index ξ_t (e.g., salinity, standardized to have a mean of zero and a variance of one by subtracting the time mean from each observation and dividing by the standard deviation of the observations), for which it is safe to assume $w_t^* = c\xi_t$. That is, suppose we can be confident that recruitment anomalies are linearly related to the environmental factor, with an unknown slope c representing the standard deviation of w_t^*. We then modify the likelihood function equation 7.2 to include c as a third unknown parameter, replacing the conditional anomalies $w_t = \acute{a} - F + bs_{t-1} - s_t$ with $\hat{w}_t = \acute{a} - F + bs_{t-1} - s_t - c\xi_t$ and adjusting the variance terms accordingly, and maximize over the three parameters $\acute{a}$, b, and c. Table 6 shows what happens to the $\acute{a}, b, F_{MSY}$ bias when we do this for the $T = 10$ and $\sigma_w = 0.3$ (estimates of c are apparently unbiased and so are not presented below), noting that c/σ_w is the proportion of the variability explained by the environmental factor.

TABLE 6: Mean parameter estimates from $n = 100$ simulated data sets where 50% and 95% of recruitment variability are explained by the known environmental index ($w_t^* = c\xi_t$).

	Mean $\acute{a}$ estimate	Mean b estimate	Mean F_{MSY} est.
(c/σ_w)	(true $\acute{a} = 2$)	(true $b = 0.5$)	(true $F_{MSY} = 0.69$)
0.50	2.42	0.32	1.91
0.95	2.10	0.46	0.86

These results are not encouraging at all. It is very uncommon in recruitment research to find environmental factors that even explain 50% of the recruitment variation, yet we see that there is essentially no reduction in bias for this case, and there is dangerous bias in the estimate of optimum F even for the 95% case. And we are being quite optimistic about assuming that we know which factor ξ_t to use in the first place, and that we know the factor has a linear effect. We cannot, in fact, know these things in the first place, and we cannot evaluate them a priori by first estimating $\acute{a}$ and b without including c, then examining the resulting w_t as a function of environmental factor choices (that research tactic would reinforce the bias in the $á$ and b estimates).

As an example of the shared-variation approach, suppose we have data from two "nearby" stocks that are thought to share environmental effects on survival, but we do not pretend to know what environmental factors cause these effects. If we calculate the conditional (on $á_i$ and b_i for stock i) total-recruitment anomalies $w_{i,t}$ for each stock as $w_{i,t} = á_i - F_i + b_i s_{i,t-1} - s_{i,t}$, the conditional maximum likelihood estimate for the shared anomaly effect w_t^* is then just the average of these $w_{i,t}$. The overall parameter estimation then proceeds by expanding equation 7.2 to sum the likelihood component terms over stocks as well as times, and with the mean of the $w_{i,t}$ subtracted from each deviation $ŵ_{i,t}$ representing the unexplained variation. Table 7 shows what happens when we do this for the $T = 10$ and $\sigma_w = 0.3$, where σ_{shared} represents the "true" (simulated) standard deviation of the shared anomaly component for each year:

TABLE 7: Mean parameter estimates for $n = 100$ simulated data sets from two stocks with shared environmental effects. Shared effects would have to be a high proportion of the total variation in order to reduce bias in parameter estimates.

σ_{shared}/σ_w	Mean $á$ estimate (true $á = 2$)	Mean b estimate (true $b = 0.5$)	Mean F_{MSY} estimate (true $F_{MSY} = 0.69$)
0.25	2.56	0.27	2.14
0.50	2.56	0.27	2.13
0.90	2.24	0.40	1.15

(If we had a large number of stocks so that w_t^* could be estimated exactly as the mean of the $w_{i,t}$, we would obtain the same results as the above table showing σ_w^*/σ_w results). As for the environmental factor example, these results indicate that shared effects would have to be an unrealistically high proportion of the total variation in order to substantially reduce the bias in the estimates of b and F_{MSY}. But the approach at least has the advantage of using only the measured recruitment variation rather than resorting to guesses about which environmental factor(s) to use in filtering the data.

7.6   CAN STATISTICAL FISHERIES OCEANOGRAPHY SAVE THE DAY?

Recruitment variation has puzzled fisheries scientists for many years, and it has been irresistible to try to explain this variation in terms of measured environmental changes. A relatively well-defined research program of statistical fisheries oceanography now guides the work (and research investments) of a quite large research community. This program involves three basic steps: (1) assembling recruitment (and other fish abundance index) time series;

(2) comparing of these series to the many environmental time series now available (temperature, salinity, upwelling indices, etc.) using methods of exploratory data analysis (trend plots, statistical regression models); and (3) elaborating of possible mechanisms to explain observed correlations.

At least three types of claims are made about the importance of this work. The first is a vague and general claim that we cannot manage fisheries successfully without understanding the causes of variation. The second is a more precise claim that improved recruitment forecasting would allow better management and economic planning, when lead time is needed for political or marketing reasons in order to effect regulatory changes. The third is that fisheries declines often seem to involve the interaction of persistent environmental and overfishing effects, and we may not be able to design effective stock-rebuilding programs without knowing how much of the problem was due to each factor (e.g., fisheries closures would be ineffective and unnecessary if the original decline were due to a decrease in the environmental carrying capacity). This last claim is a more general version of the idea presented in the last section, that we might be able to estimate mean recruitment parameters (and optimum harvest rates) more precisely and with less bias if we could "filter" the recruitment anomalies to remove known variations due to factors other than the spawning-stock size (Caputi 1988). An extreme version of the first claim is that recruitment is driven only by environmental factors, and that the spawning stock can be ignored since it explains little of the observed variation and fish have millions of eggs anyway. Proponents of this extreme claim are often very popular with fishing industries, providing arguments for continuing to fish hard despite severe spawning-stock impacts.

Most papers in the field are short on data, find only relatively weak correlations (some workers seem happy if bumps in the fisheries and environmental data occur even in the same decade), and have long explanations that mainly demonstrate how anyone with a bit of biological and physical oceanographic background can concoct a plausible story from the myriad of possible biophysical linkages that define any natural ecosystem. Most often these stories show a sad lack of recognition of the power and importance in scientific investigation of constructing and critically evaluating multiple alternative hypotheses. Most recruitment and environmental data series have fairly high autocorrelation, meaning there are many fewer independent statistical contrasts (degrees of freedom) than there are years of data, and this means that a scientist usually does not need to examine many environmental data series before finding an exciting correlation. It also means, of course, that most correlations that are found are probably spurious, as evidenced by the frequency with which correlations break down soon after publication (Drinkwater and Myers 1987; Walters and Collie 1988).

Vague claims about the need for understanding and for forecasting seem to come mainly from scientists who have never looked closely at how fisheries can and should be regulated. In fact, we can successfully design feedback policies for adjusting harvests in response to recruitment variations without knowing anything at all about the causes of those variations; more precisely, we would implement exactly the same feedback policy given expected patterns of variation due to a known environmental factor as we would if only the pattern of variation (but not its cause) were known. Indeed, there is a risk of designing an inadequate feedback regulation system if we include only the "known" components of variation, i.e., of underestimating any future variation in the policy design. Further, Walters and Parma (1996) have shown that we generally would not improve harvest management much (in terms of the total long-term catch) even if we could exactly forecast long-term recruitment variations, as opposed to responding adequately when the variations appear. As noted in chapter 3, we would make modest adjustments in response to anticipated conditions, basically by "stockpiling" fish to take advantage of future periods of high productivity and by removing fish that would not contribute to productivity just before periods of low productivity. An "omniscient" manager who made such adjustments would get into trouble (especially by harvesting harder before periods of low productivity) if the forecasts turned out to be even a bit off, so it is doubtful that any harvest policy based on trusting long-term forecasts would ever be implemented in practice.

There has as yet not been a single instance when correlative studies have allowed us to unequivocally separate the effects of fishing and environmental change during a severe stock decline. There is no doubt that the disastrous collapse of the cod fishery in eastern Canada was the result of overfishing, but there is still much debate about continued overfishing or the environmental factors that have caused persistent poor recruitment (or the depensatory predation effects by marine mammals on the current low stock) that have prevented the anticipated recovery. There is much debate about whether the decline of coho salmon in the Pacific Northwest was driven by environmental effects on marine survival, by freshwater habitat loss, by overfishing, or by the competitive (and perhaps disease) effects of releasing large numbers of hatchery coho into the marine environment. On reflection, it is quite obvious why the statistical approach cannot in principle meet the claim of teasing out interacting effects: all of the changes are going on simultaneously and are confounded in the time-series data. The confounding is made worse by time-series effects, in which poor recruitments lead to poor spawning stocks (as well as vice versa). In the end, we must hope for, or deliberately induce through management experiments, informative contrasts in the stock size, environmental factors, and other ecological factors such as predation by marine mammals.

Perhaps the best example of how environmental and stock-size effects can be confounded is in apparently cyclic populations (e.g., the Pacific halibut, box 7.1). In this case, we remain uncertain even after nearly half a century of detailed data collection about the relative importance of environmental factors, fishing, and cannibalism or competition in causing the cyclic recruitment variation that the Pacific halibut stock has exhibited. Models

BOX 7.1

PACIFIC HALIBUT AND THE THOMPSON-BURKENROAD DEBATE

Pacific halibut is one of the most valuable fisheries in the north Pacific, apparently involving one major stock distributed from Alaska to British Columbia. It has been intensively monitored and managed since the middle of the last century by the International Pacific Halibut Commission (IPHC). It has shown a striking, cyclic pattern in recruitment variation:

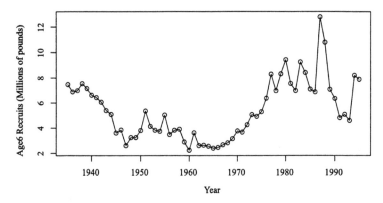

The first decline and recovery of the stock engendered the famous "Thompson-Burkenroad" debate (Skud 1975). R. F. Thompson, first head of the IPHC, maintained that the first monitored decline had been caused by overfishing, and that the first recovery was due to the sound management of harvests. His "enemy" Burkenroad, one of the first oceanographers to think critically about fish-environment interactions, maintained that the decline and recovery was not due to management at all but, rather, to the environmental effects on recruitment, and that the IPHC was taking credit for a recovery that was not, in fact, due to its management efforts at all.

In the 1980s an even more spectacular version of the debate developed, when Parma and Deriso (1990b), using longer time series, pointed out that halibut recruitment appears to have a cycle of roughly 20 years. Because halibut do not become fully mature and fecund until ages 8–10, spawning-stock peaks have occurred just out of phase with the recruitment peaks (a high recruitment in any period results in high spawning stock about a decade later, as the recruits become fully mature). Deriso and Parma pointed out that two different (in terms of fisheries policy) hypotheses equally well explain the overall stock-recruitment dynamics:

1. environmental forcing (recruitment drives spawning stock)—the recruitment variation is driven by regular oceanographic "regime shifts," and

(*Continued*)

(BOX 7.1 continued)

spawning-stock changes are a result of this (yet fishing needs to be carefully regulated during the periods of low abundance, lest poor spawning stock and poor environmental conditions combine to cause a stock collapse);

2. cannibalism/competition (spawning stock drives recruitment)—recruitment is suppressed by cannibalism or competition when the spawning stock is high (a dome-shaped or Ricker recruitment relationship; one need only look at a large halibut's mouth to see the credibility of this hypothesis—they are piscivores, with many, many large sharp teeth).

Since both hypotheses "fit" the historical data equally well, the only ways to distinguish between them are to (1) maintain existing harvest policies and hope that the environmental forcing pattern changes in some way that breaks up the negative recruitment/spawning-stock correlation (which would disprove the cannibalism hypothesis); or (2) fish the stock very hard so as to prevent the spawning stock from increasing and see if high recruitment is maintained as predicted by the Ricker stock-recruitment model. Clearly, the adaptive-management experiment (2) is a very risky one and would be unacceptable to most stakeholders (Parma and Deriso 1990b).

So this is an example in which spawning stock and environmental effects are confounded, time-series bias effects could be severe (could be creating the appearance of a Ricker recruitment pattern), and we would not be prepared to conduct the experiments needed to decide what is really going on. We could spend much research effort in finding particular environmental correlations and possible mechanisms, without proving anything. We could also look at many, many halibut stomachs without proving anything (if we do not find cannibalism, we could resort to a competition argument about how big halibut drive smaller ones out of favorable habitat; if we do find it, we cannot decide whether it is a real driving factor or a symptom of some environmental effect that makes juveniles periodically vulnerable to cannibalism). So this is a puzzle that will likely haunt fisheries science for a very long time. Probably the most serious mistake we could make would be to pretend that we can do simple environmental correlation or trophic studies to get the answer, then waste a great deal of research funding on that pretense. ■

that predict cycles due to cannibalism have been widely described in population dynamics; Ault and Olson (1996) show that similar dynamics could also be produced by more complex interactions involving competition and changes in individual metabolic states. Similar uncertainty about whether to assume such "internal" cyclic mechanisms (or instead assume periodic environmental forcing) has been well documented in the analysis of cycles in Pacific Dungeness crab populations (Higgins et al. 1997; Botsford and Hobbs 1995).

There are very few instances in which even the more modest aim of filtering data to provide better estimates of the mean stock-recruitment parameters appears to have been successful. As shown in the last section, for this aim to be successful we would need to find much stronger statistical correlations (to explain much higher proportions of the recruitment anomalies) than has yet been achieved.

Despite these objections, there is at least some promise in some of the latest work on statistical fisheries oceanography. Promising studies involve cases in which data from multiple stocks and/or species are examined simultaneously for shared or spatially patterned environmental effects, thereby much reducing the risk of spurious correlations. For example, Drinkwater and Myers (1987) show that correlations of cod variation with thermal indices do appear to hold up over time for stocks near the range limits of cod. Peterman et al. (1998) show that the marine survival rates of many Pacific sockeye salmon stocks are correlated at intermediate spatial scales of a few hundred kilometers (between stream mouths), indicating that many stocks have "seen" similar environmental conditions. Recent work with Ecosim has involved fitting the model to data from multiple trophic levels, using time patterns in apparent primary production anomalies as the parameters to be estimated; these fitting exercises have uncovered apparent ecosystem-scale responses (many stocks declining or increasing at once), though the simulated primary production patterns that "explain" such shared variation do not appear to be well correlated with known oceanographic indices.

The bottom line of this section is that statistical fisheries oceanography has not yet saved the day for recruitment analysis but offers at least a little hope of doing so if the research approach is substantially altered to incorporate higher standards of critical hypothesis evaluation and the use of multiple data series. Very large research investments are now being made in the field (many scientists, much expensive oceanographic data gathering), and there needs to be a careful and objective analysis of whether these investments are a good gamble.

PART FOUR

Modeling Spatial Patterns and Dynamics in Fisheries

CHAPTER 8

Spatial Population Dynamics Models

INTRODUCTORY TEXTS ON FISH STOCK ASSESSMENT usually encourage students to think of population processes in terms of a "dynamic pool" model, in which we visualize the "stock" as living in a box with changes due to growth, recruitment, natural mortality, and harvest "flows." It is trivially true that changes in the total size of a stock do result from such flows. Unfortunately, it is all too easy to take the stock-flow imagery too literally and to think of fish of different sizes and ages as mixed up within a single spatial box or reaction vat. This kind of thinking leads to some very misleading predictions, e.g., that catch flows ought to be proportional to the total number of fish in the box and to the fishing effort, and that size-age vulnerability differences are created by the essentially "mechanical" effects of gear-fish interactions, with rates determined by such factors as mesh size or trawl speed. In fact, most fish populations exhibit a very complex spatial organization, with at least partial segregation in space of fish of different sizes and/or ages, and catches result from the ways fishers exploit this spatial organization. Spatial organization in fish and fish-fisher interaction is both a threat and an opportunity for fisheries managers. On the threat side, changes in fisher behavior can lead to much more rapid and complex changes in the apparent vulnerabilities of fish of different sizes and substocks than would be predicted from simple dynamic pool reasoning. On the opportunity side, the spatial organization can create a variety of powerful options for selectively protecting and exploiting fish, thus enhancing our ability to conserve stocks and sustain higher yields. Such options have, in fact, long been a routine part of the practice of fisheries management even if we do not see them in most published stock assessments: there are many nursery area closures, often seasonal closures of spawning locations and other places where fish aggregate and are thus particularly vulnerable to overfishing, and protected areas designed with the sometimes vague aim of "seeding" the surrounding fishing areas.

This chapter encourages students to think of the spatial organization in fish populations in terms of the concept of "spatial life-history trajectories," the full life cycle of fish from egg to spawning, and to think of viable natural populations as collections of fish that share similar trajectories that are "closed" (spawning "ends" in the same locale where eggs "begin" the trajectory). It then identifies a number of options for modeling dynamic changes (growth, mortality, reproduction, harvest) along such trajectories, with such models having the explicit aim of helping to identify and screen spatial policy options for conservation. These options are organized around the idea that there are basically two ways to represent spatial movement processes mathematically: "Eulerian" calculations treat movements as flows that contribute to changes in variable values (stocks) at a set of fixed locations, whereas "Lagrangian" calculations treat individuals or parcels of individuals as moving over a spatial reference frame with the rates applied to each parcel dependent on the spatial characteristics of its calculated location at any time. Usually, Lagrangian calculations are numerically much more efficient, an important consideration in developing of models for policy comparison and gaming.

8.1   LIFE-HISTORY TRAJECTORIES

Most fish species do not just sit in one place or small home range and produce larvae that randomly disperse to or seed other places. Rather, they display quite complex dispersal and migration trajectories that allow individuals to exploit a variety of resource opportunities and deal with a variety of threats as they grow. In most cases, these trajectories show at least some degree of "closure," with individuals tending to return to spawn near where they were spawned; it seems that evolution has placed some broad premium on such "homing," presumably because a good predictor of where to reproduce is where you were successfully produced. The basic elements of a typical life-history trajectory are identified in figure 8.1; this figure also warns that fish movement is likely to display a fractal structure, with a complex pattern at all scales that we might attempt to examine.

Most marine fishes show at least some aggregation behavior during spawning (which exposes them to obvious fishing risks). Spawning aggregations are most often formed in locations from which passive larval drift with currents will transport larvae to or at least near favorable places for juvenile rearing, especially if larvae use vertical migration that may allow them to selectively move with particular vertical and temporal components of the current field. Larval settlement is often displaced quite a distance (up to 100 km) from favorable juvenile nursery areas, which are often defined primarily by protective cover from predation. This means that post-settlement migrations are very common and are likely times of very high predation risk. After reaching juvenile rearing areas, juveniles may be quite sedentary or may undertake quite complex seasonal dispersal/migration movements for both feeding and protection from seasonal predator risks. As juveniles reach sizes allowing them to escape some predation risks (commonly at body sizes of 100–200 mm), they most often show quite strong "ontogenetic habitat shifts" to utilize larger rearing areas and foraging opportunities. Within these rearing and foraging areas, older juvenile and adult fish commonly show regular seasonal migrations, or at least shifts in apparent depth preferences. Seasonal maturation and spawning-migration patterns vary widely in spatial extent but generally involve fairly rapid directed movement to and from the spawning locations.

One of the most misleading classifications in fish population biology has been among so-called "resident," "migratory," and "highly migratory"

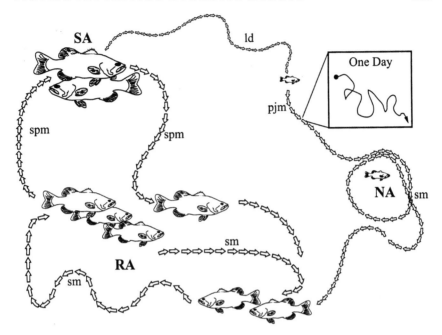

FIGURE 8.1: A cartoon representation of a typical fish life-history trajectory: spawning typically occurs in a spawning aggregation (**SA**); larval dispersal (ld) is dominated by phyiscal processes, after settlement juveniles migrate (pjm) to a nursery area (**NA**), where seasonal migrations (sm), often onshore-offshore, start to occur. As juveniles grow, they move to rearing areas (**RA**) and mix with adults. Sexually mature individuals undergo spawning migrations (spm), sometimes indistinguishable from seasonal movements. The inset box emphasizes that spatial movement patterns are "fractal": movements at various time scales mean the pattern looks just as complex no matter how closely we examine it, down to time scales of a few seconds or minutes.

stocks. Part of the reason for this distinction has been to help policy-makers recognize cases in which management may depend critically on cooperation among political jurisdictions. But politics aside, the basic structure shown in figure 8.1 is played out at many different scales, from the short distance (few kilometers) movements of supposedly resident lingcod (*Ophiodon elongatus*) of the northeast Pacific coast to the grand hemispheric migrations of the southern bluefin tuna (*Thunnus maccoyii*). We can be sure that fishers will recognize and exploit the trajectory structure at whatever scale it occurs, from a few tens of meters upward. Further, even supposedly highly migratory stocks often show a mixture of life-history trajectory patterns, with at least some pockets or substocks of "resident" fish that complete the trajectory within a relatively small area even if most fish do it over a much larger region. Commonly, early fisheries development is targeted on such short-movement substocks (they are usually easy to find, predictable, and associated with bays, islands, or other coastal structures), and fishery development "erodes" the overall stock structure by overfishing these substocks. In other cases (e.g., the cod off Newfoundland), large fisheries have virtually destroyed the long-distance migratory substocks, leaving only short-distance or "nonmigratory" pockets of fish as a basis for stock rebuilding.

There are many variants on the basic structure in figure 8.1. Semelparous species like Pacific salmon die after spawning, cutting off the return migration part of the *spm*. Freshwater fishes most often do not have pelagic larvae (but even a few river fishes such as sturgeons have larvae that drift downstream for a few days), but their juvenile dispersal to rearing sites is commonly assisted by using stream flow and other currents. Seasonal migrations, especially onshore-offshore movements, are nearly ubiquitous in temperate fish species but are uncommon in tropical reef-associated species.

Further, we suspect that many local substocks, especially of reef-associated species, do not involve a closed egg-spawning trajectory "cycle" but, rather, are seeded largely by larval dispersal from "upstream" (in an ocean current sense) aggregations or substocks. In such cases, the overall stock within a large region has a local "source-sink" structure, with some substocks or local aggregations of fish acting as source "epicenters" of recruits for sink substocks. Such a structure has strong implications for policy issues such as the siting of marine protected areas (Crowder et al. 2000).

From a mathematical modeling and computational perspective, there is a very important point to keep in mind about the dynamics implied by figure 8.1: it must usually be represented on multiple space and time scales. Simple population dynamics and ecosystem models are usually formulated as differential equations or discrete-time prediction equations with a single, fixed prediction time-step or integration interval. In general, we cannot model life-history trajectories without some nested or hierarchic time-stepping structure (figure 8.2). This is because some dynamics, particularly larval dispersal and spawning migrations, happen very rapidly in time but lead to a spreading of the effects over large spatial scales. That is, we lose the ability that physical/chemical modelers generally assume of being able to make choices between short time-step calculations for fine spatial movements to longer steps for coarser movements.

It is nice to imagine that we might someday send an army of biologists out, or develop some radical new tagging technology, to follow a cohort of fish through the full life-history trajectory of figure 8.1, and to precisely document the patterns of risk and opportunity seen by each fish along its individual development path. Such observations would doubtless help us build very realistic models of life-history trajectories. Of course, such a research investment will probably never be economical, especially considering that it might not tell us very much from a fisheries policy perspective that we cannot deduce from radically simpler observations and models. Some very detailed models have been developed for particular parts of life-history trajectories, particularly for larval dispersal and foraging movement, but these models have been aimed mainly at understanding the causes of vari-

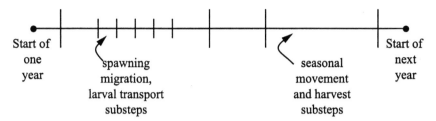

FIGURE 8.2: Computational schemes for modeling life-history trajectories generally involve a nested or hierarchic structure of calculations. Larval dispersal and spawning migrations occur rapidly and need to be represented over short computational time steps, while other processes (growth, mortality, harvest) can often be represented by integrated change over much longer steps.

ability in recruitment (e.g., Kruse and Tyler 1989; Walters et al. 1991; Wing et al. 1998) and individual growth (e.g., Mason and Brandt 1999; Tyler and Brandt 2001). While such models help satisfy our curiosity as scientists about the causes of variation, it is not clear that they are all that valuable from an applied perspective (recall that we can account for most of the overall effects on abundance of growth and recruitment variability in models for fisheries policy design by directly incorporating the observed patterns and levels of variability into policy calculations, without regard to the particular causes of such patterns).

8.2 MULTISTAGE MODELS

Some valuable policy insights have been obtained with quite simple spatial models that do not explicitly represent movements along life-history trajectories but, instead, attempt only to account for some effects of this movement at the scale of relatively large "production" (spawning, nursery) and "harvest" areas or grounds. When movement patterns are not directly modeled, the particular spatial arrangement of such areas is not used in any calculations, so a "ground" can even consist of a collection of smaller spatial areas that are not even contiguous (in a geographic sense). In such "implicit spatial" models we generally represent the life-history trajectory as a set of stages or stanzas, with the spatial habitat characteristics represented by the amounts or areas available but without regard to the particular spatial arrangement of these amounts, and we try to predict only the aggregate survival/growth rates over each stage (as an "input-output" relationship). Habitat areas may be partitioned by both ecological support characteristics (e.g., "nursery habitat") and by the economic and political characteristics important to a spatial distribution of the fishing effort, e.g., we may partition the feeding grounds into subareas with different access costs or risks, fished by different fleets from different political jurisdictions. Linkages between population dynamics and habitat change are represented by changes in the amount of habitat (e.g., the total nursery area available for juvenile rearing), and recruitment/growth/mortality rates are predicted directly from the amounts of habitat.

An example of this approach is the set of models developed by Walters et al. (1993, 1998) for managing of rock lobster stocks in South and West Australia. Here we divided the nearshore waters of each state into a set of rearing-spawning grounds representing statistical data-collection and licensing units. Because of the long larval life (>12 months) of rock lobsters, we considered it reasonable to simulate annual recruitment by adding the predicted larval production from all grounds into a single larval "pool." We calculated the survival rate to total recruitment for this pool from an overall stock-recruitment relationship, and allocated this total over the grounds in proportion to estimates of local recruitment rates based on an analysis of historical density and catch data for each ground. This approach led to some very interesting, and hotly debated, implications about how recruitments have been sustained: the models indicated that a few particular license areas and grounds have likely contributed the bulk of recruitment, while other grounds have apparently been overfished (and involve local fisheries that might not be sustainable on their own). Fishing efforts have tended to concentrate on inshore, more accessible grounds, with the bulk of the spawn then being produced by lobsters that have accumulated in less accessible grounds (see fig. 9.7). A similar approach and concerns can be found in Caddy and Seijo (1998), but with an emphasis on the local variation in body growth (recruitment being assumed independent of stock size) and the possibility of increasing yields through rotational harvesting.

Another example is the set of models used by scientific advisory committees to the International Pacific Salmon Commission for the reconstruction and prediction of stock contributions and catch allocations of chinook and coho salmon in the Pacific Northwest (Morishima and Henry 2000). These models divide the rivers of the Pacific Northwest into fairly large production areas, and divide the coastal fisheries into large harvest areas. Hatchery release policies and stock-recruitment relationships for wild stocks are used to predict the total numbers of salmon likely to enter the fisheries from each production area. But instead of trying to solve the very complex problem of predicting how fish move up and down the coast through the fishing areas, scientists use historical data from coded-wire tag recoveries (millions of tagged fish are released along the coast each year) to predict the proportions of each stock likely to be available and harvested in each fishery (gear type in a harvest area). These proportions are used to calculate the effects on meeting the conservation (spawning-stock size) and (international) allocation goals of altering the exploitation rates (or allowable catches) in each of the fisheries. It is well recognized that this approach fails to capture fully the "gauntlet" structure of the fishery; fish move from area to area along the coast, so that the numbers caught in any one area depend at least partly on the exploitation rates suffered in areas through which the fish have moved earlier the same year. But the seasonal foraging and spawning movement patterns of chinook and coho are so complex and uncertain that scientists rightly fear making larger prediction errors by trying to model these movements explicitly than by treating the movement effects as already "hidden" in the coded-wire tag data.

A promising approach for linking nursery area dynamics and recruitment to changes in nursery habitat characteristics (size, quality) and predation risk is the Beverton-Holt model derived in chapter 6. That model predicts that the net output R_s of juveniles from life-history stage s will vary with the number N_s of juveniles entering the stage according to a relationship of the form

$$R_s = \frac{N_s e^{-\alpha_s P_s / C_s}}{1 + \frac{\beta_s}{A_s}(1 - e^{-\alpha_s P_s / C_s})N_s} \tag{8.1}$$

where the parameters α_s and β_s depend on the behavioral characteristics of the juvenile fish and their prey, and habitat factors are represented by the three variables

A_s Total "foraging-arena" area over which juvenile feeding is
   distributed;
P_s Predation risk per time spent foraging;
C_s Large-scale mean food density in the nursery region (local
   food density within foraging arenas reduced by competition).

The form of equation 8.1 was derived by thinking first about very fine scale behaviors (inset box in fig. 8.1), then integrating over time (and along the life-history trajectory) to predict the net survival while explicitly retaining the effects of the nursery-scale variables A, P, and C. So far, equation 8.1 has not been fit to time-series data because we know of no data sets for which there are simultaneous, long-term measurements of R, N, A, P, and C. Its main practical application to date has been as a core prediction equation in multispecies models aimed at evaluating the possible effects of habitat restoration and exotic predators on endangered cyprinid fishes of the Grand Canyon, Arizona (Walters and Korman 1999; Walters et al. 2000). In that model, we have attempted to predict changes in habitat areas A_s and mean invertebrate food concentrations C_s as these variables might be affected by changing water-release practices from Glen Canyon Dam (diurnal and seasonal flow fluctuations, temperature). Changes in predation risk P_s are predicted from population models for both exotic fishes (mainly trout, channel catfish) and native fishes that may prey upon cyprinid juveniles. This model has made the prediction that water-release practices aimed at restoring a more "natural" hydrologic pattern may, in fact, have a net deleterious effect on recruitments of the endangered native fishes, by also enhancing the abundances of exotic predators. Surprising but reasonable predictions like this are a warning that we should not try to use simple regression models or arguments based solely on "bottom-up" habitat factors (A_s, C_s) to derive predictions about spatial effects in fish nursery areas.

8.3   EULERIAN REPRESENTATION

The next step toward more realistic models for spatial relationships in fish stocks is to develop a "Eulerian" or "box model" representation of stocks and flows among specific locations on a spatial map. The idea here is to essentially extend the dynamic pool concept to multiple spatial locations, with changes in stock within each of these locations represented by a combination of local processes (growth, mortality) and exchange processes (movements represented as flows of stock among locations). There are basically two ways to represent "locations": as habitat patches of irregular size and shape, or as regular-sized spatial grid cells or "rasters" arranged as rows and columns to form a map. Irregular areas may represent spawning, nursery, or rearing grounds and/or may be defined by fishing patterns and political jurisdictions. The key distinction between this approach and the implicit approach of the previous section is in the explicit prediction of "flows" of fish between specific boxes, at rates dependent on fish characteristics (e.g., size, age, density) and proximity (arrangement) of the boxes. Early spatial models mainly used irregular areas or patches so as to reduce the computational requirements, but at the expense of making it much more difficult to calculate or predict movement flows. Now we generally use raster-based calculations (i.e., represent the spatial area using a two-dimensional grid), admitting that we will waste some computer time (on trivial calculations for unoccupied boxes) but gaining "visual credibility" of the results and a simplification of the spatial movement "rules."

The dynamic equations for explicit spatial models can range from very elegant reaction/diffusion/advection differential equations as have been used in some tuna stock assessments (Bertignac et al. 1998; Sibert et al. 1999), to much simpler discrete-time rules that alternate between localized dynamic pool calculations (growth, mortality, effort, and catch) and calculations of the exchanges of surviving individuals among spatial cells (Guenette et al. 2000; Martell et al. 2000; Possingham and Roughgarden 1990; Walters et al. 1998).

One thing is very clear from the experience so far with explicit spatial models: by far the most difficult part of the modeling problem is to develop reasonable predictions of spatial movement patterns. The local dynamic pool calculations of growth and mortality within each spatial cell involve relatively simple, well-understood functional relationships, especially now that we appear to have good models for predicting the spatial distribution of fishing effort (see chapter 9). But when we take a dynamic pool approach to movement modeling, we often cannot get away with calculating simple

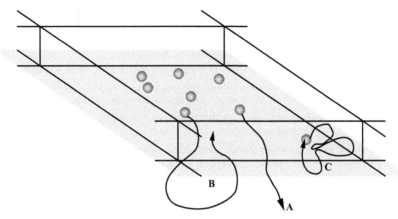

FIGURE 8.3: For species that orient to fixed topographic features (bottoms, shorelines, etc.), we may need to model at least three distinct kinds of movements of individuals across spatial cell boundaries. Spheres represent the dynamic pool or "stock" of individuals in the cell at any moment. Imagine counting movements across the near face of the spatial cell as "flow." Some flow (A) is the unidirectional dispersal/migration of individuals that have "forgotten" or forsaken their previous spatial locations, but there can also be organized "forays" (B) by individuals that will predictably return to their original residence cell. There can also be routine, frequent boundary-crossing (C) by individuals whose core home range is in the cell.

movement probabilities or instantaneous rates from tagging data or distribution shifts, then applying these in advection/diffusion or flow-between-cells calculations. Such calculations may be realistic for pelagic fishes but can be misleading for fishes that associate with physical habitat structure (and use this structure to guide their activities). For the vast majority of species, we somehow have to face the conceptual problem shown in figure 8.3 when we try to predict the "flow" of individuals between spatial areas. To understand this problem and the modeling options for dealing with it, you may find it helpful to shift the way you visualize movement: instead of thinking about following fish along life-history trajectories as in figure 8.1, imagine instead sitting at the face of one spatial cell and counting the number of fish that cross this face over some discrete (e.g., practical simulation) time step of a few days or weeks. Emerging video, acoustic, and tagging technologies are now making it possible to actually do this, to examine issues such as how many fish cross marine protected-area boundaries into areas of high fishing risk.

The conceptual problem is that at least some flows that we might observe or calculate across cell boundaries involve fish that "remember" a home range or focal point within the source cell and will return to spend much or most of their time in that source cell. This distinction has a really important practical implication when the cell boundary represents the edge of a protected area, since it implies that "resident" fish near the protected area's boundary may still be exposed to considerable fishing mortality. There are at least three distinct ways to model the spatial "memory" effect:

1. **Ignore it:** do not model foray and routine boundary crossings at all, and treat all flows as dispersal (diffusion) or migration (advection), with the calculated flow in each time step dependent only on numbers of individuals, not the origins of these individuals. This is most reasonable when calculations are restricted to large enough spatial cell sizes that return movements are small compared to the number of fish in each cell.

2. **Partition flows into two parts:** distinguish dispersive/migration versus foray, and force the "backflow" calculation from each destination cell to each origin cell to include a component of individuals that have survived some period of risk exposure in the destination cell.

3. **Follow a clever suggestion by Geoffrey Meester (Meester et al. 2001):** do not model the foray/return movements as explicit flows but, instead, calculate predation/fishing rates for the "resident" stock within each cell as partly dependent on the rates in surrounding (destination) cells.

Of these, probably (3) is the wisest. Assuming away foray movements may mean greatly underestimating the fishery impacts of nearby fishing on locally resident fish. Partitioned flow calculations are a clumsy and computationally expensive way to maintain a stock-flow representation that really does not fit the behavior being modeled. Meester (2000) suggests techniques for estimating "partial vulnerabilities" for animals within protected areas, by integrating the exposure probabilities over the individuals resident at different sites across each cell, using relatively simple data from acoustic tag tracking, or other techniques, to establish movement distance distributions for typical individual fish. So far, only Meester has applied such techniques to actual management problems (a marine-protected-area design for the Florida Keys region); his results indicate that the "SLOSS" (single-large versus many-small reserves) debate is more complex than some proponents have admitted, but that there is a minimum protected-area size below which no beneficial effects can be expected at all for a given (with particular movement patterns) fish species.

To illustrate how difficult it is to model even simple dispersive/migratory movements, consider the FISHMOD spatial modeling "shell" that is under development at the University of British Columbia and has been used for life-history trajectory and spatial-harvest analysis for a variety of fish stocks (e.g., Guenette et al. 2000; Martell et al. 2000). The development of a model under FISHMOD begins with laying out a mapped area large enough to contain full life-history trajectories for fish from at least one spawning area, and defining a grid of spatial cells over this map (cell sizes depend on expected movements and have ranged in size from 1×1 km for lingcod up to 100×100 km for cod and tuna). Next the user defines dynamic pool parameters (growth curves, natural mortality rates, age-vulnerability and catchability coefficients). Then the user must "sketch" a mean or typical life-history trajectory on the map, showing monthly mean position(s) for each age (and spawning-area substock) of fish; the monthly representation is chosen as a way to make it possible to model seasonal migrations. Finally the user must run the dispersal/migration submodel (with dynamic pool changes "turned off") and adjust a set of residence/dispersal parameters to ensure that the simulation will produce reasonable spatial patterns of movement and a spread of fish around the nominal or mean trajectory. The movement simulation itself involves dividing each monthly time step into short enough substeps to make it very unlikely that any individual fish would move more than one spatial cell per step. Then for each of these substeps, the fish in each cell (of each age and stock) are apportioned into five pools: fish that will stay in the cell over the time step, and fish that will move to each of the four adjacent grid cells over that step. The probabilities of movement in the four possible directions are calculated as functions of distance from the mean trajectory point for that month, i.e., are biased to progressively favor movement toward the point, for fish starting the time step more distant from it. Technically, the biased movement is accomplished by calculating the probabilities of movement in each direction i (i = East, West, North, South, stay in same cell) as

$$p_i = mw_i \bigg/ \sum_i mw_i \qquad (8.2)$$

where mw_i is an arbitrary index inversely proportional to the distance between the cell under consideration and the "target" or trajectory mean cell. The probabilities of staying in the same cell p_{stay} are varied by changing mw_{stay}, unusable (e.g., dry land) destination cells are assigned $mw_i = 0$, and high values can be assigned to destination (or origin) cells that represent "preferred" habitats. Note that the p_i have to sum to 1.0, i.e., all fish are accounted for at each substep.

An implicit notion in equation 8.2 to create biased movement and migration is that fish can directly detect either geographic position or "gradients" in the direction of preferred locations caused by environmental factors such as temperature or salinity. That is, saying that p_i is related to conditions in another cell in the i direction from the cell being processed is the same as saying that the p_i reflect some ability to "assess" conditions in movement directions i. The assumption that fish somehow can assess at least gradients toward preferred locations has been used in most spatial movement modeling (Ault et al. 1999; Walters et al. 1999; Bertignac et al. 1998). Without such an assumption, diffusion calculations predict the undamped spreading or dissipation of aggregations, whether or not there is directional bias that creates advection/migration.

The net effect of repeated substep reallocations of fish using equation 8.2 in spatial models is to produce both net "advection" along the life-history trajectory, and also a diffusive spread of fish away from it. The balance of "attraction-diffusion" rates defines how widely distributed the simulated fish cohort will be over time. In some versions of the calculations, we have also incorporated density-dependence in dispersal rates by reducing the residence probability at high fish densities (cod, e.g., appear to maintain near constant densities near the center of migrating shoals, and the area covered by shoals increases with increasing stock size; see Guenette et al. 2000).

There does not seem to be any way to avoid spending quite a bit of time to "tune" the movement weighting parameters (e.g., varying the mw_i in the movement submodel for FISHMOD) by trial and error so as to produce realistic distribution patterns. While this may seem like a purely ad hoc approach without strong functional justification, it must be remembered that fish movements are a very complex resultant of myriad immediate behavioral actions-reactions to factors like predation risk, gradients in environmental factors, and "urges" that apparently drive fish toward places like natal spawning areas. Hence it is difficult to model the resultant distributions of net movements among individuals in any precise or mechanistic way.

While it is easy to become lost in the details of how to simulate movement rates and directions in relation to environmental and habitat factors that obviously influence individual behavior, one main concern in developing useful models for fisheries policy design should be with two easily overlooked "emergent effects" at large space-time scales. These are density dependence in dispersal and mixture of fish with persistently different life-history trajectories.

The default assumption in spatial models should probably be that dispersal (and perhaps even larger-scale migration) rates are strongly density dependent (MacCall 1990). There is good evidence that fish make fine-scale movements and habitat choices as though they were calculating how to balance risks and rewards (Houston and McNamara 1987; Abrahams 1989; Grand 1997; Grand and Dill 1997); there is no reason to think that they are any less calculating about less frequent, larger-scale dispersal decisions. Foraging arena theory (chapters 6 and 10) warns us to expect some effects of intraspecific competition for food (and perhaps other resources) at fish densities far lower than we might expect based on a sampling of large-scale mean food abundances. The combination of choice behavior with localized competition is likely to cause density dependence in dispersal and mortality risk (Fryxell and Lundberg 1998; Hixon and Carr 1997; Sutherland 1996), whether or not we can see the density effects as obvious large-scale patterns like range collapse following overfishing. There are at least two major policy implications of density-dependent dispersal:

1. reduction in the efficacy of protected areas as a means of accumulating larger fish and hence acting as seed sources for surrounding fishing areas;
2. density dependence in catchability coefficients and hence the risk of depensatory increases in fishing mortality rates at low stock sizes even following reductions in the fishing effort (recall that catchability is defined by the ratio of the area

"swept" by each unit of effort to the area over which fish are distributed, i.e., $q = a/A$; reductions in abundance are likely to lead to reductions in A, hence increases in q).

These are obviously major management concerns, especially in programs for rebuilding depleted fish stocks. So whether or not we can document density-dependent dispersal effects from sound field data, we should certainly not assume these effects away as a modeling convenience or response to scientific skepticism about their ecological importance.

When we examine the dynamic pool of fish available to harvest in any spatial cell or ground, it is not uncommon to see a mixture of fish that were spawned in different locations and have drifted/dispersed/migrated to the ground (particularly if the ground is a major feeding area or migration bottleneck). This means that if there is any tendency for a behavioral substock structure, such that fish exhibit life-history trajectory closure (return at spawning times to near where they were spawned), the ground represents a potential "mixed stock" management problem. That is, high fishing mortality rates on the ground can potentially impact some substocks, particularly ones with relatively low productivity (growth/survival rates), harder than others. In the long run, mixed-stock fishing is likely to result in "erosion" in the overall stock structure (Ricker 1973). Spatial models should represent this possibility and risk, even when there is not strong evidence of homing to natal spawning areas, by the simple accounting tactic of keeping separate track of numbers, ages, and sizes of fish in each spatial cell/ground by spawning area origin as well as other attributes. In our experience with FISHMOD, this extra accounting is a trivial computational burden, with large potential benefits through warnings about the risk of erosion in stock structure.

8.4 LAGRANGIAN REPRESENTATION

The different movement patterns shown in figure 8.3 present difficulties about foray versus dispersal movements, and it is tempting to discard Eulerian, multiple-area dynamic pool models entirely in favor of models that create (recruit) and track the movements of individuals and parcels of fish over aquatic landscapes. In such Lagrangian movement models, each parcel or cohort or individual fish is seen as exhibiting some dynamic pool rates (growth, mortality), but its "system state" at any time also includes measures of its current spatial position (X-Y or lat-long coordinates, or map cell row i, column j). The simulation "rules for change" in system state then include rules for changes in these spatial position variables, for each individual or parcel. This approach sounds simple enough, but it is at least as difficult to implement in practice as the Eulerian calculations. Some "trivial" movement calculations, such as the passive advection/diffusion of fish larvae away from spawning areas, are quite easy to simulate using obvious techniques for following "drifters" in hydrodynamic models (e.g., Walters et al. 1991). But the more complex dispersal and migration patterns of larger fish are at least as difficult to simulate with Lagrangian calculations as with the dynamic pool approaches reviewed in the previous section.

It is not yet practical to simulate full life-history trajectories and emergent population dynamics for large numbers of individual fish, quickly enough to allow the many simulation runs that are generally needed in practical fisheries policy exploration. However, Lagrangian calculations have been extremely useful in modeling some specific relationships and patterns as part of larger dynamic pool spatial models: (1) a source-sink population structure created by larval dispersal, (2) boundary-crossing patterns for fish resident in protected areas, and (3) the space-time distribution of harvest impacts on migrating fish ("boxcar" models for movement and exploitation).

The REEFGAME model developed by Walters and Sainsbury (1990) is an example of mixed dynamic pool and Lagrangian simulations to help understand source-sink dynamics relationships. REEFGAME was developed for the Australian Great Barrier Reef Marine Park Authority as a tool to allow public exploration of possible overfishing problems and management solutions for the coral trout (*Plectropomus* spp.) that are prized by both recreational and commercial fishers of the Great Barrier Reef (GBR). To develop this model, we divided the GBR into some 800 major reef structures or nearby clusters of structures (fig. 8.4) and developed a generic dynamic pool model for post-settlement growth, survival, and harvest of fish resident on each of these structures or model "cells."

A gravity model was developed to allocate the fishing efforts from eight regional urban centers to the various structures, in relation to fish abundance and fishing cost as measured by offshore distances (see chapter 9); there is an onshore-offshore and north-south gradient in the fishing effort and possible overfishing on the GBR, with fishing concentrated in more northerly and inshore reefs, and trout density surveys by T. Ayling (pers. comm.) had indicated inverse relationships between coral trout densities and this effort pattern. There is also a worrisome link between coral trout densities, fishing, and hydrodynamics: from about Cairns southward, advective water flow through the GBR is generally southward, likely creating a "stepping-stone" or source-sink pattern in the larval dispersal of many fish and invertebrates. To model this effect, we used the current fields from a hydrodynamic model developed by M. James (pers. comm.; see also James et al. 2002), along with a (Lagrangian) larval drifter calculation, to simulate the drift of a large sample of larvae from each of the 800 modeled reefs. These sample results allowed us to construct a destination list of likely settlement locations for larvae from each reef and to assign proportions of source-reef larvae likely to settle on each destination reef in each of the destination lists (fig. 8.5). Then as the REEFGAME simulation proceeds over time in annual time steps, the recruitment calculation involves (1) predicting larval production from each

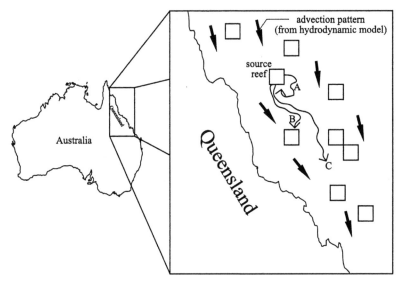

FIGURE 8.4: For each of the distinct reefs modeled in REEFGAME (fig. 8.5), a hydrodynamic model and larval drifter simulation was used to predict the proportions of larvae from the reef likely to settle on each of a list of destination reefs likely to obtain significant numbers of larvae from the reef. Sample trajectories show: A—entrainment, "self seeding" event, probably unlikely except on more northerly reefs in figure 8.5; B—successful settlement on a "downstream" reef; C—unsuccessful settlement.

reef, (2) allocating this total using the stored destination list proportions, and (3) summing these allocations to obtain a predicted total larval settlement on each reef for that simulation year. Though all these calculations appear quite complex, REEFGAME could run long-term management scenarios (3–50 yr) in just a few seconds on even the PCs available in the early 1990s. Scandol (1999) has used a similar approach to develop management games for the crown of thorns starfish on the GBR.

As noted in the previous section, Meester (2000) has demonstrated that individual or small cohort-based models of fine-scale (hourly, daily) fish movements can be used to evaluate the exposure of fish resident in marine protected areas to fishing outside the protected area's boundaries. Such models could also be used to assess the probable impacts of encroachment (poaching) by fishers into marine protected areas, a major concern particularly in remote areas where the enforcement of closures can be very expensive. Meester et al. (2001) have shown that they can also be used, in conjunction with the fine-scale mapping of habitat characteristics and preferences, to evaluate the impacts of habitat disturbances such as dredging.

Another very useful application of Lagrangian movement simulation has been in the development of so-called "boxcar" models for the exploitation of Pacific salmon during their return spawning migrations. In the boxcar

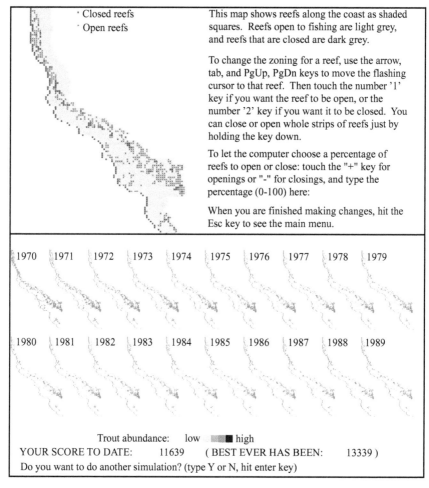

FIGURE 8.5: The REEFGAME model interface for simulating spatial organization of coral trout abundance and fishing on the Great Barrier Reef, Australia. The upper panel shows the user interface for defining a spatial closure policy, and the bottom panel shows maps of simulated coral trout densities predicted to result over time from the policy.

representation of migration, we divide a returning "run" of salmon into a "train" of parcels of fish (fig. 8.6), such that each parcel or boxcar represents the collection of fish that will pass a given point along the coast (or move through a given coastal channel or strait) in a short period (usually 1/2 day). We then simulate the movement of this train along the spawning migration path and over time (again in 1/2-day steps). Salmon fishery openings usually occur in small areas, for short times (1/2 to a few days) and typically generate very high exploitation rates (90% and higher) on the boxcars of fish that happen to be in any area when it is opened (or enter the area during the

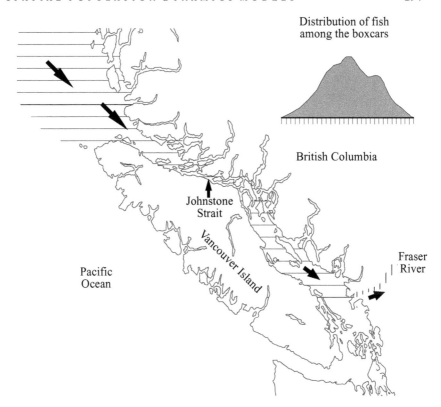

FIGURE 8.6: A boxcar representation of a run of salmon moving along the British Columbia coast and entering the Fraser River. A simulation of the effects of fishing at places like the Johnstone Strait and Fraser River mouth involves moving the train along its migration path over time and applying high exploitation rates to those boxcars that happen to be in the fishing areas on the days when the areas are opened. Note that boxcars are seen as "rubbery" with variable widths depending on topographic constraints.

opening). When we simulate various date-area opening patterns representing overall fishing plans or proposals for a fishing season, we can calculate how many (and which) boxcars are likely to be "emptied" of fish and, hence, the overall exploitation rate likely to be suffered by the whole train (representing the biological stock of concern). Indeed, short fishery openings on restricted proportions (boxcars) of a stock can be a very effective way to limit exploitation rates even when the total stock size is highly uncertain at the time when fishery openings need to be declared (fig. 8.7). We can also show how sensitive the realized exploitation rates are likely to be to unusual or unexpected run timing patterns, such as having the fish "bunched up" into relatively few boxcars (tight run timing) or moving at unusually low swimming speeds through some fishing area(s).

1991 Fraser River Sockeye Salmon Returns

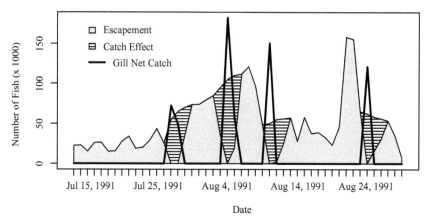

FIGURE 8.7: Estimates of the daily number of sockeye salmon passing an acoustic counting point in the the Fraser River, upstream of the Fraser River fishing area in figure 8.6. Note the "holes" that have apparently been cut in the migration timing pattern by downstream fisheries; these "boxcars" of fish suffered high exploitation rates. Note also that we can estimate (and limit) the exploitation rate by estimating (and controlling) the area in the cutouts relative to the total area under the run timing curve. Note that single one-day catches can cut holes more than a day wide in the distribution, because the fishing area is large enough that several boxcars of fish are present and can be depleted at the time of each opening.

   Notice that once we adopt a Lagrangian frame of reference as in fig. 8.6, it becomes relatively easy to add further realism to the movement calculations. For example, individual salmon migrate at different speeds, and this variability can be represented as the mixing of fish between boxcars over time. Also, changes in vulnerability/catchability due to the "compression" of fish as they pass through topographic bottlenecks (e.g., Straits) can be represented by viewing the boxcars as "rubbery," compressing and stretching over space and time; this avoids the much more complex computational problem of representing the run as distributed over a two-dimensional grid.

8.5   POLICY GAMING WITH SPATIAL MODELS

Much of the emphasis in this chapter has been on the development of ways to simplify the numerical prediction of the spatial movement patterns of fish, by examining their movements from different reference frames and scale perspectives. In part this emphasis reflects the current limitations on computer speeds and also the need to avoid "numerical diffusion" (the divergence of predicted distributions over time) and overparameterization of submodels for movement. But much more importantly, it reflects our experience that the most useful spatial (and for that matter, all) models for policy development are ones that can be easily "exercised," i.e., run many times quickly, for both a "sensitivity analysis" to see how they perform when their parameters are varied and for the exploration of policy choices. There are many, many potential ways to improve management by making use of spatial policy instruments such as closed areas, and related economic instruments like gear restrictions that make some areas too costly to exploit. Indeed, the key to the future sustainability of many fisheries may be to discover and use such spatial instruments, especially to meet complex management objectives and to avoid the impossibly high costs of more precise stock assessments.

Today, we do not yet understand the full range and probable efficacy of spatial management options. In this intellectual environment, it is absolutely critical to use models and other intellectual devices that stimulate imaginative thinking. Our imaginations work best when we can quickly get feedback on ideas, i.e., when we can quickly "screen" ideas using relatively simple models. Once we identify promising options, there is plenty of time to proceed with more complex and detailed calculations to look for flaws "hidden in the details" or to satisfy those who cling to the scientific pretense that precise predictions are both possible and necessary for effective management. One of the biggest mistakes that an inexperienced modeler can make is to assume that it is necessary to "get it right" in all details at the outset, before initiating any policy exploration or discourse.

Temporal and Spatial Dynamics of Fishing Effort

ONE OF THE BIGGEST MISTAKES in the development of fisheries science has been the broad assumption that this science is about fish, i.e., could equally well be labeled "fish science." In fact, a fishery is a complex dynamic system in which the unregulated economic dynamics of fishers are just as important to the outcome of management policy choices as are the biophysical dynamics of the fish and the ecosystem that supports them (Hilborn 1985; Wilen et al. 2002). In a sensible world, there would be just as much research on the dynamics of fishers as on the dynamics of fish. Unfortunately, there has been very little research on the dynamics of fishers. There is plenty of descriptive information on how fishers have behaved in space and time, particularly in relation to stock collapses (see, e.g., the discussions of temporal and spatial effort patterns during the Newfoundland cod collapse, by Hutchings and Myers 1994 and Hutchings and Ferguson 2000). But only a handful of researchers have attempted to develop useful predictive models for fishing-effort dynamics and the response to harvest-management policies (Botsford and Johnston 1983; Eales and Wilen 1986; Hilborn 1985; Holland and Sutinen 1999; Holland 2000; Millisher et al. 1999; Sampson 1992; Watson et al. 1993; Wilen 1979). And it is not just fisheries biologists who have failed to think carefully about the policy implications of fishing-effort dynamics; only a few economists have examined these dynamics from a behavioral perspective, as opposed to a prescriptive or normative perspective. In short, both biologists and economists have tended to focus mainly on how fisheries would or ought to look in the long term, in terms of fish abundance patterns and regulatory strategies such as MSY and quota management, without careful attention to the tactical management problems created by unregulated aspects of fisher behavior (where and how they fish, any growth in their capacity to do harm via technology improvements and, often, total fleet sizes).

As noted in chapter 4, early concepts of fishery regulation were based on "input control" of fishing effort, in hopes that the fishing mortality rate would be proportional to effort expended (as would be expected if fishers deployed their gear at random with respect to the distribution of fish). This approach was soon recognized as dangerous, given that nonrandom searching can lead to "hyperstability" in catch per unit effort, (Hilborn and Walters 1992) and that, in turn, to depensatory increases in the fishing mortality rate at low stock sizes. Also, it has proven difficult to track temporal changes in catchability q (in predictions $F = q(\text{effort})$) due to technology improvements. These concerns, along with arguments about economic efficiency when fishers are less constrained, led to a move toward "output controls" on catch, via TACs and quotas. But recently, output-control approaches have been questioned because of the "hidden costs" of providing precise enough stock assessments to avoid depensatory changes in F during stock-size declines (Pearse and Walters 1992; Walters and Pearse 1996), and there has been a resurgence of interest in both input controls and alternative ways to limit fishing mortality rates, particularly the use of protected areas. The efficacy of alternatives involving spatial controls on fishing are clearly dependent on how fishers behave in response to them, and the prediction of spatial patterns of effort distribution (and redistribution) has become an important concern.

This chapter discusses some simple models for both long-term and short-term fishing effort dynamics. It is useful to think of effort dynamics as involving three response scales: (1) long-term changes (decadal time scales) in capacity and technology, representing capital investment and decision making (development of fishing communities and infrastructures, changes in fishing fleet size, investment in GPS or other technologies); (2) short-term changes (days-months) in actual fishing activity (search and capture of fish), within limits defined by capacity; and (3) a fine-scale (hours-days) spatial allocation of fishing activity among fishing sites. Scales (2) and (3) are obviously not distinct, since decisions about whether to fish at all will generally depend on assessments by fishers of where fishing opportunities might be. Here we discuss each of these scales in turn. The chapter concludes with an example of how we might use spatial-effort-dynamics ideas to improve management and minimize the economic impact in multispecies fisheries.

9.1 LONG-TERM CAPACITY

There are obvious parallels between fishers and natural predators, and it has been irresistible to think about the long-term "population dynamics" of fishing fleets as involving recruitment (investment) and mortality (depreciation) processes linked to prey (fish) abundance. The depreciation of fishing capital (equipment and knowledge) is precisely similar to the natural mortality in a natural population. If investment in a fishing fleet comes mainly from profits earned within the fleet, so that the growth rate of investment (fleet size) is low both early in the fishery development (few investors) and after the stock size has been much reduced (so that most predators are no longer making any profits, i.e., fleet growth and increased cost to capture each remaining fish make profits "dissipate"), then the fleet-dynamics recruitment process is also precisely parallel to recruitment in natural predators. Fisheries textbooks such as Hilborn and Walters (1992) or Clark (1976) show predator-prey equations and plots of fish abundances and fleet sizes, complete with the usual predator-prey patterns that overshoot the long-term equilibrium during initial development and perhaps persistent cycles.

A central prediction of the predator-prey analogy for long-term fleet dynamics has been the idea of a "bionomic equilibrium." After some period
of transient development and perhaps over-capitalization, we expect fishing
fleets that depend on a single stock (if that stock is not so patchily and predictably distributed so that all the remaining fish can be found even if the
stock is very low) to reach an uneasy equilibrium with the remaining stock.
Near this equilibrium, the fishery should be essentially "self-regulating," in
the sense that further stock declines should trigger a reduction in the fishing effort and mortality (hence allowing recovery to start), and recoveries
should be followed by effort increases that tend to push the stock back down
toward equilibrium. The bionomic equilibrium need not even represent an
overfishing state for the stock, i.e., the equilibrium may be at a higher stock
size than would produce MSY or than would be economical to pursue if the
fishers were subsidized (and could ignore costs). Or it could be "sustainably
overfished" even absent any overt management regulations. Such equilibria
are expected to be "uneasy" because of long-term changes in fishing technology; at any equilibrium, at which profits are dissipated, there is a particularly strong incentive for fishers to invent and invest in ways to catch more
of what is still available. As technology is improved, the equilibrium may
become unstable and/or move toward a lower stock size (i.e., the technology allows fishers to drive the stock lower before costs force them to reduce
effort).

Bionomic equilibrium ideas have helped us understand why limiting total fishery catches via TACs or quotas, as conservation measures, does not
ensure healthy fisheries from a social and economic perspective. Limits on
catches do not directly prevent "overcapitalization," dissipation of profits
on "wasted" fishing costs, and poor fisher incomes. To see this, imagine a
developing fishery for which a total quota Q has been imposed in year t, and
in which b_t fishing vessels share this quota to obtain an average per-vessel
catch of Q/b_t. If this per-vessel catch is high, so that the fishers make a
good profit, additional investors are likely to enter the fishery, making the
next year's per-capita income Q/b_{t+1} smaller. Absent license-limitation programs or individual allocations of Q (Individual Transferable Quota (ITQ)
system), b will continue to grow until Q/b is reduced to the point at which
no further investment is attracted. Further, the regulatory methods used to
prevent the catch from exceeding Q, such as shortening the fishing season
as b grows (see fig. 4.3 on page 109 in Hilborn and Walters 1992 for a Pacific halibut example), will generally create "perverse incentives" for fishers
to invest in technologies aimed at getting a bigger share of Q more quickly
(faster boats, quicker net-hauling gear), rather than just harvesting fish at
the least cost.

Unfortunately, while the predator-prey (or "bionomic development")
model has helped us understand the historical dynamics of many fisheries,
it is misleading as a tool for thinking about fleet-size dynamics in modern
fisheries, for at least three reasons.

1. **Broad availability of "venture capital"**: investment in modern fisheries can occur much more rapidly than would be predicted from profits earned within the fishery, due to "speculative" investment by a much broader community of investors than were traditionally willing to consider fisheries as investment options. A modern fishing vessel is now about as likely to be owned by a doctor or lawyer as by the scion of an old fishing family.
2. **Global mobility of accumulated capital stock**: the world now has very large fishing fleets that are inactive much of the time due to regulatory measures and stock declines. These vessels can be purchased relatively cheaply and/or moved quickly in response to new fishing opportunities. That is, fishing fleets can develop almost overnight.
3. **Most often, the predator does not depend on a single prey**: few fishers are completely dependent on a single fish stock any more. Most can choose not to fish at all (and make a living elsewhere) when fish are not abundant, and many can stay out fishing (and impacting) any particular stock by continuing to take other species that are present on the same fishing grounds. The classic example is the whalers who continued going to the Antarctic long after it became uneconomical to go there in search of blue and fin whales, because other smaller baleen whale species were still available: neither fleet size nor effort responded immediately to the blue and fin stock declines.

The basic implication of points (1) and (2) is that fisheries managers can no longer assume fleet-size changes during fishery development and restoration to be "slow" variables that need not be closely regulated. Indeed, most developing fisheries today in such places as Canada and Australia are subject to strict license limitations, with new licenses issued at rates deemed safe in relation to the accumulation of information on stock sizes and productivity (Walters and Pearse 1996; Perry et al. 1999). For fisheries for which license limitation has traditionally been seen as unacceptable, particularly the large recreational fisheries of North America, management planners can (and should) reasonably assume that "fleet capacity" is effectively unlimited.

The main implication of point (3) is that fleet size or capacity is not a good predictor of realized fishing effort, for which we generally define effort as a quantitative measure (i.e., the number of vessels × the time spent fishing) in order to obtain an index that is as closely related as possible to the realized fishing mortality rate or risk of harvest. There is an important policy implication as well: fishery managers cannot generally use simple limits on numbers of vessels or fishing licenses to achieve "input control" objectives on fishing mortality rates. Perhaps the best example of this problem has been a set of very lucrative, apparently successfully managed invertebrate (prawn, rock lobster) fisheries around Australia. These fisheries have been subject to strict limits on the numbers of licenses and gear deployed per license, and additionally to restrictive fishing seasons aimed at both "capping" the fishing mortality rates and taking fish at the most economical times of the year. Australian fisheries managers point to these "input control" systems with

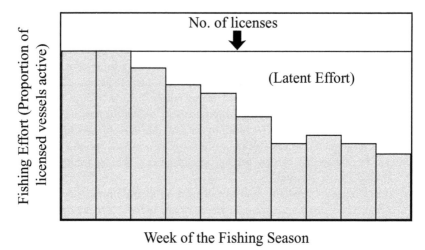

FIGURE 9.1: Australian prawn and rock lobster fisheries are subject to strict license limitations and fishing seasons, but the realized fishing effort is generally much below the limit set by (licenses) × (weeks). License limitation is not what is really limiting the fishing effort, except in the few weeks just after the annual fishery openings.

justifiable pride, noting that they have produced high and sustained yields along with prosperous fishing industries (with good individual incomes). But when we look closely at the seasonal dynamics of the actual fishing effort in these fisheries, what we generally see is that the gear is fully deployed (all fishers active) for only the early part of each fishing season. Later, many license holders give up or turn to other fisheries, so that the fishing effort actually exerted is often less than 50% of the potential represented by the product of fleet size times fishing season length (fig. 9.1). The difference between potential and realized effort is often called "latent effort." For at least some of the Australian fisheries, this latent effort is more than sufficient to cause severe overfishing, should it ever become economical for fishers to deploy it. In more vivid terms, license limitation per se is not the reason for ecological "success" (prevention of overfishing) in these fisheries; rather, the fisheries are at bionomic equilibria with respect to effort, and the equilibria are stabilized to some degree by license limitations and fishing seasons.

9.2   SHORT-TERM EFFORT RESPONSES

The "latent effort" problem (for fishery managers) shown in figure 9.1 tells us that we need to be every bit as concerned about the short-run dynamics of fishing effort—i.e., how many fishers will actually exert fishing effort—as we are about fleet capacity even when there are tight input controls on that capacity. The concern is even greater in open-access recreational fisheries (and a few remaining commercial fisheries), where fleet capacity can be treated as effectively infinite (and is, hence, useless as an effort predictor).

In trying to understand the linked dynamics of fish and fishers, the most important prediction that we need to make is about the effect of fish abundance (and distribution) on short-term fishing effort. Lots of factors affect fishing effort on fine time scales: weather, fishing prospects information (gossip, fishing news columns, landing statistics), and the availability of alternative employment and recreational opportunities. On longer time scales (i.e., months to years) we expect the total fishing effort to depend mainly on the most basic factor that drives fishers to go out, namely, the abundance of fish. It is not that fishers can ever "see" or measure that abundance directly in making their individual decisions but, rather, that they see a variety of other indicators such as catch per effort and "big-fish-stories" that are ultimately linked in some (generally complex) way to abundance. When we talk to the fishers who are no longer going out near the end of the seasons (fig. 9.1), we obtain replies ranging from "they don't enter our pots at this time of year" to "there're too few fish left on the grounds" to "my catch rates are too low to meet my operating costs"; i.e., fishers may have very divergent models for why catch rates are low after periods of high exploitation, but they still respond as though they were measuring actual abundance changes.

When we plot data on fishing effort versus total fish abundance for a given fishing area or fishery, we usually see a strong positive relationship. It is important to recognize that this relationship represents a cumulative, statistical summation of the effects of individual fisher choices, with the "functional form" of the relationship determined not by the behavioral "rules" that any individual follows but, rather, by the pattern of statistical variation among fishers in decisions about whether to fish (fig. 9.2). This is essentially the same distinction that economists use in describing the differences between rational behaviors of individual firms in an industry, and the supply and demand curves that arise when economic activity is summed across firms; the form of supply and demand curves need not bear any obvious relationship to the supply or demand curve exhibited by any one firm.

Figure 9.2 points out that there are two quite different ways to approach the problem of predicting the overall effort response (panel C) needed for an analysis of fleet-fish dynamics. The individual-based modeling or "IBM" approach would be to gather detailed information on the decision rules used by individual fishers, use these to predict individual responses, and add these predicted responses to give the total effort predictions. In large fisheries, this approach would be wildly impractical, especially considering that the B_i^* are created by complex individual appraisal processes that can vary grossly among fishers and are likely to be unstable over time due to changes in fisher perceptions and capabilities; the individual response models would have to be rebuilt every year! The alternative, much more practical approach is to treat the overall "numerical response" in panel C as a relationship to be measured directly by analysis of historical data on the total efforts and

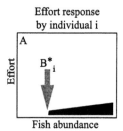

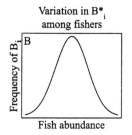

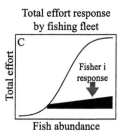

FIGURE 9.2: Individual fishers "see" fish abundance differently in deciding whether or not to fish, with any individual fisher i "turning on" to a given fishing opportunity at some fish abundance B_i^* (panel A). There is a distribution of B_i^* values among the $i = 1 \ldots n$ fishers potentially entering a fishery (panel B). The total fishing effort response to abundance is the sum of these individual responses (panel C), and the shape of this response is determined not by any one fisher's response pattern but, rather, by the pattern of variation among fishers in B_i^* values.

abundances. The big trouble with this approach is in assuming that the overall relationship is, in fact, stable (the distribution in panel B changing slowly over time) despite the movement of individual fishers within the distribution. With more research on the problem, we might well discover how to combine these approaches, with an analysis of the changing behavior of sample fishers used to provide information on likely changes in the frequency distribution of panel B.

Fishing fleets and fisheries can differ dramatically in the shape and variance of the probability distribution of B_i^* in panel B, figure 9.2, and this can change the overall form of the effort response as shown in figure 9.3. Recreational and artisanal fisheries tend to have highly heterogeneous fishers, with widely varying knowledge, motivation, and constraints on activity related to factors like variable spatial access to a given fishing opportunity (e.g., variable travel time to the fishery). This leads to a high variance in the panel B distribution, and that, in turn, leads to an overall effort response with a more or less constant slope, i.e., a roughly linear relationship (for examples, see the plot of recreational angler response to salmon abundance in Argue et al. [1983], or the sport effort responses shown in the next section). In contrast, modern industrial fisheries often have relatively homogeneous technology and highly "professional" fishers with similar knowledge, skills, and information about the stock(s), and fleets are often based in one or a few ports (where access to vessel and technology servicing is available) so that the fishers see similar costs to access a given fishery. In such cases, low variance among fishers can lead to a very sharp, highly nonlinear response of effort to fish abundance. Note also that, as shown in panel B of figure 9.3, the industrial fishery response is likely to have a much lower B^* mean, i.e., the average fisher will find it profitable to start fishing at much lower fish abundance than would be needed to attract the average recreational or artisanal fisher.

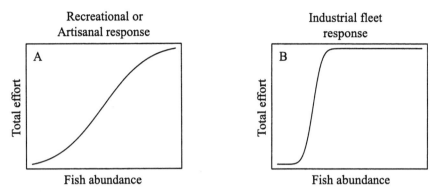

FIGURE 9.3: Fishing fleets can differ dramatically in the form of their overall effort responses. Recreational and artisanal fisheries tend to have a high variance among fishers, leading to smooth effort responses (panel A). High-technology industrial fisheries can have much less variation among fishers, leading to nonlinear effort responses (panel B).

Obviously, a fishery can evolve over time from exhibiting a smooth "type A" response in figure 9.3 to exhibiting a "type B" response, especially when regulatory structures and/or government planning encourage (or even subsidize) investment in technology improvements to meet myopic objectives related to reducing fishing costs (more "efficient" fisheries). At the same time, the relationship between fish abundance and catch per vessel is likely to evolve from nearly linear (random, spatially restricted, relatively inefficient search by individual fishers) to highly "hyperstable" (highly nonrandom search, catch per effort limited by vessel capacity/handling time rather than search time).

When we combine the effort response and the catch-per-effort responses, to predict total catch as a function of fish abundance absent quotas, we expect to see roughly the same shape of relationship as shown in figure 9.3 (whether or not the catch per effort is hyperstable). If we then recognize that industrial fisheries are likely to be "capped" by quota or TAC limits as management responses to the increased risk of overfishing, and compare the resulting catch responses to the likely patterns of biological surplus production in relation to fish abundance, we see that two very different kinds of "bionomic" (biological production minus catch) relationships can occur (fig. 9.4). In the type A case, biological production and unregulated catch are likely to balance at a single unregulated equilibrium, when the stock size is relatively high. In contrast, the type B fishery may exhibit two stable equilibria, at stock sizes B_L and B_U. If the stock size is initially high (above the unstable equilibrium or critical stock size B_C) when quota limits are imposed, biological production dynamics will tend to move the stock toward the high "happy" equilibrium around which production will tend to exceed the quota following modest natural reductions in the stock size, and

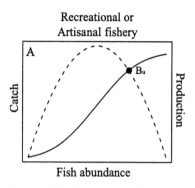

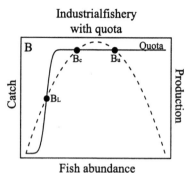

FIGURE 9.4: Combining effort, catch per effort, and mean biological surplus production responses to stock size can result in two quite different "bionomic equilibrium" patterns. Type A (recreational/artisanal) fisheries may exhibit a single, stable bionomic equilibrium even absent any effective limits on catch. Type B (industrial) fisheries may exhibit three equilibria (at which surplus production is balanced by catch): B_L is stabilized by effort responses, B_C is unstable, and B_U is stabilized by changes in surplus production relative to the catch limit (quota). The solid lines show catch (effort × catch per effort), and the dotted lines show surplus production.

will tend to be less than the quota after natural increases. But if the quota is imposed after the stock size has been reduced to below B_C, or if some natural production variation causes the stock to decline to below B_C, the system will likely move to an "unhappy" equilibrium at which the quota is no longer an effective limit on fishing and the stock size is maintained near the equilibrium by strong fishing effort responses (whenever the stock starts to rebuild, it will get clobbered by fishers going out to target it). The overall "slow" movement from single- to multiple-equilibrium dynamics over time creates the qualitative dynamics pattern known as a "cusp catastrophe" (Jones and Walters 1976).

The type B dynamic structure in figure 9.4 is troublesome for any fishery manager who might believe in the importance of taking a slow, incremental approach to quota changes so as to minimize the disruption of (and complaints from) a fishing industry. Should some collection of accidents result in the fishery's dropping toward equilibrium B_L, large quota reductions might be needed in order to destabilize B_L and send the stock on a recovery trajectory. A bitter and very apt phrase that fishers have used to describe the effects of enough incremental quota reductions to eventually do the job is "death by a thousand cuts." Systems that require changes in order to restore a desired or productive state are said to exhibit "hysteresis" in response to changes in management variables. A possible example of such pathological management dynamics is one of the world's largest fisheries, for Peru anchoveta (fig. 9.5). Following a collapse in 1971, the fishery has obviously recovered to some degree, but the effort was incompletely regulated so that effort responses may have prevented recovery for more than a decade.

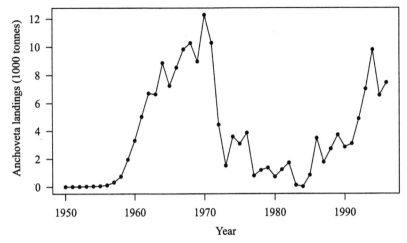

FIGURE 9.5: The catch history for the Peru anchoveta fishery, as summarized by the International Research Institute for Climate Prediction (data available at *http://ingrid.ldgo.columbia.edu/SOURCES/.IRI/.EPPF/.catch/.yearly/.anchoveta/*). Following the collapse driven by El Niño in 1971, the stock apparently began to rebound several times but may have been prevented from doing so by fishing-effort responses (B_L equilibrium in fig. 9.4). Since 1990, it appears that the stock has finally "escaped" B_L and is recovering to a high level that might be stable if management controls are effective in preventing future collapses involving the interaction of El Niño and fishery effects.

Usually, this fishery is discussed by fisheries scientists mainly in terms of ocean-fish dynamic interaction and El Niño impacts; it would seem that to understand the historical dynamics of the fishery, it is equally important to understand the economic and regulatory dynamics that have followed the effects of environmental variation.

Figure 9.6 shows a lovely example of how important temporal effort responses can be in determining the efficacy of policy options. The pompano (*Trachinotus carolinus*) is a prized sport and commercial fish in Florida. In the mid-1990s, commercial gill netting was banned in inshore waters because of concerns about overfishing and bycatch of nontarget species and because of a powerful sport-fishing lobby that saw conservation arguments as an opportunity to promote the reallocation of fish to the sport fishery. But following the ban, the sport-fishing effort roughly doubled so as to prevent the sport catch per effort (an index of the quality of sport fishing) from increasing at all. A rough estimation of fishing mortality rates using a simple stock-assessment model indicates that there has been no net change in the total fishing mortality rate at all, due partly to the sport response and partly to the development of an "offshore" (outside the 3-mile limit) commercial fishery intended to "compensate" commercial fishers for lost inshore fishing opportunities.

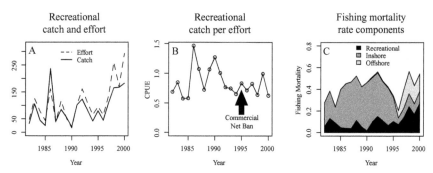

FIGURE 9.6: Recreational effort responses, catch per effort, and calculated changes in fishing mortality rate components for the pompano fishery on the west coast of Florida, following a ban on commercial gill netting in inshore waters. Data kindly provided by Behzad Mahmoudi and Bob Muller, Florida Marine Research Institute (pers. comm.).

9.3 SPATIAL ALLOCATION OF FISHING EFFORT

Just as temporal effort responses can make conservation policies appear more effective than they are, as in figure 9.1, or prevent conservation policies from working as in figure 9.6, the spatial allocation of fishing effort can be a critical reason for historical sustainability or a cause of failure for closed areas policy options. The spatial equivalent of figure 9.1 is the pattern shown in figure 9.7: in many fisheries, some spatial locations have received much lower fishing pressure because of high access costs (most often related to distance offshore or along shore from fishing communities), and these areas have acted as "economic refuges" or buffers against overfishing. Very probably, the famous Newfoundland cod fishery that persisted for centuries did so only because most of the stock was concentrated offshore much of the year, where it was economically inaccessible to the fishing gear used by Newfoundland fishers. The high costs of fishing far offshore in the mid-Pacific may be an important mechanism that helps stabilize the major Pacific tuna fisheries. An important reason to engage in modeling spatial effort patterns is that such economic refuges can be unstable in the long term, due to technological innovations and other changes in economic factors that reduce costs so as to destroy the refuge effect.

Four approaches have been used to predict short-term changes in the distribution of fishing effort among locations or grounds. These approaches vary in their complexity of explicit assumptions and data requirements similar to how population models vary from simple surplus production calculations to complex individual-based calculations.

1. "Gravity" models distribute a predicted total effort E_t among $i = 1 \ldots n$ sites or grounds using some index of attractiveness for each ground, $\ddot{A}_i$. For example, $\ddot{A}_i$ might be the ratio of fish density to fishing cost on ground i. Effort

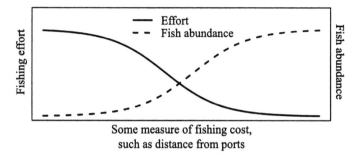

FIGURE 9.7: Many fisheries operate across multiple spatial sites or grounds, which are differentially costly to access. The spatial concentration of effort in the least costly grounds can result in the more costly grounds' acting as economic "refuges" preventing overfishing.

E_i on ground i is then taken to be $E_i = E_t \ddot{A}_i / \sum \ddot{A}_i$. Caddy (1975) first suggested this approach for fisheries modeling. Recent application examples are groundfish trawling (Walters and Bonfil 1999) and the Ecospace software in Ecopath/Ecosim for spatial ecosystem modeling (Walters et al. 1999).

2. Value-leveling or "ideal free distribution" (IFD) models use the idea that fishers end up distributing (and redistributing) their activity so that no ground stands out from any other in profitability (Fretwell and Lucas 1970; Kacelnik et al. 1992; Tregenza 1995). These models can also be used to derive gravity models (see Walters and Bonfil 1999). Since they appear to be very powerful, we discuss them extensively in this section.

3. Sequential effort-allocation models were suggested by Hilborn and Walters (1987) basically as a way to numerically calculate the results of IFD assumptions. The idea was to divide the total effort E_t into a set of small numerical components or "chunks," then sequentially allocate these chunks among grounds using a rule that each chunk goes to whatever ground is most profitable (and the allocation of chunks across grounds then occurs as each ground is fished down to the level at which it is no longer the most profitable). Advances in the solution methods for IFD models have made this approach obsolete.

4. Complex statistical models of individual fisher decision processes can explicitly represent variations among fishers in factors such as experience, perceived costs and expected catch rates, and home-base spatial cost considerations. With such models, the effort on each ground is calculated as the explicit sum of individual fisher contributions (Holland and Sutinen 1999; Wilen et al. 2002; Smith 2002). Like individual-based population models, this approach can be used to gain further insights about the mechanisms involved in creating large-scale patterns, but it is not yet clear whether it provides any practical advantage over the gravity and IFD methods (Wilen et al. 2002) when the predictive need is just for spatial effort distribution changes in relation to management policies (the more detailed models may be critical in evaluating the efficacy and subtle impacts of other regulatory tactics such as gear restrictions and fuel taxes).

The remainder of this section will discuss the development of type (2) IFD models, show one of the remarkable predictions that these models make about relationships between catch and effort, and discuss how to incorporate them into larger spatial models for long-term stock and effort dynamics. Gravity models are used today mainly in situations like Ecospace when we need "quick-and-dirty" predictions of effort distributions over a very large number of fishing sites (map grid cells), possibly for several fleets or gear types, and when we are simultaneously doing complex ecological calculations. Detailed IBM models have been developed for only a few fisheries because they require rich and detailed logbook information on historical fishing patterns and factors that have affected past fishing decision choices.

In deriving gravity and individual-based effort-prediction models, we reason "forward" in time about how factors such as fish abundances and travel costs *will* affect probabilities or proportions of potential effort directed to alternative fishing sites. In deriving IFD predictions, we begin instead by first hypothesizing what the net effects of many individual decisions will be on some critical site measure such as profitability, then reason "backward" to calculate how much effort *will have* been allocated to each site after enough effort has been allocated so that no site is likely to stand out from any other in attractiveness. The basic justification for IFD prediction methods is that fishers rarely seem to miss (fail to detect) unusual fishing opportunities, so that if any site temporarily stands out in profitability from others (e.g., due to high local fish abundance), at least some fishers will detect this and "home in" on the site to knock it back until it no longer stands out as more profitable. Note that the development of an IFD prediction always involves looking not only at the dynamics of fisher choice but also simultaneously at the effect of choice on local abundances/availabilities of fish; that is, we do not attempt to separate the effort allocation and fish depletion-dynamics calculations into two "submodels."

Here we illustrate the IFD reasoning with two derivations that have been useful in interpreting both recreational and commercial fishing data. The examples differ in their specific assumptions about fine-scale spatial search processes and the availability of fish within each site (or ground or lake or reef or stream reach or ...). For both examples, we will use the following notation and assumptions.

1. We predict the allocation of effort over a short enough fishing "season" or simulation time step that natural mortality, growth, and recruitment during the season can be ignored. Time t within the season is scaled so that t varies from 0 to 1.0 over the season.

2. Over such a season, instantaneous fleet sizes or fisher numbers $f_{i,t}$ working on sites i at times t within the season result in total effort $E_i = \int_{t=0}^{t=1} f_{i,t} dt$. Total fishery-scale effort $E_T = \sum_i E_i$ may or may not be constrained to some "known" or policy-limited total; when it is not constrained, the prediction will be of both site efforts E_i and total effort E_T for the season.

3. On each site i, there is an initial total abundance $N_{i,0}$ of fish; this total abundance may be measured in either biomass or numbers units depending on which is most appropriate for profitability calculations. As effort accumulates on site i, abundance $N_{i,t}$ decreases to an end-season value $N_{i,T}$ that depends on E_i.

4. The instantaneous average profitability $pr_{i,t}$ of fishing on site i at time t within the season can be represented as a simple difference between a "price" P times catch per effort, minus a site-dependent cost c_i of exerting a unit of effort on site i: $pr_{i,t} = P(CPUE_{i,t}) - c_i$. We assume that the effort allocation tends to drive all of these $pr_{i,t}$ toward a single "flat" or total fishery-scale value $pr_{o,t}$ due to the reallocation of effort toward any site i that "stands out" as momentarily having $pr_{i,t} > pr_{o,t}$.

An important thing to keep in mind about assumptions such as these is that the precision of the predictions based on them will depend critically on how the sites i are defined: if each site i represents a very small area (or lake or whatever) that would never likely see high effort E_i, the actual time series of efforts received by the site may be grossly unpredictable with any model. That is, the modeling will fail if we are too "ambitious" about predicting the details of the effort distribution. On the other hand, if the sites i represent relatively large grounds or regions, the total efforts may be quite predictable due to statistical averaging effects over variable microsites within the grounds. There is no "right" or "wrong" way to define sites, and site definition for IFD prediction should be based on specific policy questions/options and the availability of data for estimating response parameters.

Continuous Leveling of Profitability When Fish Exchange between Vulnerable and Invulnerable States within Each Fishing Site

Here we show the development of an IFD model that was originally derived for effort prediction in two grossly different British Columbia fisheries: commercial groundfish trawling (Walters and Bonfil 1999) and recreational fishing for rainbow trout in lakes of the British Columbia interior (Cox et al. 2002; Cox and Walters 2002). We begin the derivation by examining the fine-scale dynamics of interaction between fish and fishers within any site i, from which we derive a seasonal depletion submodel. That submodel is then used in the backcalculation of a fishing effort for each site. In both these cases, we believe that only a small proportion of the target stock at each site is actually "vulnerable" to fishing at any moment, and that fish exchange between vulnerable and invulnerable states occurs in much the same way that is assumed in the derivation of foraging arena equations for recruitment and trophic interaction processes (chapters 6 and 10). Trawlers fish only a small proportion of the bottom, concentrating their efforts at microsites and along "trawl lines" that have produced high catch rates and have low risk of gear fouling from boulders and such. Fish appear to exchange between safe places and these heavily fished microsites via movement processes (small-scale dispersal and migration). Rainbow trout fishers also concentrate their effort in areas where fish aggregate (e.g., lake shoals or stream pools), but there also appears to be some other behavioral exchange process at work as well; as is typical for most fish pursued by sports anglers, only a small proportion of the fish are "reactive" to gear at any time.

To model the vulnerability exchange process and its effect on catch (and effort distribution), we treat the stock $N_{i,t}$ in area i at any moment as divided into two components: $V_{i,t}$ fish that are vulnerable to fishing (are in trawlable areas or are reactive to fishers besides Carl Walters), and $N_{i,t} - V_{i,t}$ fish that are safe for the moment. We then treat $V_{i,t}$ as a "fast variable" with dynamics $dV_{i,t}/dt = v(N_{i,t} - V_{i,t}) - \acute{v}V_{i,t} - q_i f_{i,t} V_{i,t}$, where v is an instantaneous rate of movement of fish from an invulnerable to a vulnerable state, $\acute{v}$ is rate of movement of vulnerable fish to an invulnerable state, and q_i is the catchability coefficient or proportion of $V_{i,t}$ caught by a unit of fishing effort $f_{i,t}$. Except for a short transient period at the start of the fishing season, singular perturbation arguments (box 6.2) tell us that $V_{i,t}$ should vary over most of the season as $V_{i,t} = v N_{i,t}/(v + \acute{v} + q_i f_{i,t})$, i.e., it should remain near equilibrium with respect to changing $N_{i,t}$ and $f_{i,t}$ (to obtain this equation, set $dV_{i,t}/dt = 0$ and solve for V). Then at any moment we expect the catch per unit effort $CPUE_{i,t} = q_i V_{i,t}$ to depend on $V_{i,t}$ and $f_{i,t}$, as

$$CPUE_{i,t} = \frac{q_i v N_{i,t}}{(v + \acute{v} + q_i f_{i,t})} \tag{9.1}$$

leading to instantaneous profitability $pr_{i,t} = Pq_i v N_{i,t}/(v + \acute{v} + q_i f_{i,t}) - c_i$. That is, the profitability for fishing in site i should depend directly (and "immediately") on the abundance $N_{i,t}$, the relative catchability q_i of fish at site i, and the local cost of fishing c_i.

The IFD argument is, then, that fishers will move about ($f_{i,t}$ will vary) so as to level $pr_{i,t}$ among sites, i.e., to drive $pr_{i,t}$ toward a single value pr_o that is independent of both site and t. At this leveling point, the condition $pr_{i,t} = pr_o = PCPUE_{i,t} - c_i$ along with equation 9.1 implies the relationship

$$pr_o = \frac{Pq_i v N_{i,t}}{(v + \acute{v} + q_i f_{i,t})} - c_i \tag{9.2}$$

from which we would predict (by solving eq. 9.2 for $f_{i,t}$) that $f_{i,t}$ should vary among sites and over time t as

$$f_{i,t} = \left[\frac{Pq_i v N_{i,t}}{(pr_o + c_i)} - (v + \acute{v}) \right] / q_i \tag{9.3}$$

Further, $CPUE_{i,t}$ should level out at the base value $CPUE_{io} = (pr_o + c_i)/P$, independent of $N_{i,t}$ (any increase or decrease in CPUE should result in a compensatory effort change, a short-term and spatial version of the CPUE-flattening example shown in fig. 9.6). Note that this equation can predict negative $f_{i,t}$; in that case, the predicted effort is taken to be zero (site i should attract no fishing).

To complete the IFD prediction of E_i as a function of pr_o, all we then need to do is integrate $N_{i,t}$ and $f_{i,t}$ over the season $t = 0$ to 1. To perform this integration, we first note that $N_{i,t}$ will decline over the season according to $dN_{i,t}/dt = -f_{i,t}CPUE_{io}$, absent natural mortality, growth, and recruitment. Substituting equation 9.3 for $f_{i,t}$ into this differential equation and integrating the resulting linear dynamical equation, we get that $N_{i,t}$ should vary over the season as

$$N_{i,t} = N_{i\infty} + (N_{i,0} - N_{i\infty})e^{-vt} \tag{9.4}$$

where the "asymptotic stock size" parameter $N_{i\infty}$ is a function of the original dynamics parameters:

$$N_{i\infty} = \frac{CPUE_{io}(v + \acute{v})}{(vq_i)} = \frac{(pr_o + c_i)(v + \acute{v})}{(vq_iP)} \tag{9.5}$$

$N_{i\infty}$ can be interpreted intuitively as the "residual," stable abundance that would not attract any further fishing effort if the seasonal harvest process were allowed to continue indefinitely without new recruitment, growth, or natural mortality. As we would hope, equation 9.5 predicts higher residual abundances in areas with higher fishing costs c_i and lower catchabilities q_i, and if either the rate of becoming vulnerable v is low or the rate of becoming invulnerable $\acute{v}$ is high. We can now calculate E_i as the total catch $N_{i,0} - N_{i,T}$, divided by the mean catch per effort $CPUE_{io}$. Using equation 9.4 with $T = 1$ to predict $N_{i,T}$, we get the apparently complicated predictive relationship

$$E_i = \frac{(N_{i,0} - N_{i,T})}{CPUE_{io}} = \frac{(N_{i,0} - N_{i\infty} - (N_{i,0} - N_{i\infty})e^{-v})P}{(pr_o + c_i)} \tag{9.6}$$

If you examine this relationship closely, you will see that it actually predicts a quite simple linear response of E_i to space/time changes in abundances $N_{i,0}$, of the form $E_i = a_i + b_iN_{i,0}$, where a and b are somewhat complicated functions of the "structural" (site but not short-time dependent) parameters $v, \acute{v}, q, c, pr_o$, and P. Cox and Walters (2002) have shown that such linear predictions do in fact describe the effort variation in British Columbia trout fisheries quite well; Cox also obtained estimates of the v parameter in equation 9.4, showing that asymptotic exploitation rates are well below 1.0 (eq. 9.4 predicts a maximum possible exploitation rate of $1 - e^{-v}$ even if $f_{i,t}$ is very large and $N_{i\infty} = 0$).

End-Season Leveling of Profitability When Instantaneous Effort on Each Site Is Unpredictable and When Fishers Search for and Exploit Aggregations of Fish within Each Site

This model was originally derived for the analysis of spatial-effort patterns in the Australian northern prawn fishery (NPF) that takes place mainly in the Gulf of Carpentaria. This fishery is particularly interesting in terms of its spatial-effort dynamics, because it begins each year with a fishing "season"

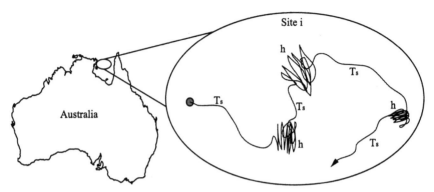

FIGURE 9.8: Search processes by fishers can very often be modeled as a two-stage process, such that fishing time is divided between a random search for aggregations (times marked T_s along the sample search track) and the time spent "mopping up" each aggregation by tactics such as systematic trawling (times marked h).

on banana prawns, a schooling species that fishers search for by day, followed by a second season during which fishers trawl mainly at night for tiger prawns. Some fishers switch to tiger prawn trawling well before the second formal season, in response to the declining chances of finding banana prawn schools. As in many fisheries, NPF fishers search for and capture prawns in a roughly two-stage process. More or less random search tactics are used to find the aggregations (or schools) of prawns, then highly concentrated and systematic fishing is used to fish down each aggregation (fig. 9.8). In the tiger prawn fishery, random search or "scratching" involves towing a small trawl ("trigear") that is checked every few minutes to detect whether an aggregation has been encountered that would warrant switching to systematic "mopping up." In the banana fishery, the mop-up process usually involves a quick few trawls that take (or disperse) each school.

   Search processes like that shown in figure 9.8 are likely to lead to hyperstability in the catch per unit effort. To model local profitability and the depletion process within each season in site i, we need to account for this effect. A simple model for CPUE as a function of the overall abundance at site i can be derived by following the same basic arguments that Holling (1959, see box 10.1 on page 241) used to derive the disc equation. Think of the total fishing time T_t along the trajectory in figure 9.8 as being divided into two components, search time T_s and "handling" or "mop up" times $T_h = h\widehat{NA}_s$, where $\widehat{NA}_s$ is the number of aggregations encountered (three in the figure example) and h is the time needed to systematically fish each aggregation. Since $T_t = T_s + h\widehat{NA}_s$, the search time T_s has to decrease when many aggregations are encountered. Suppose the total stock abundance $N_{i,t}$ is divided into n_s aggregations, so that the mean exploitable biomass per aggregation is $B_s = N_{i,t}/n_s$, and ignore minor catches that might be made while "scratching" during T_s. If the search is random during T_s, the encounter rate with aggregations should be proportional to the number of aggregations (or schools) at risk to encounter, i.e., $\widehat{NA}_s$ should vary as $\widehat{NA}_s = qT_s n_s$, where the rate of effective search q represents the area covered per time searching divided by the total area over which the n_s schools are distributed. This implies that the search time should vary as $T_s = \widehat{NA}_s/(qn_s)$, and substituting this relationship into the time budget results in $T_t = \widehat{NA}_s/(qS) + h\widehat{NA}_s$. Solving for $\widehat{NA}_s$ and rearranging terms then gives the familiar disc equation:

$$\widehat{NA}_s = \frac{qn_s T_t}{(1 + ahn_s)} \tag{9.7}$$

Then converting back to total abundance units using $N_{i,t} = B_s n_s$, with catch per aggregation attacked being B_s, this functional relationship predicts the total catch per effort (per total time fishing T_s) to be

$$CPUE_{i,t} = \frac{B_s\widehat{NA}_s}{T_t} = \frac{q_i N_{i,t}}{\left(1 + (q_i h/B_s)N_{i,t}\right)} \tag{9.8}$$

Note that this relationship does not include $f_{i,t}$, i.e., it does not recognize the possibility of direct, short-term "interference competition" among fishers who might at times try to simultaneously fish the same aggregation. Note further that equation 9.8 reduces to the familiar random search assumption $CPUE = qN$ if the denominator term $q_i h/B_s$ is very small, i.e., if fishers take very little time h to mop up each aggregation.

When we do not assume localized, instantaneous effects of interference competition as was done above in deriving equation 9.3, we cannot directly predict instantaneous fishing efforts $f_{i,t}$ for all times and locations during each fishing season (eq. 9.8 cannot be solved directly for $f_{i,t}$). Instead, we can make the weaker (and possibly more robust) assumption that each $f_{i,t}$ may vary in complex and unpredictable ways, and that the only predictable IFD outcome of this variation is that profitabilities will have been leveled across fishing sites by the end of each fishing season. That is, the end-season catch rates $CPUE_{i,T}$ should approximately satisfy the equal-profitability relationship $pr_o = PCPUE_{i,T} - c_i$ for all sites i, implying $CPUE_{i,T} = (pr_o + c_i)/P$ for each i. The analytical task is then to predict how much effort it would take to drive $CPUE_{i,t}$ down to $CPUE_{i,T}$ by $t = T$, given the initial abundance $N_{i,0}$ and a model for the dynamic decline of $N_{i,t}$ as local effort accumulates in i. Note that $CPUE_{i,T}$ does not depend on $N_{i,0}$, warning us to expect a positive relationship between E_i and $N_{i,0}$.

Under the closed-population assumption (effort being calculated over a season short enough to neglect mortality, growth, recruitment) we would expect $N_{i,t}$ to vary according to the following differential equation if $CPUE_{i,t}$ varies as predicted by equation 9.8:

$$dN_{i,t}/dt = -f_{i,t}CPUE_{i,t} = \frac{-f_{i,t}q_i N_{i,t}}{\left[1 + (q_i h/B_s)N_{i,t}\right]} \tag{9.9}$$

This equation can be readily solved using elementary calculus, resulting in the integral relationship (and recalling that E_i is the time integral of $f_{i,t}$)

$$(q_i h/B_s)(N_{i,0} - N_{iT}) - \ln(N_{i,T}/N_{i,0}) = q_i E_i \qquad (9.10)$$

Note here that $N_{i,T}$ appears in transcendental form, i.e., the integral equation cannot be solved directly for it unless $q_i h/B_s = 0$, in which case the integral reduces to the exponential removal equation $N_{i,T} = N_{i,0}e^{-q_i E_i}$ that has been commonly used to predict seasonal depletion patterns in fisheries. The transcendental form forces an annoying numerical calculation when we need to predict $N_{i,T}$ as a function of arbitrary effort E_i.

Fortunately, we do not need a direct analytical solution for $N_{i,T}$ in order to complete the IFD prediction of E_i using equation 9.10, since the IFD proposal is that $N_{i,T}$ will be driven to a particular value dependent only on pr_o and the parameters that determine $CPUE_{i,T}$. The IFD effort prediction is obtained by dividing the left-hand side of equation 9.10 by q_i, after first finding the value of $N_{i,T}$ that will make $CPUE_{i,T} = (pr_o + c_i)/P$. Substituting the equation 9.8 prediction of CPUE as a function of N in this relationship and solving for $N_{i,T}$, we get the IFD prediction of $N_{i,T}$:

$$N_{i,T} = \frac{[(pr_o + c_i)/P]}{\left[q_i - (q_i h/B_s)(pr_o + c_i)/P\right]} \qquad (9.11)$$

For readers that find it helpful to think in terms of computational algorithms, the IFD effort prediction for the dynamic depletion model equation 9.9 proceeds in three steps for each modeled fishing season (assuming that the time-independent parameters q_i, c_i, h, B_s have been entered):

1. set a value of pr_o and values for the local abundances $N_{i,0}$ for all sites i;
2. calculate $N_{i,T}$ for each site i using equation 9.11;
3. calculate E_i for each site using equation 9.10;

We could put all of these steps into one big fat equation just to impress the reader, but we suspect that the IFD calculation is confusing enough as it is.

A Prediction about the Relationship between Catch and Effort When Both Vary Dynamically with Stock Size

One way to test the IFD predictions from equations 9.6 and 9.10, and to obtain parameter estimates for larger models, would be to plot observations of the historical efforts E_i against the estimates of historical abundances $N_{i,0}$. Equation 9.6 predicts linear relationships, while equation 9.10 predicts relationships that are somewhere between linear and logarithmic depending on the handling-time parameter $q_i h/B_s$. Regression analysis might then be used to fit the models to give values for aggregate parameters, and these might be disaggregated using independent assessments of at least some of the economic (P, c_i) and technological (q_i, h, B_s) parameters. Unfortunately, there are two big practical problems with this approach: (1) most often, we do not have good historical estimates of abundances $N_{i,T}$ from localized stock assessments; and (2) it is difficult or impossible to directly estimate what fishers assess the costs c_i to be, since these perceived costs include subjective time and risk factors that can be large compared to direct, measurable operating costs (e.g., Prince [1989] has shown that Tasmanian abalone fishers avoid dangerous diving areas except when prices are high, and this results in a positive correlation between CPUE and price because CPUE is higher in the more dangerous areas).

An alternative approach to testing IFD predictions is to directly compare the observed variation in catches C_i with efforts E_i. Plots of catch versus effort have long been used as a stock-assessment tool in fisheries. The IFD models predict a striking and surprising departure of such catch-effort patterns from the predictions of classical fisheries models (fig. 9.9). For fixed $N_{i,0}$, the classical "catch equation" predicts a saturating relationship between catch (as an "output") and effort (as an arbitrary "input"). When $N_{i,0}$ is negatively affected by effort, classical population dynamics models predict a dome-shaped equilibrium relationship between catch and effort (the classic "Gulland method" for estimating optimum effort by regressing catch or CPUE on effort is based on assuming that $N_{i,0}$ remains near equilibrium over time with respect to effort; see Hilborn and Walters 1992 for warnings about why this method should never be used). Now, consider the predictions of equations 9.6 and 9.10. Under the closed-population assumption, these equations can be rewritten in terms of catch by noting that catch is predicted to be just the change in N over the fishing season, i.e., $C_i = N_{i,0} - N_{i,T}$. For fixed economic and technical parameters among seasons ($P, pr_o, c_i, q_i, h, v, \acute{v}$ constant), the IFD relationships reduce to equations of the forms

$$E_i = k_i C_i \quad \text{for model 9.6} \tag{9.12}$$

$$E_i = [\ln(k_{2i} + C_i) + k_{3i} C_i - \ln(k_{2i})]/q_i \quad \text{for model 9.10} \tag{9.13}$$

(Here the k's are just constants that depend on site i). Obviously, equation 9.6 predicts a simple linear relationship between catch and effort, which in field data could easily be misinterpreted as implying no effect of effort on abundance! For a large handling time per aggregation (large k_{3i}), equation 9.13 reduces to equation 9.12. For small handling times (or a random search with no time cost at all per fish captured) equation 9.13 predicts an exponential relationship of the form $C_i = k_{2i}(e^{q_i E_i} - 1)$, i.e., a relationship of exactly the opposite form as would be predicted from the classic yield equation $C_i = N_{i,0}(1 - e^{-q_i E_i})$. To understand why these surprising reversals are predicted, we need to recognize that equations 9.12 and 9.13 are not predictions about how effort causes catch but, rather, about the pattern of covariation that should occur in both catches and efforts when both of these dynamic-response variables are "driven" by changes in abundance $N_{i,0}$. At the scale of local fishing grounds, it is likely common for an interannual

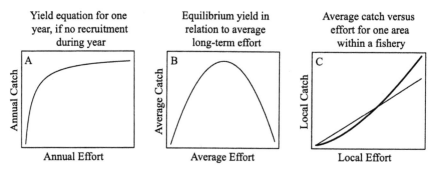

FIGURE 9.9: Three qualitatively different predictions of the relationship between fishing effort and catch. Model A is the classic fisheries "catch equation," B is the classic equilibrium yield relationship, and C is the predicted local relationship for one area when fishers can redistribute their efforts among areas and can target fish aggregations within each area; a straight line is predicted when the handling time per aggregation is high, while a curve is predicted when it does not take much time to harvest or disperse each aggregation. Points along the curves in C are generated by the interannual variation in abundance, driving changes in both effort and catch.

variation in $N_{i,0}$ to be driven by factors entirely independent of the past local fishing efforts E_i; equations 9.12 and 9.13 predict how this variation should be expressed through combinations of E_i and C_i, entirely independent of why $N_{i,0}$ varies among years in the first place. Also, we need to be careful not to confuse plots of E_i versus C_i data collected over years from a single area i, with plots of catches versus efforts measured across sites (within a single year or averaged across years); a variation in the average efforts among sites that are differentially costly to access can provide valuable data on the equilibrium relationship between yield and effort, as in Marten and Polovina (1982) and Munro and Thompson (1983).

The IFD models point out that there are two quite different reasons why we might see a nearly linear relationship between catch and effort when we plot data from different years for a single fishing ground. A linear C versus E relationship implies a constant CPUE from year to year. One mechanism that can cause a constant CPUE is hyperstability due to an efficient search combined with large handling times (a high q in eq. 9.8). A quite different mechanism is through "effort flattening" such that any change in CPUE causes an immediate effort response to allow CPUE either to rebound (if it starts to fall) or to be cut back (if it starts to rise), as assumed in deriving equation 9.3. There is actually a third possibility not considered in the simple IFD derivations, i.e., "effort sorting," in which catchability coefficients differ greatly among fishers (common in recreational fisheries) and low-q fishers tend to quit sooner (or not go out at all) in years of low abundance, which causes the mean q of active fishers to be higher in years of low abundance.

Figure 9.10 shows plots of interannual covariation in C_i and E_i for different grounds i in the Australian northern prawn fishery. For the nonschooling

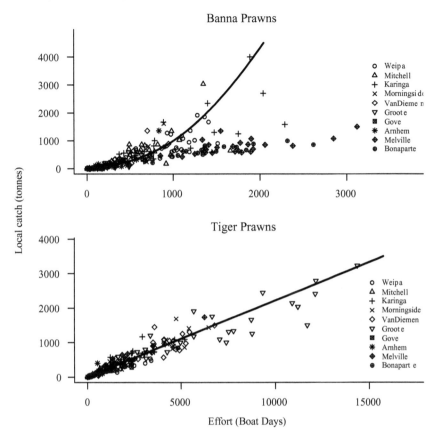

FIGURE 9.10: A test of prediction C in figure 9.9. Catch-effort relationships for local areas within the Gulf of Carpentaria prawn fishery, northern Australia. Each point represents an annual effort and catch for one of the areas listed, for the years 1970–1996. Banana prawns (top panel) are strongly schooled (except the species taken in the Bonaparte and Mitchell areas), while tiger prawns are more widely distributed. Data kindly provided by C. M. Robins (pers. comm.); see also Robins et al. (1998), Somers (1994), Somers and Wang (1997).

tiger prawns, these plots are obviously close to linear. But there is a distinct curvature in the plots for the schooling banana prawn, as we would expect if the handling time per school encountered is low. This seems counterintuitive, because we would ordinarily expect schooling to cause hyperstability in CPUE and hence a linear catch-effort relationship due to strong handling-time effects. What appears to be happening in this fishery is that the handling times per aggregation (h) are effectively much lower for banana prawn schools than for tiger prawn aggregations. That is, banana prawn schools get mopped up or dispersed through just a few quick sets, while it takes much longer to systematically mop up a more dispersed aggregation of tiger prawns. There may also be some vulnerability exchange dynamics with tiger prawns, due to the movements of prawns into fishable aggregation sites from deeper waters and/or areas that are dangerous to fish because of reef structures. But one thing is very clear: IFD models appear to be very good predictors of the C versus E relationship, and hence probably of the future effort-distribution patterns in the fishery. Further, they make it very clear that any attempt to simplify grounds-scale stock assessments by using simple catch-effort relationships (assuming equilibrium, case B in fig. 9.9) would be misleading.

Incorporating Effort Predictions from IFD Models into Larger Simulations

There are two ways to use effort-response models like equations 9.6 and 9.10 in larger simulations of multiseason dynamics that include processes that cause variation among years in the abundances $N_{i,0}$ and in the bioeconomic parameters P, q_i, c_i, and pr_o. The simplest approach from a computational perspective is to treat base profitability pr_o as a "known" parameter, then compute the E_i directly from the models, being careful to assign zero efforts whenever the equations predict negative values. Unfortunately, this approach can result in a violation of known or proposed (as policy) constraints on the total effort E_T. To assure that such constraints are met, we must resort to a second, iterative calculation of the E_i to force them to sum to known E_T wherever appropriate. The iterative calculation proceeds by choosing a reasonable starting value for the base profitability pr_o, then calculating the E_T that this would imply, then adjusting pr_o upward (to reduce E_T) or downward (to increase E_T) until the E_T constraint value is matched. Various line-search methods can be used to perform the iteration efficiently (see Press et al. 1996).

   In fact, it can be very interesting from a policy perspective to "map" the expected relationship between the total effort E_T and the predicted average profitability pr_o, for a range of overall stock abundances and spatial-distribution patterns. The general form of the E_T and pr_o relationship (fig. 9.11) is $1/X$, which just means that high efforts are associated with low profitability, and high profitabilities are associated with low total effort.

   For models of large regional fisheries, where fishers from many ports or population centers can fish in many grounds, the IFD (or gravity model) calculations of efforts should generally be disaggregated by port or center. That is, the total predicted effort for each center ought to be allocated over spatial sites using information on the average costs and observed effort patterns specific to fishers from that center. Otherwise, the model will make some silly hidden assumptions about how fishers are willing to travel great distances to exploit particular local opportunities. Also, the model will fail to capture the possible inequities or disadvantages to fishers from particular centers caused by spatial management options (e.g., an MPA squarely in front of one fishing port).

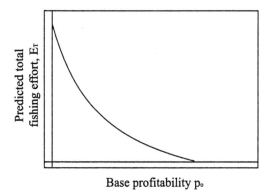

Base profitability p₀

FIGURE 9.11: The general pattern of variation in the predicted total (summed over sites) fishing effort E_T as a function of the location-independent profitability pr_o parameter used in IFD models for spatial effort prediction. Note that the Y-intercept of the curve is the total effort needed to "dissipate" the average profit, and the X-intercept of the curve is the maximum profitability that could be achieved if only one fisher were active.

9.4 MOSAIC CLOSURES

Most fisheries capture multiple stocks/species of fish that differ in productivity and exploitation-rate goals (including zero goals for bycatch species that should be protected entirely if possible). There has been much research on how to improve gear and fishing technique so as to increase selectivity and reduce bycatch. Beyond such changes in the technical details of fishing, there are generally options for improving selectivity by using information on larger-scale spatial and temporal patterns in the vulnerabilities of various stocks, so as to target fishing to the most productive places/times while avoiding places/times where the bycatch is worst. Partial separation in spatial distributions (and/or timing of vulnerability to fishing) is created in most fish communities by the same variety of factors that drive biological niche differentiation in such communities; e.g., species often show differential distributions along depth gradients (fig. 9.12), creating a spatial mosaic of relative abundances when combined with typically complex bathymetry.

Suppose we have mapped the distributions in figure 9.12 from historical information based on harvest and survey data, and then ask what distribution of fishing effort along the spatial (or temporal) gradient would maximize total fishing profit, subject to constraints on the total fishing mortality rate suffered by each of the stocks/species over some specified period (like one fishing season). This is a formal optimization problem that is quite easy to set up and solve using nonlinear optimization procedures. In general, we want the optimization procedure to find the distribution of fishing efforts E_i

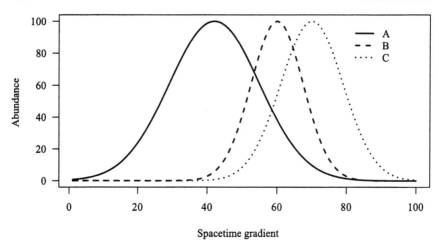

FIGURE 9.12: The hypothetical distribution of stocks or species along a space-time gradient such as depth or timing of migration through a fishing area. For the example in the text, we assume stock A is widely distributed and valued for fishing, while stocks B and C have low target-exploitation rates and/or bring low prices to fishers.

over space/time locations i, for a set of species j, that will maximize the total profit V from fishing predicted as the sum

$$V = \Sigma_i E_i \left(\Sigma_j P_j q_{ij} \bar{B}_{ij} - c_i \right) \tag{9.14}$$

where P_j is the unit price of species j, q_{ij} is the catchability of species j in this fishing location i, c_i is the unit cost of fishing in location i, and $\bar{B}_{ij}$ is the mean biomass of species j in area i over the period for which the value V is to be optimized. We want the optimization to be constrained so that the fishing mortality rates by stocks j and F_j do not exceed target values:

$$F_j = \frac{\Sigma_i q_{ij} E_i \bar{B}_{ij}}{\Sigma_i \bar{B}_{ij}} \leq F_{target,j} \quad \text{for every species } j. \tag{9.15}$$

For such formulations to have sensible solutions, the model(s) for mean biomass must represent the "diminishing returns" effect of increased fishing effort on the catch obtained from each location i. One possible model for this effect is the short-term (short season) catch equation, when fishing on each location takes place over a short enough season to partially deplete the local stock before there is enough time for significant natural mortality, recruitment, migration, or growth to occur. In that case, $\bar{B}_{ij}$ can be modeled as

$$\bar{B}_{ij} = \frac{B_{ij} \left(1 - e^{-q_{ij} E_i} \right)}{q_{ij} E_i} \tag{9.16}$$

where B_{ij} is the biomass of stock j present in location i at the start of the fishing season (and catch over the season is predicted to be just $C_{ij} =$

$B_{ij}(1 - e^{-q_{ij}E_i})$). An extreme alternative assumption would be to treat each location as having an independent carrying capacity biomass K_{ij}, an intrinsic biomass growth rate r_{ij}, and a logistic surplus production equal on average to $r_{ij}B_{ij}(1 - B_{ij}/K_{ij})$, implying the long-term average biomass

$$\bar{B}_{ij} = K_{ij}(r_{ij} - q_{ij}E_i) \quad \text{if} \quad q_{ij}E_i \leq r_{ij} \quad \text{else} \quad \bar{B}_{ij} = 0 \qquad (9.17)$$

Other alternatives can be constructed by considering more complex spatial dynamics, such as the exchange of fish among locations using movement submodels (chapter 11) to calculate $\bar{B}_{ij}$. The critical thing about the choice of a model for $\bar{B}_{ij}$ is that the model recognize the competitive effect of fishing on profitability; constraints such as equation 9.15 can be used instead of long-term dynamic models like equation 9.17 to represent assessments of long-term productivity and exploitation-rate goals.

Absent fishing mortality rate constraints (eq. 9.15), solutions for the maximum V in equation 9.14 are typically effort distributions (E_i) similar to or identical to the distributions of effort predicted by IFD models for effort allocation. In both types of models, the effort is "added" to each location until the marginal profitability of putting more effort into that location is no greater than for other locations. But when we add fishing mortality rate goals or constraints to the optimization (and/or capacity constraints on the total effort), the optimum solution typically changes so as to predict that no effort should be allocated to locations where species that have low target fishing mortality rates are concentrated. An example solution for the three stocks in figure 9.12 is shown in figure 9.13, where low fishing mortality rate targets $F_{target,j}$ have been set for stocks B and C. Note that this solution involves low fishing effort near the center of the stock C distribution, and a reduced or zero effort over much of the distribution of stock B.

The interesting point about the fishing-effort distributions predicted by combining profit maximization with fishing-rate target constraints is that the optimum is likely to involve avoiding fishing entirely in a "mosaic" of locations where species that need special protection are concentrated. This mosaic can easily be far more complex (when mapped in space rather than with respect to a simple habitat gradient measure) than would typically be considered when designing protected areas, and might well need to be moved from year to year in response to changes in fish distributions.

An example of the possible mosaic of closed areas produced by multi-species optimization with harsh constraints on fishing mortality is shown in figure 9.14. In this example, we used AD Model Builder (Otter Research 1994) to maximize equation 9.14 for the trawl fishery off the British Columbia coast, for the 16 most important species caught in that fishery, subject to varying target fishing rates (eq. 9.15) and assuming seasonal depletion dynamics for diminishing returns to increased effort (eq. 9.16). Using mandatory logbook information from 1996, Walters and Bonfil (1999) estimated the average catch per effort and density for these 16 species for 6,000 1-nm^2 spatial cells ($i = 1 \ldots 6000$ in eqs. 9.14–9.16) where most

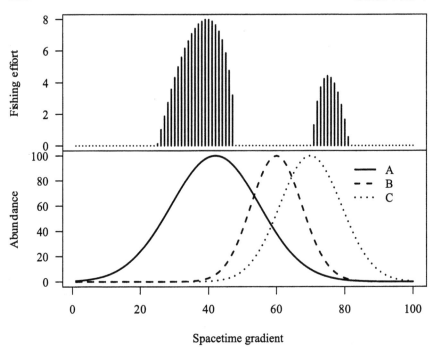

FIGURE 9.13: The optimum distribution of fishing effort (for maximizing the total profit from fishing) for the three stocks shown in figure 9.12, where the "target stock" A has a high price, B has zero price, and C has an intermediate price. Target-exploitation rates are low for stocks B and C.

of the trawling has been concentrated. The species differ widely in depth and north-south distributions (see fig. 1 in Walters and Bonfil) and in target fishing-mortality rates. In particular, target fishing rates $F_{target,j}$ are very low (< 0.05 to 0.1) for a collection of long-lived rockfish (*Sebastes*) species, while target rates are much higher (0.2–0.4) for Pacific cod and flatfish. In this case, the optimization warns that it may eventually be necessary to close a large number of 1-nm^2 cells to fishing because there are concentrations of at least a few "weak" rockfish stocks in most fishing locations, and that effort should be shifted toward (concentrated in) particular areas where cod and flatfish are most abundant.

It is an entirely open question whether there is any practical way to implement a complex mosaic of closures as is likely to result from formal optimization. Such closure patterns are actually in routine use in one particular situation, harvest management for Pacific salmon, because the habitat gradient variable is time and the distributions represent migration timing patterns. In other cases, there are serious issues about the practicality of enforcing complex closures, especially for fishing fleets where VMS (vessel monitoring system) and on-board fishing observers are not yet required. But

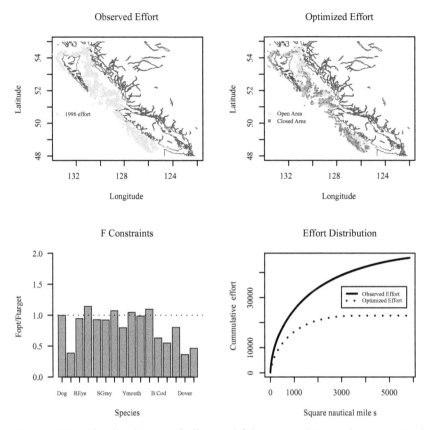

FIGURE 9.14: The distribution of effort and fishing-mortality rate by species in the British Columbia groundfish trawl fishery, 1996 (from Walters and Bonfil 1999), and the change in this distribution (optimized effort) that might have maintained the maximum possible profit subject to stringent constraints on fishing-mortality rates (F constraints) for some of the 16 main species in the fishery. Roughly 90% of the total observed fishing effort was distributed over 5,000 square nautical miles, whereas 90% of the optimized fishing effort is distributed over 2,500 square nautical miles.

such regulatory innovations are spreading rapidly across fisheries in general. Also, it is possible at least in principle to create economic incentives for fishers to voluntarily avoid recommended no-fishing locations, e.g., by imposing a threat of overall fishery closure at whatever time of year the target fishing rate $F_{target,j}$ is reached or exceeded for even one stock j.

PART FIVE

Food Web Modeling to Help Assess Impact of Fisheries on Ecological Support Functions

CHAPTER 10

Foraging Arena Theory (II)

THERE IS A LONG HISTORY of trying to construct useful models for trophic interactions in aquatic food webs. In the late 1960s and early 1970s, there was much work done on methods for ecosystem modeling, mainly centered on what we called "compartment models" for material and energy flows. Some of these models were specifically aimed at understanding the bottom-up control of production processes in marine ecosystems, attempting to link hydrodynamic factors (upwelling, spatial water flow), nutrient cycling, and plankton production. Others tried to provide mass-balance accounting for food webs (e.g., DYNUMES II, Laevastu and Favorite 1988), using information on diet compositions, food-consumption rates, and abundances of various creatures to back-calculate biomass flows and mortality rates.

Most of the early dynamic models were built upon the so-called "mass action principle" borrowed from elementary chemistry. The idea behind this principle is that if there are two "well-mixed" (randomly distributed) chemical species in a reaction vat (or modeled spatial area), then the number of encounters between these species ought to be proportional to the product of their densities. That is, the interaction rate (feeding, predation rate) ought to vary as encounters/time = k(species 1 density)(species 2 density). Such encounter rates were thought to be reasonable predictors of predation-interaction rates, especially if corrected for predator satiation and handling-time effects (the so-called type II functional response by predators). Simple versions of using the mass-action principle were multispecies biomass-dynamics models built as generalizations of logistic growth and Lotka-Volterra rate equations. Apparently more realistic but functionally identical variations on the mass-action theme have included adding terms to instantaneous natural mortality rate predictions in single-species models so as to represent predation impact, e.g., to predict the impact of predators on survival rates by using survival = $\exp(-M_o - M_1 \text{predator abundance})$, where M_o represents the nonpredation mortality risk and M_1 represents a per-predator increase in the prey mortality rate.

The relatively simple early models were partly successful at explaining some trophic interaction effects, when the analysis was restricted to adding a few (relatively weak) interaction terms to single-species logistic or age-structured models. But some serious problems occur when modeling the dynamics of whole ecosystems and complex food webs. The mass action or reaction vat structure for predicting interaction rates, especially when combined with type II functional response limits on predator feeding rates, generally results in predictions of (1) very strong "top-down" or "trophic cascade" control of abundances by predators; (2) predictions of dynamic instability (predator-prey cycles), especially in more productive ecosystems (the "paradox of enrichment," Rosenzweig 1971); and (3) unstable community structure, involving predictions of the loss of biodiversity through competitive interactions and through predators' overexploiting some prey species. All of these model predictions fly directly in the face of most of the field data. Strong trophic cascades are much less common than are mixed bottom-up and top-down control effects, involving apparent "ratio-dependence" in the predation rates (Berryman 1992; McCarthy et al. 1995; Scheffer and De Boer 1995; Brett and Goldman 1996). Regular predator-prey cycles appear to be relatively uncommon and are not more common in more productive lakes and ocean regions. Aquatic communities maintain diversity despite obviously strong competitive and predator-prey interactions.

It was not until the 1990s that we realized that these basic model failings might be due to a common bad assumption, and that this might be with the core assumption that species interactions are similar to random encounters of chemical species in reaction vats. Every biologist with even the slightest field experience knows that random distributions of organisms never occur in nature. Almost every metazoan animal has more or less highly organized behaviors for limiting its predation risk (feeding at particular times of day, spending most of the time in relatively safe refuge habitats or in the midst of a school of conspecifics, etc.). We used this idea in chapter 7 to explain why single-species recruitment relationships most often have a Beverton-Holt shape, and why recruitment "limits" rarely seem to coincide with the simple limits calculated from the total food supplies as we see (measure) them (but not as they are seen by juvenile fishes).

In this chapter we develop the "foraging arena theory" argument further, to show how it can be used to build ecosystem models that overcome the failings (stability, loss of structure) of earlier models. The basic idea of this theory is that most organisms exhibit spatial habitat-choice behaviors aimed at moderating their predation risk, and this behavior in turn limits access to prey resources. Trophic interactions then take place mainly in spatially and temporally restricted "foraging arenas," where the competition for food resources can be intense but where only small proportions of the total food population may be at risk to predation at any moment. Further, these arenas of feeding activity generally expose feeding animals to predation risk, so that the processes of eating and of being eaten are closely linked.

10.1 Understanding Foraging Arena Theory

Perhaps the easiest way to understand foraging arena theory and its ecosystem-scale implications is to imagine how an arena structure would evolve in a primitive ecosystem. Consider a pond with only three types of creatures, with densities: *Al*-algae, *Zo*-zooplankton, and *Pr*-predator on zooplankton. If these three creatures drifted or swam about in the system at

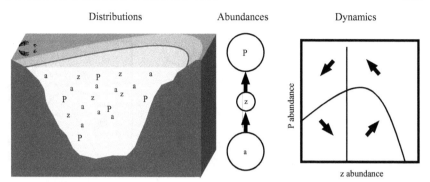

FIGURE 10.1: A primitive three-level ecosystem with random distributions of a = algae (*Al* in text), z = grazer (*Zo* in text), and P = predator (*Pr* in text) on grazer. Predators limit the abundance of grazers, and algae is abundant (cascade). The grazer-predator interaction leads to a stable limit cycle depending on algae productivity (the vertical predator isocline implies a stable limit cycle when the prey isocline has a positive slope at the equilibrium). Increased productivity causes instability, and harvesting grazers causes predators to decline rather than grazers.

random (fig. 10.1), encounter (grazing, predation) rates would follow mass-action rules; *Zo* would eat *Al* at rates proportional to the product $aAlZo$ (a = volume searched per time per z), or to $aAlZo/(1 + ahAl)$ if the zoo-plankton require a handling time h for each *Al* captured. This ecosytem would exhibit strong "top-down control," with *Pr* building up until the *Zo* density was reduced enough so that each *Pr* individual was barely getting enough to eat to replace (by growth, reproduction) itself. That is, *Zo* would be limited by *Pr*. This limitation of *Zo* would free *Al* to build up to densities set by environmental productivity factors (nutrients, light). Under increasing fertility, systems of this kind would exhibit the "paradox of enrichment": when/where *Al* is very productive, the $Al - Zo$ and/or $Zo - Pr$ predator-prey interaction becomes cyclic rather than having a stable equilibrium point. Any increase in the mortality rate of *Zo*, e.g., by human harvesting, would lead to reductions in the equilibrium abundance of *Pr* but not of *Zo* (unless the *Zo* were harvested hard enough to drive it toward extinction).

Next, suppose that a heritable phenotypic variation appears in the *Zo* population, for a behavioral preference to rest in edge habitats that are physically inaccessible to the predator *Pr* (fig. 10.2). Selection would favor this risk-avoidance behavior, leading to (1) an increase in *Zo*, at least in the edge habitats; (2) a possible reduction in predator abundance *Pr*; (3) the depression of algal density *Al* near refuges, and a possible increase in *Al* in offshore areas where the algae would be "protected" from *Zo* by the predators *Pr*. At this point (fig. 10.2), the ecosystem has already developed foraging arena structure, with both the $Al - Zo$ and $Zo - Pr$ interactions concentrated in "foraging arena" areas near the edge habitats. Food available to *Zo* in these arenas would be limited partly by grazing effects and partly by the physical

Distributions   Abundances   Dynamics

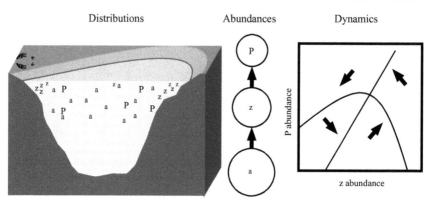

FIGURE 10.2: A spatial organization of the same three-level ecosystem as the previous figure, but after the grazer (z) species has evolved near-shore hiding behavior that is favored by selection because it reduces predation risk. Note the qualitative change in the predator-prey isocline structure, the predicted responses to eutrophication.

mixing processes that determine the delivery rates of Al to the arenas. Food available to Pr would likewise be limited in nearshore arenas by exchange rates and "spillover" of Zo individuals from safe sites to the arena areas. To model the interaction rates for this system, we would need to divide both the Al and the Zo populations into "invulnerable" and "vulnerable" density components, with exchange rates between these components that might be low enough to severely limit the interaction rates. This organization would immediately destroy the simple, relatively unstable top-down control structure of figure 10.1, resulting in "ratio-dependent" predator-prey interactions that are much more likely to exhibit stable equilibrium points even under increasing productivity. Under the harvesting of Zo, both predator and Zo equilibrium abundance would decrease with an increasing harvest rate on Zo.

Now the development of the ecosystem would really get interesting (fig. 10.3) because none of the players (Al, Zo, Pr) would be able to utilize the system's trophic resources fully. This means that the system would be vulnerable to invasion by new Al forms that are able to utilize light/nutrient supplies inaccessible to the initial Al because of grazing effects, of new Zo forms able to use the "offshore" Al resource by having other behavioral strategies for reducing the risk of predation by Pr (e.g., vertical migration), and perhaps of new Pr forms able to pursue Zo (and the new, invading Zo forms) into their behavioral refuge areas (e.g., shallow-water ambush predators). Depending on the behavioral strategists involved in accidental invasions/colonizations (and mutations of existing forms), the ecosystem might develop considerable interesting (and apparently unpredictable) biodiversity (Hixon and Menge 1991; Hixon and Carr 1997). In particular, it would start to exhibit violations of the "one resource–one consumer" competitive exclusion rule of simple mass-action models; e.g., two Zo populations might share the same Al resource, one in nearshore feeding arenas and one in vertical-migration

Distributions   Abundances   Dynamics

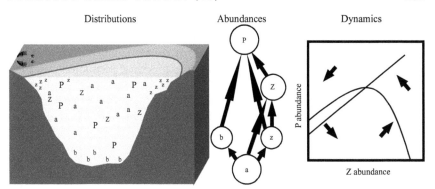

FIGURE 10.3: Further complication of the ecosystem in the above two figures, due to an invasion of new species capable of utilizing resources not accessible to the original species due to risk-management behaviors, etc.

(diurnal pulse feeding) arenas, even if neither of these Zo forms is strictly "limited" (except in spatial distribution) by the predator Pr.

Still further evolution of the behavioral characteristics of the species in the pond might involve selection for actively managing time so as to gain partial access to the larger system, to balance foraging gain with predation risk (by varying the timing of diurnal migrations, etc.). These more complicated behaviors would generally preserve the fine-scale spatial structure of foraging arenas as localized areas of intense feeding and predation risk near refuges but would alter the quantitative effects of both predator and prey density on interaction rates.

For prey species with large adult body sizes (Pr, perhaps Zo), selection might favor a delay to maturity, i.e., maturation at large enough body size so that adult prey can forage widely with impunity. Such species would then start to exhibit complex "trophic ontogeny" and "life-history trajectories." Biologists observing the dynamics of this complex ecosystem might now be puzzled by (1) a recruitment variation apparently unrelated to measurable large-scale habitat changes; (2) apparently unpredictable and perverse effects of fishing on particular species; and (3) apparent nonadditivity in mortality components, so that mortality rates are not just proportional to the abundance of predators.

What has foraging arena theory allowed us to explain? It can apparently help us understand a wide range of "puzzling" observations about patterns at several levels of biological organization, from individuals to ecosystems. Here is a short list of observations that the theory apparently helps explain: (1) lots of empty stomachs when we do field diet studies (low "Pr" values—proportions of the maximum ration actually achieved—in bioenergetics models, Schindler and Eby 1997), because most animals spend relatively little time feeding so as to manage their predation risk; (2) high search efficiencies (volumes swept per time feeding) but low rates of attack success when measured relative to the overall (system scale rather than arena scale) prey densities; (3) the widespread occurrence of complex trophic ontogeny; (4) lack of proportional dependence of natural mortality rates M on predator abundance; (5) apparent conservation of the total mortality rate Z (M decreases as fishing F increases, because the predator component of M is impacted by fishing); (6) "bottom up" control patterns and responses to enrichment along productivity gradients (lack of "paradox of enrichment" effects); (7) high sensitivity of top predators to fisheries; and (8) a higher biodiversity than predicted from diet overlaps (e.g., suites of euphausiid feeders).

10.2   PREDICTING TROPHIC FLOWS

It does not appear practical to measure and model all of the microscale behavioral and habitat factors that are likely to define the particular foraging arena structure of any large aquatic ecosystem. But we need some way to predict how spatially and temporally restricted encounter patterns are likely to limit trophic interactions, beyond just the simplistic notion that there are refuges from predators. In particular, we need some way to represent the idea that in any trophic interaction, there is likely to be a fast dynamic (compared to the rate scales for population change) exchange process of prey between relatively safe and vulnerable (in the foraging arena) behavioral states.

One simple way to model arena-scale exchange dynamics for any link between prey i and predator j in a food web is to imagine that at any moment in time the prey abundance, measured by, say, the total prey biomass B_i, is divided into two components:

- biomass V_{ij} that is vulnerable to predation by predator j;
- "safe" biomass $B_i - V_{ij}$ that is safe at the moment from predator j.

Next, imagine that organisms exchange between these pools or components at instantaneous rates v_{ij} and $\acute{v}_{ij}$, so that there is a flux rate $v_{ij}(B_j - V_{ij})$ into the vulnerable pool over time, and a flux rate $\acute{v}_{ij}V_{ij}$ back from the vulnerable pool into the safe pool. Finally, assume that the encounter rates between vulnerable prey and predators satisfy a mass-action relationship, so that biomass flow rate Q_{ij} from prey i to predator j satisfies the rate relationship

$Q_{ij} = a_{ij}V_{ij}B_j$ where B_j is a measure of predator abundance.

Combining the behavioral exchange flux and predation rate assumptions results in a rate equation for V_{ij},

$$dV_{ij}/dt = v_{ij}(B_i - V_{ij}) - \acute{v}_{ij}V_{ij} - a_{ij}V_{ij}B_j \qquad (10.1)$$

We can easily verify, by solving this equation under time-varying total abundances B_i and B_j, if the exchange coefficients v and $\acute{v}$ and the per-V predation mortality rate $a_{ij}B_j$ are large compared with the rates of B_i and P_j change (the most likely field case), V_{ij} is likely to stay close to the moving equilibrium defined by setting $dV_{ij}/dt = 0$ in equation 10.1 (a variable speed splitting approximation as discussed in box 6.2). Performing the simple algebra of setting the left side of equation 10.1 to zero and solving for V_{ij}, we end up predicting that V_{ij} will vary with B_i and B_j as

$$V_{ij} = \frac{v_{ij}B_i}{v_{ij} + \acute{v}_{ij} + a_{ij}B_j} \qquad (10.2)$$

If we then substitute this equilibrium prediction into the rate prediction above for Q_{ij}, we obtain a basic "work-horse" prediction relationship for foraging arena theory, namely,

$$Q_{ij} = \frac{a_{ij}v_{ij}B_iB_j}{v_{ij} + \acute{v}_{ij} + a_{ij}B_j} \qquad (10.3)$$

This model says that the total consumption rate Q_{ij} ought to vary as a mass-action product (avB_iB_j), but modified downward by a "ratio-dependent" effect $(v + \acute{v} + aB_j)$ representing localized competition (within foraging arenas) among predators for the prey that are vulnerable at any moment. High mixing rates of the prey (high v_{ij}) imply that Q_{ij} will approach the Lotka-Volterra prediction $a_{ij}B_iB_j$, while very low mixing rates imply that Q_{ij} will approach the "donor-controlled" limit $v_{ij}B_i$ as the predator abundance B_j increases.

We can now use equation 10.3 to see two of the most important qualitative predictions of foraging arena theory. Dividing equation 10.3 by the prey abundance B_i results in a prediction relationship for the instantaneous natural mortality rate component M_{ij} caused on prey i by predator j:

$$M_{ij} = \frac{Q_{ij}}{B_i} = \frac{a_{ij}v_{ij}B_j}{v_{ij} + \acute{v}_{ij} + a_{ij}B_j} \qquad (10.4)$$

Likewise, dividing equation 10.3 by predator abundance results in a prediction relationship for the instantaneous food-intake rate qb_{ij} of prey type i per predator j, $qb_{ij} = Q_{ij}/B_j$. The qualitative forms of these relationships are shown in figure 10.4. They are strikingly different from simpler reaction-vat predictions in at least two ways:

1. the predation mortality rate M_{ij} should not be proportional to predator abundance and may, in fact, be quite flat across a wide range of predator abundances (i.e., the rate may be largely independent of predator abundance except at very low predator abundances);
2. there can be strong "compensatory" increases in the per-capita predator food-intake rates qb_{ij} with decreases in predator abundance, even if there is no change in the overall prey abundance B_i that we might measure by various large-scale spatial sampling methods.

There is at least some empirical support for both of these predictions, like the examples shown in figure 6.5 on page 139 (e.g., Post et al. 1999).

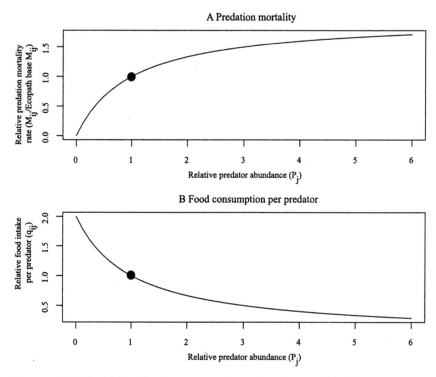

FIGURE 10.4: The relationship between predator abundance and predation mortality (panel A), and the relationship between predator abundance and food-intake rates (panel B) relative to baseline rates specified through an analysis of historical data using mass balance models like Ecopath (black circles).

Note further about point (2) that compensatory qb_{ij} effects are expected to be greatest at low predator abundances, i.e., the qb_{ij} relationship is a $1/X$ form; we have already noted the importance of this form to recruitment limitation in chapter 6. A more general theory for forms of functional response in relation to predator and prey clumping and interference can be found in Cosner et al. (1999); this theory also results in the basic prediction that responses may vary from prey-dependent (type II) to ratio-dependent as a function of the localized spatial organization of interactions.

From the standpoint of general trophic interaction and ecosystem modeling, the simple form of equation 10.3 is both a blessing and a curse. On the blessing side, it can represent a wide variety of behavioral and physical mixing processes. But this is also its curse: we cannot expect to "construct" reasonable estimates of the parameters $v_{ij}, \acute{v}_{ij}$, and a_{ij} from first principles and/or a careful analysis of the biological details of particular interactions (the v and a parameters are highly context dependent). The "rate of effective search" (Holling 1959) parameter a_{ij} can be interpreted as an area or volume searched by each predator in B_j per unit of time, divided by the area or volume of foraging arena over which the vulnerable prey biomass V_{ij} is distributed. For this parameter, the difficult (and location/context dependent) thing to measure would be the volume of the foraging arena; further, for ambush predators it may be the prey that "search," making even the per-predator search area difficult to predict. The v and $\acute{v}$ parameters can represent a much wider variety of processes, e.g.:

1. movement of stream insect larvae into the water column as drift, and back into the substrate (v = drift rate, $\acute{v}$ = settlement rate);
2. emergence of pupating insects from a lake or stream (v = pupation rate, $\acute{v} = 0$);
3. physical mixing of pelagic zooplankton into lake shallows or along/across reef structures (v and $\acute{v}$ set by water advection/mixing rates);
4. diurnal behavioral "emergence" of organisms (juvenile fish, plankton vertical migrations) from hiding places (reefs, schools, dark abyssal waters) to feed (v = rate of initiation of feeding, $\acute{v}$ = rate of return to hiding sites);
5. growth of individuals through a narrow size-range of vulnerability to a specialized predator (v, $\acute{v}$ = growth rates);
6. dispersal of juvenile fish during ontogenetic habitat shifts (v = dispersal rate, $\acute{v}$ = settlement rate into new habitat).

Notice that for at least some of these cases, it would be unsafe even to assume $v_{ij} = v_{ik}$, i.e., to hope that the rate at which type i prey become available to type j predators is the same as the rate at which these prey become available to some other type k predators. For example, the rate at which zooplankters become vulnerable to reef predators could be completely different from the rate at which they become vulnerable to schooling pelagic fishes. This emphasizes what we call the "three to tango" argument about foraging arenas: when a species (dancer one) restricts its activities or distribution in response to a predator (dancer two), it will also restrict its access to its own food organisms (dancer three). In other words, foraging arenas are defined by the joint behaviors of predators and their prey, not just by one or the other player.

Along the same lines, we need to be aware that v and $\acute{v}$ must be treated as mean-exchange-rate and vulnerability predictors, averaging exchange rates that are almost always highly variable over fine time scales of hours to days. In particular, all the processes listed in the previous paragraph are likely to have a very strong diurnal variation. This means, of course, that at very fine space-time scales, the fast equilibration or variable speed-splitting assumption used to derive equation 10.3 must fail. So on close inspection, we are forced to resort to the claim that (at the longer time scales of interest for ecological dynamics), the fine-scale or high-frequency variation ends up causing the same effect (a strong ratio-dependent departure from mass-action predictions) as if the flow between vulnerability states had been smooth in the first place. This is certainly not a very satisfactory argument, but it is surely better than pretending global mass-action interactions, i.e., that vulnerability exchange dynamics do not happen in the first place.

An important point to notice about equation 10.4 is that it does not recognize the possibility that the mortality rate M_{ij} may depend on prey abundance as well as predator abundance. It does not represent depensatory predation impacts, by which we mean an inverse dependence of M_{ij} on the prey abundance B_i. Likewise, it does not represent so-called "Type III" predation effects, involving a compensatory decrease in M_{ij} with decreases in prey density B_i due either to predators switching behaviors or to prey's being less active (or the remaining prey's being only in places where they are less vulnerable) at low prey densities. To include these effects in models, we need to add some additional realism, as discussed in the next two sections.

10.3 ADDING REALISM (I): FORAGING TIME ADJUSTMENTS

There are at least two ways to improve the basic consumption-rate model by accounting for behavioral time-budget factors that may limit predator feeding rates and prey vulnerability. The first is to include handling time (or "Type II functional response") effects, so as to recognize that each prey capture while foraging may cost the predator a significant amount of time in pursuit, manipulation, and nonreactive resting time. The second is to include adjustments in the overall predator foraging time associated with "satiation" (reaching a desired ration, then resting) and/or direct fright response to predation risk while foraging. Handling time and predator foraging time adjustments can lead to depensatory predation effects, while prey foraging time adjustments can lead to compensatory mortality effects.

Box 10.1 shows Holling's (Holling 1959) classical derivation of the type II functional response form for predation rate per predator as a function of prey density. It is worthwhile to understand this derivation both in terms of its useful result about the effective time spent searching for prey, and as a demonstration of how time-rate predictions can sometimes be constructed by looking first at the time budgets of individual animals. The basic result of this analysis is the suggestion that we should modify the arena foraging prediction for the consumption rate of type i prey by type j predators $Q_{ij} = a_{ij} V_{ij} B_j$ by modifying the effective search rate a_{ij} downward to account for time lost to handling prey while foraging:

$$Q_{ij} = (a_{ij}/H_j) V_{ij} B_j \qquad (10.5)$$

where $H_j = 1 + \sum a_{kj} h_{kj} V_{kj}$ and the sum is over all prey types k eaten by predator j. Unfortunately, this complication means that we can no longer obtain an analytical solution for V_{ij}, except for predators that eat only one prey type, since a_{ij}/H_j now appears in the denominator of equation 10.2. To deal with this problem in simulation models, the simplest strategy is to make an initial estimate of H_j using "dummy" V_{kj} estimates calculated by ignoring H_j. Then in successive simulation time steps (or iterations within a single time step), use the (a_{ij}/H_j) estimates from previous steps to update (improve)

BOX 10.1

DERIVATION OF HOLLING'S DISC EQUATION

Holling (1959) demonstrated how to include handling-time effects in predation models, with his derivation of the "disc equation" from time-budget considerations. We use his original notation below to explain this demonstration. Holling's basic argument is that the total time T_t that an animal spends foraging (in foraging arenas) can be partitioned into the time spent reactive to prey, T_s, and the time spent handling prey, T_h:

$$T_t = T_s + T_h$$

If the average time spent handling (lost from searching) for each prey captured is "h," and if the predator captures NA prey during the total time T_t, then the handling-time component of T_t can be expressed as

$$T_h = hNA$$

Suppose then that encounter rates with prey during the active search time T_s are proportional to the prey density N_o. More precisely, suppose that the predator searches an area or volume "a" per time searching so as to encounter all the prey (at density N_o per area or volume) that are in the total searched area or volume aT_s. Then NA should vary as the mass-action product:

$$NA = aT_s N_o$$

Substituting this prediction into the prediction for handling time, and that prediction in turn into the basic time-budget equation, we get

$$T_t = T_s + haT_s N_o$$

From this we can solve for the predicted time spent searching T_s as a function of prey density:

$$T_s = \frac{T_t}{(1 + ahN_o)}$$

Substituting this into the mass-action product prediction for NA, then rearranging the terms into a more convenient format, we obtain Holling's disc equation:

$$NA = \frac{aT_t N_o}{(1 + ahN_o)}$$

To model multiple prey types, we modify the handling-time (denominator) component of the prediction, $T_h = hNA$, to be a sum over prey types i of type-specific handling times h_i, type-specific search rates a_i, and type-specific prey attack rates NA_i:

$$T_h = \sum_i a_i h_i NA_i$$

(*Continued*)

(BOX 10.1 continued)

and the prediction for the attack rate on type i prey becomes just

$$NA_i = \frac{T_t a_i N_i}{(1 + \sum_i a_i h_i N_i)}$$

Note how this equation predicts that attack rates on each prey type i will decrease with increasing abundance of other prey types j, just due to the effect of those other prey types on the predator's search time. Such changes in the diet composition are commonly observed and are often incorrectly called "switching." In ecology, we usually reserve the term *switching* to refer to cases in which the rates of effective search a_i for particular prey types change with the abundance of those types or of other types. ∎

V_{ij} estimates and H_j estimates (calculate V_{ij} using a_{ij}/H_j rather than just a_{ij} as in eq. 10.2). In this approach, H_j for each predator type becomes a dynamic state variable that is simulated along with predator and prey abundances.

Foraging time adjustments (changes in the Holling T_t of box 10.1) may occur in response to animals failing to achieve (or exceeding) the "desired" per-capita food-consumption rates $qb_j = \sum_i qb_{ij}$, or in direct response to the total predation risk as measured by $M_{pi} = \sum_j M_{ij}$. A simple way to model these adjustments is to include the relative foraging time factors τ_i and τ_j in the prediction of trophic flows Q_{ij} from equation 10.3, where the relative factors are scaled to 1.0 when qb_j and M_{pi} are at some baseline or initial values. The handling-time effects Q_{ij} at any moment in time can then be calculated in the hierarchic steps:

1. predict adjustments in τ_i, τ_j, and H_j using the q, M_p, and V_{ij} values from last time
2. calculate a_{ij} and v_{ij} for the current time step from

$$a_{ij}(t) = a_{ij}\tau_j/H_j$$
$$v_{ij}(t) = v_{ij}\tau_i$$

3. calculate

$$Q_{ij} = \frac{a_{ij}(t)v_{ij}(t)B_iB_j}{v_{ij}(t) + \acute{v}_{ij} + a_{ij}(t)B_j} \tag{10.6}$$

Note in step (2) that changes in the predator's rate of effective search are seen as occurring due to both handling time and relative feeding time effects, and that changes in the prey's relative feeding time also affect the rates of becoming vulnerable to predation. Note further that changes in H_j, τ_i, and τ_j are represented as species or biomass pool-scale effects on i and j, so that changes in the abundance of other prey types and in the predation risk are recognized to affect the i–j trophic linkage.

In this hierarchic structure, the key problem then becomes to predict the variation in τ_i and τ_j over time. There does not appear to be any simple analytical model for accomplishing this because of the cross-prey and cross-predator linkages involved in the calculation of the time-dependent food intake and risk measures $qb_i(t)$ and $M_{pi}(t)$. The approach that we have taken in Ecosim (Walters et al. 1997, 2000) has been to use an iterative, discrete-time updating (at each monthly simulation time step) of τ_i as

$$\tau_{i,new} = \tau_{i,old}(1 - \zeta_i) + \zeta_i qb_i(0)M_{pi}(t-1)/[qb_i(t-1)M_{pi}(0)] \qquad (10.7)$$

where ζ_i is a numerical rate-adjustment factor $(0 < \zeta_i < 1)$ chosen so as not to cause an unstable oscillation in the calculated $\tau_{i,new}$. Here, the $qb_i(0)$ and $M_{pi}(0)$ parameters represent baseline $(t = 0)$ feeding and predation-risk rates, and the ratios $qb_i(0)/qb_i(t-1)$ and $M_{pi}(t-1)/M_{pi}(0)$ represent recent (time $t - 1$) relative-feeding and predation-risk ratios. These risk ratios are arranged so as to drive $\tau_{i,new}$ upward whenever qb_i drops below $qb_i(0)$, and to drive $\tau_{i,new}$ downward whenever M_{pi} exceeds $M_{pi}(0)$.

The foraging time adjustment equation 10.7 is a simple way to implement an argument from evolutionary ecology that animals ought to vary feeding times so as to optimize a fitness measure proportional to the predation-risk/food-intake ratio commonly called "μ/g" (Abrams 1984; Werner and Gilliam 1984; Abrams 1993; Anholt and Werner 1998; Gilliam and Fraser 1987; McNamara and Houston 1994; Werner and Hall 1988). There is much experimental evidence that such adjustments do, in fact, occur and can lead to spectacular habitat shifts when predator abundances are manipulated. There is debate about whether the best fitness measure is some simple ratio of risk to feeding rate but not about the qualitative notion that there should be inverse relationships between the feeding time and the components of the ratio.

The linkage of equations 10.6 and 10.7 can result in some very complex forms for the apparent or "emergent" functional relationship between prey density and predator feeding rate. In particular, foraging time adjustments by prey in response to the effects of increasing prey density on prey feeding rates (qb_i) can result in a basic reversal of the form of the overall functional response, from Type II to Type III, as shown in a hypothetical experiment in figure 10.5. This example, removing all habitat structure from the experimental arena results in a classic Type II response pattern, dominated in shape by handling-time limits for the predator. But adding a bit of habitat complexity, in the form of a refuge where prey can hide, can lead to the experimental result being dominated by the responses of prey to localized competition within and near the refuge.

Reaction vat model   Foraging arena model

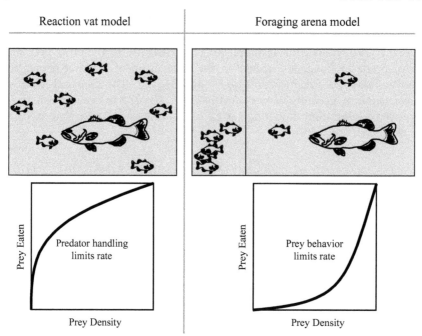

FIGURE 10.5: Risk-sensitive foraging behaviors by prey fish imply a reversal of the predictions about predation impact. Small increases in the space-time scale of experimentation and modeling can result in a reversal in the form of the functional response observed, from a type II response for a reaction-vat experiment to a type III for an experiment in which prey can hide from predators unless the prey density is high enough to force the prey to spend more time foraging. Redrawn from Walters and Kitchell (2001).

10.4   ADDING REALISM (II): TROPHIC MEDIATION

The functional response models of the previous sections invite us to think of the foraging arena vulnerability exchange rate v_{ij} and search rates a_{ij} as constants arising from the way organisms use characteristics of the physical environment (hiding places, water clarity, etc). Unfortunately, this can be a very dangerous oversimplification, in at least the following three ways:

1. some organisms use other organisms as protective cover from predation, e.g., juvenile fishes commonly use macrophyte or sea-grass beds as hiding places;
2. some organisms can directly impact the search rates of other organisms, e.g., high algal concentrations can reduce the search efficiency of visual predators;
3. some foraging arenas may be created by response to predation risk, e.g., pelagic birds (e.g., albatross) may depend on piscivores such as tunas to drive small baitfishes to the surface.

These "indirect effects" (Dill et al. 2002) or "trophic mediation effects" can alter, and even reverse the qualitative direction of, predictions about the impacts of fishing and habitat disturbances related to fishing. For example, in one simple model of pelagic food webs in the Pacific Ocean, we predicted that tuna purse-seine fisheries should enhance bird populations by reducing competition between tunas and birds for bait fishes. But the opposite effect may have actually occurred, due to effect (3). In an analysis of fish community changes in trawled versus untrawled *Posidonia* (marine macroalga) beds, Rodriguez-Ruiz (Ph.D., UBC, ms in prep) found that her Ecosim model would not correctly predict even the direction of biomass response for several fish species unless she included the effects of the macroalga biomass on the vulnerabilities of these species to predation (see fig. 12.5 on page 304). Day and Branch (2002) provide excellent field experimental results to demonstrate how a very valuable abalone fishery in South Africa depends on the protection of sea urchins, which provide a spatial refuge for juvenile abalone (effect [1]).

The incredible variety of trophic mediation effects implies that there is unlikely to be any simple, general functional form or equation for predicting them. We have tried to include mediation effects in Ecosim models by allowing model users to specify arbitrary functions $f(B^*)$, where B^* is a weighted sum of model component biomasses and $f(.)$ is scaled to 1.0 at the simulation initial biomass state. Model users may sketch the form of $f(.)$ with a computer mouse or choose from a variety of functional forms (e.g., linear, exponential, logistic, increasing or decreasing with increasing B^*). These functions are then used as multipliers on user-selected vulnerabilities v_{ij}. That is, the v_{ij} calculation in equation 10.6, step (2), is modified to $v_{ij}(t) = v_{ij}\tau_i f(B^*(t))$.

The existence of complex trophic mediation effects as part of the "assembly" of aquatic communities through colonization and succession processes (e.g., Yodzis 1984, 2000; Moyle 1996) does not bode well for any hope that we might eventually be able to make useful a priori predictions about community structure and invasions. There is much more going on than competition and predation processes. Very likely we will uncover the importance of particular mediation effects largely by seeing predictive models fail, i.e., by seeing predictions that go in the opposite direction of the observed dynamic patterns. That is fine if we view modeling and model testing against data as an adaptive learning process, in which we learn most from failures rather than apparent successes. But it is catastrophic for scientists and managers who think we can build good enough models to "get it right" in the first place.

10.5 ECOSIM

To understand the implications of foraging arena interaction dynamics for exploited ecosystems, we need somehow to link the predictions of time-varying flows Q_{ij} as gains and losses to the creatures involved, and we need to include fishery and perhaps environmental changes in the calculation of system state over time. There is currently a popular software for accomplishing this, called Ecopath with Ecosim "EwE" (Walters et al. 1997). The Ecopath part of this software helps users to enter the baseline state-rate information needed to set initial states and calculate some parameter values for simulations (see chapter 11), and Ecosim then combines the rate information by solving the set of differential equations

$$ dB_i/dt = g_i \left[\sum_k Q_{ki}(t) \right] - \sum_j Q_{ij}(t) - M_{oi}B_i - \sum_f F_{if}(t)B_i \qquad (10.8) $$

for species or biomass pools $i = 1 \ldots n$. Here the first sum represents the food-consumption rate summed over prey types k of species i, and g_i represents the growth efficiency (proportion of food intake converted into biomass available to flow to mortality processes). The second sum represents predation loss rates over predators j of i. All Q's in these sums are calculated from equations 10.6 and 10.7 using predator biomass as an index of predator abundance. M_{oi} represents the instantaneous natural mortality rate due to factors other than modeled predation. The final sum represents the instantaneous fishing mortality rate, as a sum of fishing rate components caused by fishing fleets f.

Most previous biomass-dynamics models for multispecies fisheries interactions were based on trying to avoid at least part of the food-consumption-rate prediction $g_i \left[\sum_k Q_{ki}(t) \right]$, replacing some or all of the consumption-rate-gain terms with the surplus-production function $r_iB_i(1 - B_i/K_i)$, which may represent competition for food resources implicitly through the density effect $(1 - B_i/K_i)$. Ecosim can be configured to do the predictions this way, but the default option is to use a surplus-production-gain rate only for biomass pools i representing primary producers; in that case, the density-dependent term presumably represents self-shading and competition for nutrients. This encourages EwE model users to take a "whole ecosystem" (the whole food web including primary producers) approach.

Including primary producers in the biomass-dynamics calculations opens a door for using the model to examine temporal changes in whole-ecosystem productivity due to regime changes, upwelling, or other physical factors. EwE users can "sketch" or use statistical estimation procedures to set time-varying patterns of r_i for $i =$ primary producers, similar to the idea of estimating an unexplained variation in single-species recruitment rates by estimating a sequence of "recruitment anomalies" (see chapter 12). Note that the idea here is not to pretend that we can make successful, bottom-up a priori predictions of potential fish production from estimates of primary production; exercises of that sort have not been particularly successful or useful in fisheries management. Rather, the idea is to take baseline data on the observed productivity, checked for mass-balance consistency using Ecopath (which checks to make sure that production rates are large enough to support the baseline estimated mass flows up the food web), then examine how deviations from this baseline might affect future fisheries potential.

Numerical integration procedures must be used to obtain predictions of biomasses over time from equation 10.8. For most applications, the equation system is very "stiff" (includes both fast and slow dynamic variables), and so must be integrated over quite short numerical time steps (one month or less). We have had good success with two numerical procedures, "4th order Runge-Kutta" and "Adams-Bashforth"; see box 6.1 on page 134.

For very fast variables like phytoplankton and small zooplankters that can turn over on time scales of hours to weeks, Ecosim uses a speed-splitting argument that such variables are likely to remain near equilibrium with respect to slower variables. We ignore fine-time-scale dynamic oscillations caused by predator-prey interactions among fast species. Moving equilibria for fast variables are approximated by treating the food consumption (or primary production) rates for each monthly time step to be constant, which results in the approximation $dB/dt = I - ZB$, where I is the (constant) input or production rate $g_i \left[\sum_k Q_{ki}(t) \right]$ calculated at the start of the time step and Z is the sum of all instantaneous loss rates M_{ij}, M_o, and F evaluated at the start of the month. Then setting $dB/dt = 0$ for a fast variable, the approximate solution for that variable's equilibrium is $B_i = I/Z$. Such B_i's then change over time scales longer than one month due to changes in the I and Z rates due to changes in both B_i and the impacting species B_j. This speed-splitting tactic can reduce computing time for long-term simulations by orders of magnitude, and we have found that it generally gives solutions quite close to those obtained by grinding out full numerical solutions with time steps of days or less. In cases for which the fine-scale numerical solution predicts predator-prey oscillation among fast variables (e.g., phytoplankton and small zooplankters), the moving equilibria are generally good approximations for the mean abundances over the cycles. Presumably, such means are what slower, longer-lived creatures must "see" anyway, given that they cannot respond quickly to the high-frequency variation.

As a warning to prospective modelers, it can be grossly misleading to bypass the numerical integration problems of equation 10.8 by approximating the dynamics with difference equations over long (one-year) time steps as we typically do with single-species assessment models. In that approach, we need to calculate survival rates using some exponential approximation for mortality effects, e.g., survival $= exp(-M_o - \sum_k F_k - \sum_j Q_{ij}/B_i)$, where the instantaneous mortality rates $M_{ij} = Q_{ij}/B_i$ are predicted from biomasses at the start of each survival time step. Calculating survival rates this way results in an implicit time delay (rates over whole time steps are dependent only on states at the start of the time step), which can cause gross (and incorrect) numerical instability in the predicted states over time. The effect is essentially the same as when we assume a Ricker stock-recruitment relationship, which is dome-shaped because the density-dependent mortality rate is assumed proportional to the initial (parental) abundance, rather than a Beverton-Holt relationship derived by integrating density-dependent rates that change over the juvenile life. A great deal of consternation was caused a few years ago when Wilson et al. (1991, 1991, 1994) argued using such discrete-time survival calculations that fisheries ecosystems might commonly exhibit extremely complex, chaotic behavior. Such an assertion obviously has policy consequences. There certainly are time-delayed interactions that can cause cyclic or chaotic dynamics in fish populations, e.g., cross-age cannibalism as discussed in chapter 6. But available time-series data do not support the predictions of chaotic behavior that arise from poor numerical procedures and the hidden assumptions that these procedures imply about time delays in how interaction rates vary.

10.6   REPRESENTING TROPHIC ONTOGENY IN ECOSIM

When we use overall biomasses in interaction-rate predictions (eq. 10.8, e.g.), we are implicitly assuming that there is compositional stability within each pool. That is, the actual flux rate Q_{ij} in nature is in fact a sum of component flux rates $Q_{ki,kj}$ from biomass components (sizes, species) ki within prey pool i to biomass components kj within predator pool j, where the component rates can be grossly different due to prey-size selectivity or other factors. Consider the rate prediction $Q_{ij} = a_{ij}V_{ij}B_j$, and recognize that we might divide this into a set of size-species component rates:

$$Q_{ij} = \sum_{ki}\sum_{kj} a_{ki,kj}V_{ki,kj}B_{kj} \tag{10.9}$$

Then consider representing $V_{ki,kj}$ and B_{kj} as proportions $p_{ki,kj}$ and $\acute{p}_{kj}$ of the prey and predator populations:

$$V_{ki,kj} = p_{ki,kj}V_{ij}$$
$$B_{kj} = \acute{p}_{kj}B_j \tag{10.10}$$

Substituting equation 10.10 into the aggregate prediction of equation 10.9, we see that this prediction can be written as

$$Q_{ij} = \left[\sum_{ki}\sum_{kj} a_{ki,kj}p_{ki,kj}\acute{p}_{kj}\right]B_jV_{ij} \tag{10.11}$$

So the a_{ij} of the aggregate model is, in fact, a weighted sum of size-species selective rates $a_{ki,kj}$, i.e., $a_{ij} = \left[\sum_{ki}\sum_{kj} a_{ki,kj}p_{ki,kj}\acute{p}_{kj}\right]$. In order for the a_{ij} to

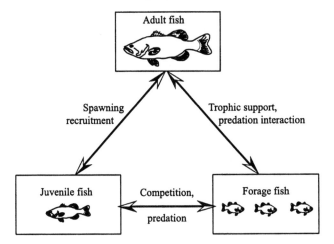

FIGURE 10.6: Recruitment depensation can occur at low stock sizes when prey organisms of adult fishes respond positively to reductions in adult fish abundance resulting in reductions in juvenile survival via competition and/or predation interactions with the juveniles.

be relatively stable over time, i.e., to be useful as predictive parameters, either the $a_{ki,kj}$ values have to be similar across ki, and kj, or else the proportions $p_{ki,kj}$, and $\acute{p}_{kj}$ have to be relatively stable over time. There are three choices for dealing with this concern:

1. restrict predictions to short enough time spans and to narrow enough ranges of system states for the assumption of constant p and $\acute{p}$ to hold as a safe approximation;
2. construct a model with many more species-size biomass (or numbers) components, so that each is relatively more homogeneous with respect to parameters like a_{ij};
3. discard the whole notion of biomass pools as state measures, construct individual-based models that can realistically represent the rich variety of ways that organisms may differ from one another, and simulate/count up all individual interaction events.

We will discuss the third of these options in chapter 11.

There is one particular feature of aquatic ecosystems that cries out for moving to choice (2), and at least dividing species pools into juvenile and adult components: trophic ontogeny. Most fishes undergo complex changes in habitat use, feeding habits, and predation risk as they grow, and many pass all the way up the food web from algal or microzooplankton feeding to piscivory as they grow. For such creatures, it is dangerous to assume compositional stability in the total population biomass, especially considering that ontogeny implies the existence of so-called "trophic triangles" (fig. 10.6; Bax 1998; Carpenter and Kitchell 1993; De Roos and Persson 2001; Rudstam et al. 1994; Ursin 1982; Walters and Kitchell 2001). In such triangles, the competition and predation effects on juveniles from other species may interact strongly with changes in the adult abundance; in particular, adults may "cultivate" the ecosystem in favor of their own juveniles by preying upon the other species. This can, in turn, result in some perverse, depensatory changes in the juvenile-adult composition and in the juvenile survival rates when the adult abundance is reduced through fishing.

We cannot solve the problem of modeling trophic ontogeny effects by dividing the biomass models into more pools and linking juvenile and adult pools through simple biomass flows representing reproductive additions to the youngest pool and "graduation" between older pools. Graduation biomass flows are often very small compared with flows associated with food consumption and predation. So adding and subtracting such small flows in biomass-rate budgets does not properly account for the limits on dynamic change in any age-stage-stanza pool caused by the fact that it is numbers of individuals that do the graduating. Pools of older individuals cannot vary in biomass independent of the numbers of individuals that recruit to them, and pools of young juveniles cannot vary in biomass independent of the numbers of young fish supplied by reproduction.

That is, in order to deal with issues that arise from trophic ontogeny, we are forced to consider numbers of individuals as well as biomass for each life-stage or life-stanza pool. The approach taken in Ecosim to this problem has been to replace the biomass-dynamics differential equation(s) for selected species with age-structured numbers and body-growth accounting (Walters et al. 2000). This single-population, discrete time accounting (for recruitment, growth, and survival) is done on the same relatively short (monthly) time steps as used for a numerical integration of the remaining differential equations. The delay-difference accounting approach (see chapter 5) is used to avoid unnecessary age-structure accounting for older animals (for more information about this accounting procedure, see Walters et al. 2000 and box 10.2).

When the calculations in box 10.2 are carried out in conjunction with the basic foraging arena rate-prediction equations for trophic flows Q_{ij} (equations 10.6 and 10.7), the combined dynamics generally produce "emergent" stock-recruitment relationships much similar to those derived in chapter 6. That is, foraging time adjustments (and/or density effects on growth and time spent in juvenile pool) by juveniles (changes in $\tau_{juvenile}$ in eq. 10.8) lead to density-dependent changes in time exposed to predation risk and hence to changes in $Z_{juvenile}$. Further, there is the added "bonus" of linking changes in $Z_{juvenile}$ and Z_{adult} per time spent foraging to changes in the abundance of predators, and the predation effects may include cannibalism by adults on juveniles so as to produce dome-shaped or Ricker-like stock-recruitment relationships.

Multiple stanza versions of the split-pool accounting described in box 10.2 have recently been implemented in Ecosim. An improvement in the representation of time-varying body growth has been included by calculating body-growth dynamics using a discrete approximation to the (monthly) von Bertalanffy growth pattern, as

BOX 10.2

<small>EMBEDDING POPULATION AGE STRUCTURE ACCOUNTING IN ECOSIM BIOMASS DYNAMICS PREDICTIONS</small>

The following method for embedding an age-structured population-dynamics calculation in biomass-dynamics models has been very useful as a relatively simple way to study the effects of trophic ontogeny on trophic interactions. As motivation for trying to understand the messy accounting described here, note that it has helped us, e.g., to uncover the possibility of cultivation-depensation effects that involve pathological recruitment changes when large, dominant fish species like Atlantic cod are overfished.

"Split pools" are treated as two life-history stages or stanzas (two biomass pools to be predicted over time, "juveniles" and "adults"). The age at transition from the juvenile to adult pool is defined by food habits, predation risk, and/or fishing impacts. The dynamic-accounting procedure in Ecosim has the following steps:

1. if the current simulated population age structure has $N_{a,t}, a = 1 \ldots k$ juvenile individuals in monthly age cohorts $1 \ldots k$, and total $N_{k+1,t}$ individuals in the adult (over k-month-old) pool, set the predator search abundances $B'_{juvenile}$ and B'_{adult} for Ecosim interaction calculations to $B'_{juvenile} = \sum_a w_{a,t}^{2/3} N_{a,t}$, $B'_{adult} = N_{k+1,t}$, where $w_{a,t}$ is the current body weight of an age-a juvenile;

2. use these B'_j estimates as the predator-abundance indices in the Ecosim prediction equations for Q_{ij} (eqs. 10.6–10.7), to obtain the impact rates on prey for the time step and the total-food-intake rates by the juveniles and adults ($Q_{juvenile}, Q_{adult}$), and to calculate the total instantaneous mortality rates Z_{juv} and Z_{adult} for the two population components;

3. update the adult numbers and biomass to the start of the next month using delay-difference accounting equations of the form

$$N_{k+1,t+1} = e^{-Z_{adult}} N_{k+1,t} + N_{k,t} \quad \text{and}$$

$$B_{adult,t+1} = e^{-Z_{adult}} [gQ_{adult} + \rho B_{adult,t}] + w_{k,t} N_{k,t}$$

4. update the juvenile numbers and body weights for each age-a by $N_{a,t+1} = e^{-Z_{juv}} N_{a,t}$, $N_{1,t+1} = \beta B_{adult,t}^{\gamma}$ (recruitment may be assumed to be density-dependent in early "larval" life by setting $\gamma < 1.0$), and $w_{a,t+1} = w_{a,t} + g' Q_{juvenile} N_{a,t} w_{a,t}^{2/3} / B'_{juvenile} / N_{a,t}$ (this last term partitions the total food consumption rate by juveniles, $Q_{juvenile}$, into a per-capita rate proportional to the relative-predation-rate index $N_{a,t} w_{a,t}^{2/3}$ for each age of juveniles).   (*Continued*)

$$w_{a+1}(t+1) = (1 - 3K)w_a(t) + G_a Q_s(t)/P_s(t)$$

where t is in units of months, K is the monthly von Bertalanffy metabolic coefficient, Q_s is the calculation of the total food intake Q, predator searching $P = \sum_a w_a^{2/3} N_a$ (as in box 10.2) represents the stanza s search area of age-a animals, and G_a is a scaling constant set to ensure that growth from a to $a+1$ will match a von Bertalanffy growth curve and will give a baseline growth curve when the food-consumption-rate index Q/P is at its baseline value. This matching is achieved by setting $G_a = [w_{a+1} - (1 - 3K)w_a] Q_s(0)/B_s(0)$ for baseline $Q, S,$ and w_a values for stanza s. The link back to Ecosim biomass dynamics is accomplished by summing $w_a N_a$ over ages within stanzas after each monthly update, and the reverse link from Ecosim rates to the size-age structured calculations over ages within each stanza is accomplished by passing $Q_a(t)$ and $Z_s(t)$ (with the total mortality rate assumed equal over each age a in stanza s) to the stanza-scale size-age accounting.

The multiple-stanza version of Ecosim allows us to test the effects of changes in biomass pool composition on aggregated Q_{ij} predictions, enter more precise information on size-dependent interaction rates, and eventually provide better guidelines about whether (or for what policy predictions) much biomass disaggregation into smaller pools is worthwhile. It allows us to better represent some "perverse" ecological interactions, such as the cannibalism of older juveniles on younger ones, and some important policy problems such as growth overfishing and the effect of stocking juveniles of different sizes in marine enhancement programs.

10.7 SINGLE-SPECIES DYNAMICS FROM ECOSIM RATE EQUATIONS

Early in the development of Ecosim, we noticed that the biomass rate equation 10.8 often gives quite good fits to the relative-abundance time-series used in single-species assessments. This happened even before we added the improved realism of delay-difference population modeling as described in box 10.2, or the more recent multistanza age-structure representation. That is, we often saw that Ecosim acted similarly to a set of almost independent surplus-production models, at least for fish populations near the tops of simulated food webs.

To see why this happens, consider pulling the rate equation for one fish population j out of the set defined by equation 10.8, and aggregating all of its prey types i into a single biomass pool B_{prey} that is not highly variable over time (e.g., has low vulnerability exchange rates $v_{prey,j}$ in equations 10.6 and 10.7, and/or a low total impact $M_{prey,j}$ caused by population j). Assume further that vulnerabilities $v_{j,preds}$ to its predators are relatively low, so that $M_{j,preds}$ is only weakly variable with predator abundances (see fig. 10.4, panel B). Under these assumptions, the biomass rate equation for population j can be closely approximated by the single-species model

$$dB_j/dt = \alpha B_j/(\beta + B_j) - M_j B_j - \sum_f F_{j,f} B_j \qquad (10.12)$$

where α, β, and M_j are aggregated "trophic parameters" defined in terms of foraging arena parameters by

$$\alpha = g_j v_{prey,j} B_{prey}$$

$$\beta = \frac{(v_{prey,j} + \acute{v}_{prey,j})}{a_{prey,j}}$$

$$M_j = M_{oj} + \sum_{preds} Q_{j,preds}/B_j$$

$$= M_{oj} + \sum_{preds} M_{j,preds}$$

Equation 10.12 states that the surplus production available to support predators, fisheries, and other mortality ought to vary as a Beverton-Holt–like function of biomass, $\alpha B_j/(\beta + B_j)$, which might well be approximated by a simpler logistic function of the form $r_j B_j(1 - B_j/K_j)$. It says that the maximum of this production function for high B_j and α ought to be positively proportional to the total prey abundance and to the prey-vulnerability–exchange characteristics. And it says that the steepness of the production function at low B_j, measured by the slope α/β, ought to be proportional to the predator's search efficiency as measured by $a_{prey,j}$ as well as to the total prey abundance.

We have used equation 10.12 as a "reduced"' model in comparisons of Ecosim to time-series biomass data, as a check on whether more complex models with variable prey and predator abundances actually do explain much of the observed variation. The results from such checks have been quite variable, but in many cases involving piscivores (near the top of the food web) equation 10.12 performs as well as a model with full trophic interactions. This should not surprise anyone with experience in single-species stock assessment. We have never had consistent luck at explaining deviations between single-species models and data, especially on recruitment rates, by using the abundances of prey and predators as explanatory variables (there are some exceptions; see the cod-herring analysis in Walters et al. 1986, and the Collie and Spencer 1993 model for the Georges Bank system). The story becomes considerably more complex for midtrophic-level species such as shrimp and herring, for which changes in trophic interactions do appear to play a much bigger role in causing abundance fluctuations (Worm and Myers 2003).

Just because we can often approximate the dynamics of single species, especially piscivores, by using single-species models that ignore changes in trophic resources and predators, does not mean that it is a waste of time to model trophic interactions. There are many policy concerns and options, such as impacts on nontarget species and the impacts of changes in ecosystem fertility, that cannot be addressed at all with single-species calculations except through largely arbitrary changes in the single-species dynamic parameters. A central part of moving from single-species to ecosystem management is, in fact, to recognize those policy options and concerns and to try to make useful predictions about them.

10.8   ECOSYSTEM-SCALE VARIATION

One of the main applications of Ecosim has been to explore how more selective fishing practices might help alleviate the apparent "cascades" of overfishing impacts that often appear to accompany regime-shift changes in marine productivity. Such cascades likely involve two types of effects: (1) "switching" by fishers to target alternative species/locations when a regime shift starts to cause reductions in the productivity of preferred species (e.g., switching to shrimp fishing following a cod collapse, and other examples of "fishing down marine food webs" sensu Pauly et al. 1998), and (2) changes in trophic interactions so that some species actually benefit from the regime change (e.g., increased shrimp production following finfish declines). Ecosim does not pretend to model the complex dynamics involved in switching by fishers, but it can at least be used to evaluate some of the ecological interaction effects under alternative policy options for improving the selectivity of fishing practices.

It is simple in Ecosim to "force" the productivity and vulnerability parameters with time multipliers representing various scenarios for regime change (see above), and it is simple to change the fishing-rate matrix $F_{i,f}$ (the mortality rate caused on each species i by each fishery or fishing gear f). Fishing rates can be changed either by arbitrarily varying $F_{i,f}$ for each gear type f, to represent improvements for that gear in reducing the undesired bycatch of some i, or by varying the overall effort by gear type f while assuming proportional changes in $F_{i,f}$ for all groups i that it impacts.

Two approaches have been used in such explorations. The first has been a simple "gaming" approach (Walters 1994), in which various productivity scenarios are suggested and evaluated using "sketchpad" methods in the software for changing the $F_{i,f}$ over time. Such gaming can at least help to screen out policy options that could not possibly improve future management, and to obtain consensus among stakeholders about what future state changes would be most desirable or important. Given some formal measures of the management objectives and performance identified through the gaming discussions, the second approach has been to apply nonlinear optimization (search) procedures to seek optimal future patterns in $F_{i,f}$. More precisely, the search procedures seek optimal relative future efforts by gear type, given fixed future selectivity but possibly including new gears f that are more selective; it is not practical to search for full optimal matrices $F_{i,f}$ by species. Essentially, the numerical search procedures do a very large number of simulation "trials" while varying the fishing efforts by gear type so as to improve the overall performance measures from trial to trial.

The main thing that has been learned from explorations to date, besides the obvious notion that fishing rates ought to be reduced in general during periods of low ecosystem productivity, is that there is no obvious "best" way to deal with the optimization of yields across interacting species. For example, a common outcome of optimizations for models that include shrimp-cod interactions has been the recommendation that cod should be deliberately overfished, since shrimp are usually more productive and valuable (and during low-productivity periods, protect the shrimp but not the cod). Such "degenerative" optimization scenarios essentially involve the prediction that we should farm marine ecosystems so as to simplify food webs to enhance the production of the most valued species while eliminating natural competitors and predators. These are very frightening scenarios, because in most cases we cannot offer firm, objective reasons (and sound examples from past management experience) for avoiding the obvious risks that such ecosystem simplification would entail.

So it appears that the most important need today in terms of bringing formal optimization methods to bear on ecosystem harvesting questions is not better models for ecological interactions but, rather, much more work on "objective functions" and optimization criteria. In particular, there is a need to develop criteria that explicitly recognize, and value in terms of potential future benefits and costs, the dangers of driving ecosystems into simplified states for which we really have no predictive experience or capability.

CHAPTER 11

Options for Ecosystem Modeling

WHENEVER ECOLOGISTS ATTEMPT TO DEVELOP and test a policy model for ecosystem management, we are faced with a confusing variety of trends and information, with a severe confounding of the effects of natural and human-induced changes. Most of us would make lousy detectives: we tend to search for correlations, then dream up ecological explanations (and models) for those we find. We are seldom critical about identifying alternative hypotheses that might equally well explain the data but would imply very different policy choices. Unfortunately, the very complexity of ecological systems assures that any fool can cook up a reasonable explanation, since there are many, many biological and physical connections, delays in response, etc. Worse, as biologists, we often assume that we can somehow bypass the confounding of effects by looking at the ecology in more detail, modeling processes "bottom up" so as to get the right answer by the brute-force approach of accounting for everything. This does not work either, because important dynamics can occur at just too many scales. As witness to the fact that serious omissions are almost certain to occur, we note that until a few years ago when the foraging arena theory of the last chapter started to be used, some apparently very complex and realistic ecosystem models failed because they did not include the limiting effects on trophic interactions of arena-scale behavioral processes. These models accounted beautifully for finer-scale, physiological ecology relationships, and some also did very well at accounting for the large-scale spatial structure caused by factors like advection; but they missed the crucial impact of intermediate-scale structure.

In trying to develop useful models for ecosystem management, we are faced with same difficult choices as in single-species assessment, and more: simple versus complex, few versus many variables, forms of relationships, where to set boundaries (and treat variables beyond those boundaries as constants). We have generally made the same mistakes as in single-species assessment: (1) being preoccupied with adding spatial and biological detail (e.g., size structure), looking downward into the system structure rather than more broadly at factors; and (2) concentrating on variables and relationships for which we have the strongest data, rather than those likely to be most important to the policy problems at hand. There are various pronouncements about the "right way" to model aquatic ecosystems, based on such preoccupations and concentrations, but we do not hear such pronouncements from people with much modeling experience. There is also much discussion about the "optimum" model complexity when we admit that trying to model too much can lead to huge cumulative errors in predictions (Bartell et al. 1999; Isawa et al. 1987; O'Neill and Rust 1979; Rasteller et al. 1992; Fulton

2002a), but that discussion has focused mainly on particular, scientific measures of model performance (as opposed to measures of the ability to provide useful policy insights).

To make matters still worse, many biologists assume that what looks bad to us must be bad for the fish, and this bias creeps into our selection of model variables and our judgment calls about interactions that are difficult to measure. When faced with the apparent failure of models that have such bias, we often ignore the evidence or find some contorted explanation for why the model did not work. An excellent example of these warnings is a model developed in the early 1970s to predict the impacts of logging on Pacific salmon populations in Carnation Creek, B.C. (Walters 1975). There was to be an experimental logging program on the watershed, and we (an adaptive management modeling team from the University of British Columbia) were challenged to see if we could predict the experimental outcomes in advance. This seemed at the time like an easy opportunity to "sell" a modeling approach, on a problem for which the negative ecological impacts appeared pretty obvious. A team of biologists worked with us to develop quite a complex model, with everything from a spatially mapped forest vegetation and water runoff submodel to a series of physical and biological linkages between water conditions and juvenile salmon (fig. 11.1). The team did not have much data then on the impacts identified in figure 11.1, so we tended to choose parameter values that made the negative effects stronger (after all, we could all see how horrible streams usually look in logged watersheds in the Pacific Northwest). So when we ran the model, we predicted the "conventional wisdom" with some confidence: logging should cause siltation and flooding that would reduce egg survival in spawning gravels, and this should result in a reduced abundance of fry and smolts and in lower adult population sizes. We got a very big surprise when the experimental data began to arrive: egg survival declined as expected for coho salmon, but the net yield of juvenile coho from the watershed (smolt production) actually doubled rather than declined as we had expected. Something that we had not modeled correctly led to a fourfold improvement in the survival rate from the fry to smolt stage, despite the obvious degradation in the apparent habitat conditions, and this was enough to completely reverse the impact from what we had predicted. The mechanism for improved survival is still not fully understood and, apparently, has something to do with increased water temperatures and improved juvenile fish growth after logging. Similar results have been obtained in other watershed experiments in the Pacific Northwest. To this day, the authors have yet to hear a professional salmon "habitat biologist" talk about the possibility that the Carnation Creek reversal might be a general effect, representing an opportunity for cooperation between forest and fisheries management.

What can we do to minimize the risk of really embarrassing qualitative model failures, given that it is absurd to pretend that we can model everything? One very obvious answer is to stay away from quantitative predictive

Factor	Bad Effects	Good Effects
Silt	Reduces egg survival Triggers emigration	Smaller juvenile territories Reduced visibility to predators
Temperature	Can be lethal in summer	Faster egg development and faster growth
Runoff	Carrying out eggs and smaller juveniles	More juvenile rearing space
Nutrients	Algae blooms and low night time oxygen	More food
Bank cover	Loss of hiding places, insect food sources	Predatory birds more visible
Instream Debris	Block migration (especially adults)	Creates cover and resting areas

FIGURE 11.1: Factors identified that would have either positive effects or negative effects on salmon production for an experimental logging program in the Carnation Creek watershed, British Columbia.

models in the first place and focus, instead, on the design of diagnostic management experiments (adaptive management) to reveal the best policy options for ecosystems (Larkin 1996; Tegner and Dayton 1999). Unfortunately, this can take too long, and can be too risky, to be economically or socially acceptable. We are stuck with the prediction problem in many, many cases. Yet another answer that appears to avoid the need for modeling is to admit a wide range of possible dynamic responses to policy, then shift the emphasis from the identification of such responses to the design of "robust" policies that should give good performance over a wide range of response possibilities. Unfortunately, this approach does not actually avoid the modeling problem, because we must use models of some sort to judge and compare the candidates for a robust policy choice.

But the idea of searching for robust and adaptive policies does suggest one thing that we can do to manage, if not minimize, the risk of prediction failure: to always deliberately develop a whole set of models, varying in complexity and emphasis on different processes and variables, so as to at least understand the range of possible outcomes that are consistent with our available experience and understanding as represented in the models (Whipple et al. 2000). This chapter suggests three alternatives that could be included in such model sets, ranging from simple, qualitative analyses of the dominant trophic interactions to very detailed, individual-based models. Further alternatives can, of course, be developed within a single general framework like the Ecosim equations of the previous chapter, by disaggregating and omitting trophic groups and rate-modifying processes like handling times and trophic mediation effects. But there is always a risk when the alternatives are constrained to a single framework that the framework itself is flawed (structurally incapable of representing some key effect, no matter how it is parameterized). The alternatives suggested below are deliberately chosen to encourage thinking about the dynamics in very different ways.

11.1 QUALITATIVE ANALYSIS OF DOMINANT TROPHIC INTERACTIONS

Even in quite complex models and historical data sets, we often see what appear to be relatively simple, dominant dynamic interactions like predator-prey cycles of the most abundant species or "flips" in community structure that appear to be associated with a few important species. There is quite a well-developed methodology in ecology for making qualitative predictions about dynamic stability and the response to management of two-species predator-prey (or two trophic–level) relationships. Backing up the methodology is basic theory suggesting that many predictive problems can be addressed by examining just a few key variables that change at speeds much faster, about the same, and much slower than the variables for which prediction is most important in the problem (Carpenter 2002; Gunderson and Pritchard 2002; Gunderson and Holling 2001; O'Neill et al. 1986). Predator-prey methodology can be used to examine policy issues ranging from the effects of changes in ecosystem fertility/productivity through to the impacts of fishing on the ability of ecosystems to support nontarget species like marine mammals. It must be used with great care, because it treats interactions as chains rather than webs, and a weblike interaction structure can result in the opposite patterns from those predicted assuming a chain structure (see Yodzis 2001 for several examples related to fishery impacts on top predators).

The basic tool in the qualitative examination of predator-prey interactions is "isocline analysis," first clearly introduced to ecologists by Rosenzweig and MacArthur (1963). The basic idea in this tool is to focus on the "state-space" of abundance combinations of predators and prey, and to partition this state-space into qualitative domains of dynamic behavior (prey increasing while predator decreasing; prey decreasing while predator increasing; etc.). These domains are separated by "isoclines," which are lines showing combinations of predator and prey abundance such that one or the other species is just holding its own (in mathematical terms, has a zero time derivative), i.e., is just about to change from one of the qualitative behavior types to another. Isoclines are usually derived by writing specific models for rates of change in predator and prey abundances, then setting the rates to zero and solving for the combinations of abundance that will satisfy the zero-rate condition. But they can also be derived by "thought experiments" in which we imagine being able to control the abundance of each species while allowing enough time for the other species to reach equilibrium (box 11.1).

The simple isocline analysis as outlined in box 11.1 can be used to explore several important policy issues in ecosystem management, mainly by asking how policy interventions of various kinds may impact the positions of the isoclines and hence long-term average (equilibrium) abundances and ecosystem stability. Here is a sample of such explorations.

Effects of Changes in Prey Productivity through Enrichment, Environmental Change

Suppose some management action like fertilization or habitat improvement for a prey species causes its isocline to move upward (in terms of the thought experiments in box 11.1, higher maximum prey numbers are supported at each experimental predator abundance). In this case, the system equilibrium will move from point A to point B in the isocline diagrams shown in figure 11.2.

There are three important features of the prediction in figure 11.2. (a) The change in prey productivity is most likely to be "felt" in an abundance increase of the predator, not of the prey (unless the predator isocline is strongly tipped). (b) The equilibrium point B is less likely to be stable, i.e., there may be a "paradox of enrichment" effect if the predator isocline is nearly vertical. (c) Density-dependent mortality in the predator (so-called "model closure") predicts a limitation of predator response as prey productivity increases, whereas foraging arena models predict a stronger response over a wider range of changes in prey productivity.

Notice in the left-hand case in figure 11.2 that changes in prey productivity can result in large changes in predator abundance, but with little or no corresponding change in prey abundance (if the predator isocline is steep). This is a very counterintuitive result; how can a system support more predators without more prey? What is happening here is that our intuition leads us to confuse prey abundance with prey productivity. When prey productivity increases, that increase may support more predators whether or not the prey abundance increases at all; in the more productive situation, the same density of prey may be present, but with the prey population turning over more rapidly due to supporting a higher abundance of predators.

This point about the distinction between changes in prey abundance and changes in prey productivity has major implications for where in ecosystems we should look for the effects of persistent "regime shifts" that affect basic productivity (e.g., a persistent decrease in upwelling rates). Predator-prey theory tells us that unless there is strong density-dependence in predator mortality rates, we should expect the biggest regime responses to be in the abundances of top predators in the ecosystem, not in the abundance of creatures at lower trophic levels. This is very likely why we commonly see very large changes in populations of large fishes in conjunction with persistent

BOX 11.1

PREDICTING PREDATOR-PREY ISOCLINES BY THINKING ABOUT THE OUTCOMES
OF CONTROLLED EXPERIMENTS

Imagine doing two sets of long-term experiments with a two-species predator-prey system. In one set, the abundance of a prey species is held constant (by stocking or removal as necessary) at a range of levels, and at each of these levels the predator population is allowed to grow/decline until it reaches equilibrium. The line connecting the experimental predator equilibria obtained this way is called the "predator isocline." In the other set, the abundance of the predator is held constant at a range of levels, and at each of these levels the prey population is allowed to change until it reaches equilibrium. The line connecting these prey equilibria is called the "prey isocline."

   The simplest prediction we might make for such an experiment would be for a type of predator that searches randomly for prey and is not affected in its feeding or reproduction by the presence of other predators. In this case, the predator isocline would be just a vertical line:

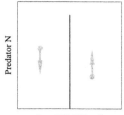

Prey N (held fixed)

   In this case, there is a minimum prey density (where the isocline intersects the prey N axis) needed for the predator to reproduce just fast enough to re-place its natural mortality rate. For any prey density less than this minimum, the predator declines toward a zero equilibrium; for any prey density above the minimum, the predator population just keeps increasing until it is imprac-tical to keep adding prey to the experimental universe fast enough to keep up with the predation rate. Corresponding to this predator outcome, the simplest prey result that we might predict would be a monotonic decrease in the prey abundance with increases in the experimental predator abundance:

(Continued)

(*BOX 11.1 continued*)

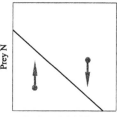

Predator N (held fixed)

Combining these experimental results into a single state-space graph, we would predict the qualitative behavior domains shown below:

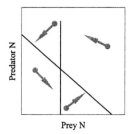

Prey N

If you track the predicted qualitative behaviors in this picture over time, you will see that successive changes will move the system through the four qualitative types of change, but in each "cycle" the system will move closer to the overall equilibrium point where the two isoclines intersect. That is, this simplest model predicts a "global stability" of the predator-prey equilibrium point.

A more complex prey isocline is predicted if the predator is admitted to have a limit on its maximum feeding rate, i.e., a "type II" functional response to prey density:

(*Continued*)

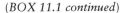
(*BOX 11.1 continued*)

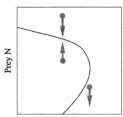

Predator N (held fixed)

In this case, there are two outcomes of experiments when the predator is held fixed at intermediate abundances: the prey may move up to an upper equilibrium or may crash to zero due to the depensatory effects of the type II predation. Low initial prey densities lead to the depensatory crash. The predator-prey long-term dynamics can also show more complex dynamics and, in particular, are expected to show a stable "limit cycle" rather than a stable equilibrium if predators are efficient enough at finding prey so that the system equilibrium occurs to the left of the "hump" in the prey isocline:

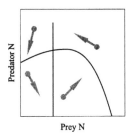

Prey N

(The easiest way to see that the equilibrium point will be unstable in this case is manually to track a set of arrows representing stepwise dynamic changes, starting near the equilibrium point; you will see that it is difficult to draw the arrows so as to maintain the qualitative domain directions but still move toward the equilibrium point in successive cycles).

In order for a predator-prey model involving an efficient predator to (possibly) have a stable equilibrium point at low prey densities absent some refuge for the prey when they are rare, there must be some mechanism that causes the predator isocline to tip to the right, i.e., for predators to "see" one another rather than seeing just the prey density:

(*Continued*)

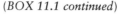

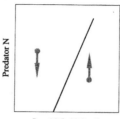

Prey N (held fixed)

At least three mechanisms can produce such tipped isoclines: (1) density-dependent mortality in the predator due to some factor like cannibalism or type III predation responses by its own predators; (2) direct interference among predators in feeding, so that more predators means each predator is less likely to see each prey that is available; or (3) foraging arena limitation on the density of available (as opposed to total) prey, with prey exchanging between vulnerable and invulnerable states (in which case predators appear to interfere with one another because they reduce the density of the available prey whether or not they impact the total prey density). The second and third mechanisms are often called "ratio-dependent" predation. When the first mechanism is used in models as an arbitrary way to prevent instability and/or large predator increases with changes in prey productivity, this modeling tactic is sometimes called "closure" of the dynamic system (Steele and Henderson 1992; Edwards and Brindley 1999; Murray and Parslow 1999b). ■

environmental changes, despite the lack of a correspondingly large change in the abundance of some lower trophic groups. In essence, the effects of small but persistent changes in productivity are amplified up the food chain (as we would expect from thermodynamic efficiency considerations even without using predator-prey models). This argument is very robust to model complexity; strong effects at the top of food webs due to a reduction in basic productivity are a general prediction of Ecosim food-web models; as a rule, we expect to see something like a 50% reduction in sustainable abundance at the top of a marine food web if the primary production is reduced by something like 20%.

Effects of Harvesting on Predators of Harvested Species

Suppose we think about the prey species in figure 11.2 as representing a harvested fish species, and the predator as being a species valued for other reasons (e.g., a marine mammal). In this case, the "natural" system equilibrium might be at point B (high predator abundance). We would expect

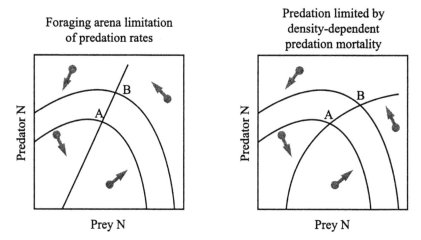

FIGURE 11.2: The effect of an increase in prey productivity on prey and predator abundances.

fishing to lower the net productivity of the prey species by "appropriating" (diverting) part of that productivity as (to) harvest. The initial impact of harvest development may be to reduce the prey population and cause the predator to exert depensatory mortality effects on the prey (by trying to maintain its feeding rate despite a decreased prey abundance). But if the predator cannot then maintain its abundance by feeding on alternative prey, or if some alternative prey does not become more abundant due to the reduced competition/predation effects from the fished prey, then ultimately the reduction in net prey production available to support the predator must result in a predator population decline (movement to point A in fig. 11.2).

In the long term, the movement from B to A in figure 11.2 may mean that the effects of fishing are felt most strongly in the abundance of non-fished predators, rather than their harvested prey. Thus it is incorrect to make management claims of the form "provided fishing is restricted enough to maintain healthy fish populations, healthy populations of their natural predators should also be maintained." Arguments of this form are seen in documents that attempt to reconcile sustainable fisheries with other ecosystem-management goals, inviting the reader to suppose that unhealthy natural predator populations should occur only where fishing is too high to be sustainable (e.g., NRC 1999). What the response dynamics in at least the left-hand case of figure 11.2 (steep predator isocline) tell us is that the appropriation of prey productivity should have deleterious effects on the natural predator abundance at all levels of fishing, including sustainable fishing at "maximum sustainable yield" points at which there can even be *no* net production left to sustain natural predators. If we want to argue that there should not be a large effect of fishing on the capacity of an ecosystem to sustain some large predator that is not fished, we must resort to arguments about the "indirect effects" (ability of the predator to switch to alternative prey, increases in alternative prey with increased fishing).

It should be noted that the field evidence is entirely equivocal about this prediction. In some cases, fishery developments have been correlated with severe declines in birds (Peru upwelling system, Pauly et al. 1989) or marine mammals (Bering Sea Steller sea lions, NRC 1996, 2002). Yet in other cases, at least some bird species appear to have prospered (Bering Sea), and there have been marine mammal increases during periods of fishery collapse (harp seal during the Newfoundland cod collapse). In the case of Steller sea lions in the Bering Sea, it has been argued that the negative correlation is not likely due to a simple mechanism like food shortage caused by the fishery (NRC 2002). Part of the problem with such observations is one of time scale (Yodzis 2001) such that we could easily confuse transient response patterns with a long-term, sustained abundance change. Another problem is that fisheries and climate-regime effects can be confounded in historical data (Shima et al. 2000). But it also appears that indirect effects can be important, and arguments for continued fishing based on such effects cannot be discounted out of hand.

Mechanisms That Can Cause Multiple Equilibria and Abundance "Flips"

Beginning with Holling's (1973) seminal paper on the resilience and stability of ecological systems, there has been much concern that ecosystems may have multiple stable states with "domains of attraction" that can be altered through human activities and deliberate management. If an ecosystem has such multiple stability domains, an "accident" that drives the system into an undesirable domain cannot be reversed by correcting whatever problem originally caused the movement into that domain; the ecosystem may be "trapped" in the domain by strong dynamic interactions that may not even have been important in the original "normal" domain of behavior. There is growing evidence that we should be very much concerned with such dynamics in aquatic ecosystems, mainly from long-term data on the failure of stocks to recover after overfishing (Hutchings 2000), from dramatic examples of persistent flips in ecosystem structure (Anderson and Piatt 1999; Scheffer et al. 2001; Nystrom et al. 2000; Nystrom and Folke 2001; Jackson et al. 2001), and from arguments and data about relationships between fishing and the spatial organization of dynamics (Cury et al. 2000).

At least three types of multiple equilibrium dynamics have been identified in aquatic ecosystems by using simple, qualitative models: clear/turbid states in relation to nutrient loading, "vampires in the basement" explosions of normally rare species, and "cultivation/depensation" effects. The following paragraphs briefly introduce each of these.

As nutrient loading increases in an initially clear-water aquatic ecosystem, the increased load may contribute mainly to the development of a benthic plant (macrophytes, seagrasses, macroalgae) biomass and to trapped sediment nutrient loads. But if loading is great enough to cause an algae bloom, the bloom may (1) reduce the benthic plant biomass (and nutrient absorptive capacity) through shading effects and (2) reduce the search efficiency of visual piscivores enough to allow increases in planktivores and, hence, reductions in zooplankton abundance. Thus once established, the "phytoplankton-dominated" system state may persist even after some reduction in nutrient loading. Qualitative models of these interactions (Carpenter 2001; Janssen and Carpenter 1999; Scheffer 1990; Scheffer et al. 2001) have been used to warn water-quality managers that simple control of eutrophication inputs may not be sufficient to restore more desirable clear-water ecosystem states, and that active biomanipulations may be needed as additional corrective measures.

Probably the most common multiple-equilibrium dynamics in aquatic ecosystems is the "vampires in the basement" phenomenon. Whenever we strongly manipulate an aquatic ecosystem, particularly through changes in nutrient loading, we can count on seeing explosive population growth of at least some species that were previously very rare (e.g., algae blooms following fertilization). Often such bloom species have highly undesirable characteristics (like toxin production by blue-green algae). We presume that these species were rare in the first place because of the competition/predation effects from the "normal" dominant species. But think about this presumption a bit more carefully: if the bloom species are persistent in the system, then they must find some low-density "escape" from the effects of the dominant species; i.e., even if they would normally decrease in the face of competition and predation, once rare, they must exhibit balanced birth-death rates in order to be persistent. In terms of the predator-prey isocline structure, they must not be strong determinants of the abundances of their predators (the predator isocline must be flat or only slightly increasing with prey density) and must show reduced depensatory mortality effects when rare so that the prey isocline shows an "S" or "fold catastrophe" (Thom 1975) shape (fig. 11.3).

A reduction in the depensatory predation effects might be caused by predators "switching" to other prey, or by the bloom species being restricted to particularly favorable or safe microhabitats when rare, or by the bloom species spending very little time foraging (and exposed to predation risk) when there are few intraspecific competitors. Note that in plots like figure 11.3, it is generally best to think of the prey abundance as a logarithmic scale, such that abundance at the lower equilibrium A may be several orders of magnitude less than during bloom developments. Unfortunately, this means that we have very little data about precisely what mechanisms cause or allow persistence at low A equilibrium points, since dynamic rates are virtually impossible to study in species that are so rare. This means, in

Rare species persists   Rare species released by
at low density A   increase in productivity

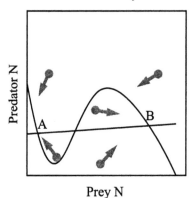
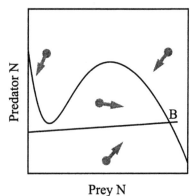

FIGURE 11.3: The predator-prey isocline structure for a prey species that is normally very rare (equilibrium A), is not a strong contributor to the abundance of its predator, but is capable of explosive growth (toward equilibrium B) should the lower equilibrium point disappear.

turn, that it is unpredictable which rare species might exhibit bloom dynamics under a given manipulation or disturbance regime (we have to find out about such things the hard way).

"Cultivation/depensation" models involve so-called "trophic triangles" (fig. 11.4) in which the juveniles of a dominant species can be impacted negatively by competition or predation from a third species that is normally kept in check by adults of the dominant species (e.g., Swain and Sinclair 2000).

In situations like this, the prey of the adult dominant has both positive (trophic support) and negative (recruitment impact) effects on the predator (dominant species) isocline and can, in principle, cause that isocline to bend down, representing a net decrease in the equilibrium abundance of the dominant at high abundances of the prey (fig. 11.5).

Figure 11.5 was drawn for an NRC panel report on the impacts of groundfish fisheries on the Steller sea lion population of the Bering Sea. We were struggling to explain an apparent "flip" in the Bering Sea ecosystem (Anderson and Piatt 1999) involving a decline in several small, "fatty" pelagic fish species that are the favored prey of Steller sea lions, while at the same time several large benthic and piscivorous fish species increased considerably. In this case, we viewed the B point in figure 11.5 as the "natural" situation that had supported a large sea lion population, and the A point as a recent development that has supported very productive fisheries (pollock, cod, flatfishes) but has been associated with the sea lion decline.

Cultivation/depensation arguments have been proposed to explain the failure of large stocks like the Newfoundland cod to recover after overfishing

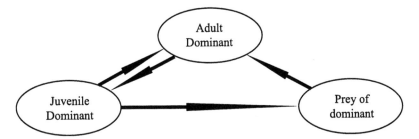

FIGURE 11.4: "A trophic triangle" interaction, in which the juvenile survival of a dominant "benthic" species may be negatively impacted by a second "pelagic" species that is a prey of the adult of the dominant species.

(Walters and Kitchell 2001). In these cases, the argument is that the "natural" state in figure 11.5 is A, with an abundance of large predators and low densities of the species labeled "pelagic prey" (which could just as easily be another benthic species; the only requirement is that the "prey" have a negative impact on juveniles of the predator while being vulnerable to predation by the adults). Then overfishing of the predator may cause the A equilibrium point to vanish (the predator isocline moves down so that only the B equilibrium exists), but the system will not quickly recover from B to A even if the original isocline structure is restored by reduced fishing. A variation on this argument is the worry often expressed by ecologists that reductions in abundant predators by fishing may result in increases in less valued competitors so that these "take over the niche" originally occupied by the abundant species. At the very least, the argument casts suspicion on simpler predator-prey analyses (e.g., Collie and Gislason 2001) that do not recognize how trophic ontogeny (the triangle structure in figure 11.4) may cause complex dynamics in marine predator-prey interactions.

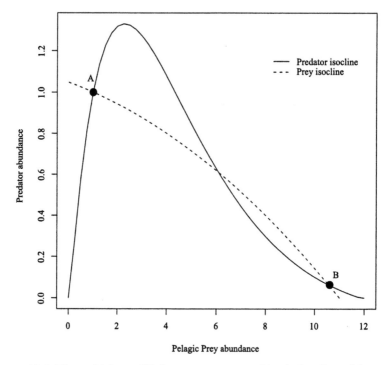

FIGURE 11.5: The multiple-equilibrium structure caused by the bending of the predator isocline in a trophic-triangle situation, representing a depression in predator abundance due to the negative effects of the prey on predator recruitment when the prey is abundant. Redrawn from NRC (2002).

11.2 QUALITATIVE ANALYSIS OF MORE COMPLEX LINKAGES

Very often we notice that when scientists and stakeholders are asked to examine alternative policy scenarios from complicated ecosystem simulations like Ecosim, there is a tendency to ignore the details of predicted time transients and to focus instead on qualitative patterns of long-term change. That is, people tend to look for qualitative indices of performance, measured by directions (up, down) and broad magnitudes of change in ecosystem indicator variables like biomasses.

Very often such qualitative predictions can be obtained without ever bothering to run temporal simulations, by examining the sensitivities of equilibrium values to (the derivatives of equilibrium values with respect to) changes in forcing inputs such as fishing mortality rates. A very powerful technique for doing this examination is "loop analysis," developed by Levins (1974). Loop analysis begins by identifying all feedback loops and connections among variables, with only the directions (+ or −) of the impact of variables on one another being initially specified; some directions of response can be predicted knowing only these signs. (Good reviews of the general ideas and assumptions can be found in Bodini 2000; Puccia and Levins 1985; Lane 1986; Lane and Levins 1977). It has been used to make general predictions about ecosystem responses to disturbances ranging from eutrophication (Lane and Levins 1977; Bodini 1998) to exotic species introductions (Li and Moyle 1981).

One of the most powerful potential applications of loop analysis is to pinpoint those system variables for which the qualitative direction of response *cannot* be predicted just by knowing the directions (signs) of interaction effects. For instance, it is easy enough to deduce that marine mammals are likely to be negatively affected by fish harvesting in simple food chains of the form plankton $\Rightarrow$ fish $\Rightarrow$ mammals. But in more complex webs, e.g., where there are several kinds of plankton-feeding fish and only one or a few of these kinds are harvested, increases in the other kinds when a few kinds are harvested may result in mammals having about the same (or even an increased) food supply. In such cases, we cannot predict the direction of response in variables like mammal abundance without making difficult quantitative calculations of competition and predation interaction changes. Loop analysis helps identify and highlight such difficult variables, essentially warning us that these are the variables to watch most closely for "surprising" responses if/when management policy changes are actually implemented.

11.3 Models That Link Dynamics with Nutrient Cycling Processes

Foraging arena theory now appears to provide quite a good framework for making detailed predictions about the impacts of trophic interactions, especially when coupled with population-dynamics models to deal with issues arising from trophic ontogeny and with trophic mediation relationships to deal with various indirect effects related to trophic interactions. But this ability to represent food-web relationships does not by any means provide a "complete" representation of ecosystem behavior and response to human policy actions. In particular, it has not (so far) dealt at all with issues related to nutrient cycling and physical limiting factors like light penetration. There is a rich history of modeling such issues, mainly aimed at providing a "bottom up" understanding of how basic production processes must be limited by physical and chemical factors in the aquatic environment, and how human activities and climate change may impact these factors, especially in coastal environments (see, e.g., Baretta-Becker and Baretta 1997; Bartell et al. 1988; Murray and Parslow 1999a; Nihoul 1998; Fulton 2001; Fulton 2002a).

Bottom-up models for nutrient cycling and basic production processes have become quite sophisticated about representing physiological mechanisms that limit productivity, spatial transport, and arrangement, especially

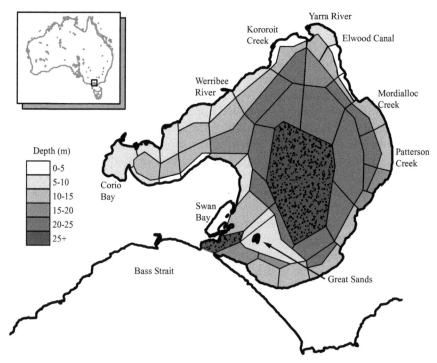

FIGURE 11.6: Spatial boxes used in a detailed model of nutrient and biomass dynamics in Port Phillip Bay, Australia (from Fulton 2001). Each box defines a coastal habitat area with a relatively homogeneous nutrient concentration, depth (the limiting factor for benthic algae and seagrass), bottom sediment type, and nutrient exchange rate.

in coastal ecosystems where nutrient loading and mixing processes are very complex, along with nutrient recycling-storage mechanisms. Spatial relationships and flows are now routinely represented by linking hydrodynamic and nutrient/biomass flux equations over detailed spatial grids and/or spatial boxes like the ones used by Fulton (2001) to partition habitats in Port Phillip Bay, Australia (fig. 11.6).

These models are standing up very well in terms of their ability to reproduce observed temporal patterns in the abundance of lower-trophic-level organisms and, more importantly, in their ability to reproduce broad spatial patterns in productivity in relation to nutrient loading (fig. 11.7).

Further, they appear to be quite capable of reproducing some of the multiple-stable-state and hysteresis behavior mentioned in the previous section, that can arise from light-mediated competition/predation interactions between benthic and pelagic components of aquatic ecosystems. They have not yet shown much capability to predict "vampire in the basement" bloom dynamics for species that are normally rare but may exhibit damaging impacts under particular conditions, but that may be because modelers have

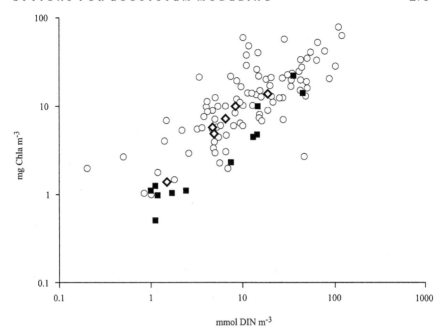

FIGURE 11.7: The ability of detailed nutrient cycling/biomass models to reproduce broad spatial patterns in algal abundance. Each open point represents observed values for a microtidal coastal ecosystem (worldwide data), while the diamonds and black squares represent predicted values from two models (from Fulton 2002b). The black squares are from a model with a less detailed representation of physiological processes.

not yet focused on this issue. As we sit writing this section, there is a horrific smell from windrows of dead fish on the beach just outside the Mote Marine Laboratory near Sarasota, Florida. Such die-offs are thought to occur in Florida coastal bays mainly after prolonged droughts, when rains can flush accumulated nutrients from shoreline areas into the bays and trigger blooms of toxic algae. Such blooms and the associated fish kills have obvious policy importance but are rare enough to be very difficult to study and model (and to predict, given that they are apparently weather/climate driven).

Fulton (2001, 2002a, 2002b) has carried out a series of careful comparative studies using nutrient/biomass models of differing functional and spatial complexity and has also compared their predictions to a biomass-numbers food-web model that does not include nutrient accounting (Ecosim). These comparisons have provided extremely valuable insights about issues ranging from optimum model complexity to conditions under which the explicit representation of nutrient cycling is "necessary" for useful prediction. Her key overall finding is that we do need to marry the food web and cycling approaches if we are to understand developing human impacts in coastal ecosystems. Another important observation concerns whether to simplify models by aggregation or the selection of key variables:

> Thus, when simplifying a food web already aggregated to the level of functional groups, judicious choice and retention of the most important functional groups in a system appears to be a much more reliable method of constructing simplified webs than aggregating across functional groups in an effort to represent everything. (Fulton 2001)

Fulton's analyses are also very clear about the importance of including at least some explicit representation of spatial variation, at least by major depth-productivity habitat type, in models that attempt to say anything useful about the biodiversity impacts of human activities. Much biodiversity rather obviously arises through spatial niche specialization and physiological niche limits (e.g., depth-light limits on seagrass development), and some human impacts on key niche dimensions (like O_2 concentration and light) cannot be seen in models that deal only with trophic interactions.

It is important to recognize that detailed modeling of bottom-up physical-chemical linkages with basic production processes is neither necessary nor sufficient to guarantee successful predictions about dynamic responses at higher trophic levels, at which most fisheries policy problems reside. Fulton showed that Ecosim outperformed her detailed nutrient cycling models in relation to some policy questions (mainly about harvesting), i.e., she showed that bottom-up modeling detail was not necessary for making useful predictions.

A good example of how care in bottom-up modeling is not sufficient to guarantee a successful prediction is the prediction failure that we encountered in adaptive management planning for Kootenay Lake, British Columbia (Ashley et al. 1997; Thompson 1999). This very large (105 km long) lake had supported valuable recreational fisheries for kokanee salmon and giant "Gerrard" strain rainbow trout. But nutrient loadings to the lake declined dramatically during the 1970s due to the closure of a fertilizer plant and hydroelectric dam developments in the lake's watershed. Further, a freshwater shrimp (*Mysis relicta*) that had been introduced as a food source for trout was discovered to be a competitor with kokanee for zooplankton but to be virtually invulnerable to predation (due to strong diel vertical-migration behaviors). By the mid 1980s the kokanee salmon population had declined to the point at which both it and its major predator the Gerrard rainbow were considered threatened. Fertilization was suggested as a possible mitigation measure, and a team of physical scientists and biologists developed a detailed model of nutrient loading, transport, and utilization by the plankton and fish community as a basis for planning the fertilization regime. This model included detailed population models (bioenergetics, growth, age structure, mortality factors) for the *Mysis*, kokanee, and trout. The team's recommendation was to fertilize near one end of the long, narrow lake so as to measure the fertilization responses along a spatial gradient as well as over time. Along the way to this recommendation, the detailed model had made a highly counterintuitive prediction that had prompted the insistence that the fertilization be done as a careful and easily reversed experiment in the first place. The modeling team predicted (confidently) that kokanee salmon and trout would actually *decrease* after the fertilization, due to the rapid response to fertilization by the *Mysis* competitor that would then lead to a net reduction in zooplankton available to kokanee. When the fertilization experiment was actually carried out, the kokanee salmon showed a strong positive response (growth and survival rates) to cladocerans that "reinvaded" the system (a "vampire in the basement" response; cladocerans had been common before the reductions in the nutrient loading but nearly absent from zooplankton samples during the 1980s), while the *Mysis* abundance remained stable or even decreased a bit.

So, as in the Carnation Creek example mentioned above, the detailed modeling lulled the modeling team into a sense of confidence in the model predictions, and these predictions proved the opposite of the experimental results. It is not understood why *Mysis* failed to respond as predicted, since the *Mysis* response dynamics were predicted from quite good information on feeding, growth, fecundity, and survival. The best guess is that the model underestimated the importance of cannibalism on first instar *Mysis* larvae by adult *Mysis* as a population-regulating factor. In hindsight, it would have been better to have developed a simple qualitative model for the possible joint responses of kokanee and *Mysis* to the fertilization, expressed in terms of isoclines of predicted equilibrium (longer-term mean) abundance of each species as a function of the abundance of the other species (fig. 11.8).

This model would at least have warned that the net outcome of fertilization might be a positive response by both species or a net gain by one at the expense of the other, depending on the details of how each species responds to fertilization (the details of how each isocline moves under fertilization). In trying to predict the experimental response isoclines, the modeling team might at least have seen the sensitivity of these isoclines to all sorts of details in the ecology of the species (like cannibalism) that the team may or may not have been able to specify precisely in the large simulation model.

This example is not meant to discourage the reader from detailed modeling but, rather, to emphasize that no single model or modeling approach should be trusted. In particular, a detailed ecosystem analysis should be accompanied by a qualitative analysis of any dominant interactions suggested by the development of the detailed model, and simpler models of the dominant interactions should then be used to evaluate the possible missing elements and sensitivities of the more detailed model.

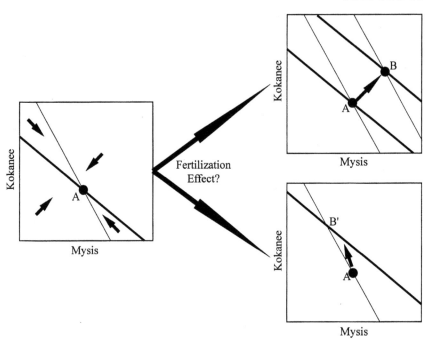

FIGURE 11.8: Possible qualitative outcomes of a lake-fertilization experiment that might favor one or both of the two competing zooplanktivores, *Mysis relicta* and kokanee salmon. Point A represents a persistent abundance combination prior to fertilization, while the lines represent experimental equilibria for each species expected if the other were held constant at various levels (the darker line represents the kokanee response to *Mysis* density). In the top graph on the right, both species are predicted to prosper under fertilization (move to abundance combination B). In the bottom right graph, kokanee become more abundant while *Mysis* "lose" (move to B′). The bottom graph could equally well have been drawn to show a *Mysis* increase but a kokanee decrease, as incorrectly predicted by a detailed Kootenay Lake simulation model.

11.4 REPRESENTATION OF MESOSCALE SPATIAL-POLICY OPTIONS

The title of this section was chosen with care. Much detailed modeling of marine ecosystems has been aimed at representing the spatial organization of production and consumption processes at intermediate spatial scales (mesoscales) created by advective divergence (upwelling) and convergence patterns, light penetration and mixed-layer depth effects, and benthic habitat structures (e.g., soft vs. hard bottoms). The motivation for pursuing such calculations has most often been the naïve presumption that we need to do them in order both to "understand" the dynamics of marine ecosystems and to make "accurate" predictions about trophic flows and production limits. In fact, when our main policy concerns are about spatially aggregated abundances (overall harvest-rate management, impacts of fishing on nontarget species), detailed bottom-up calculations are an invitation to pathological error propagation, complacency about an understanding that we do not really have, and delays in policy exploration due to wasteful computing effort. We are often better off basing the calculations on spatially averaged data and rate processes, counting on the vulnerability exchange dynamics of foraging arena equations (chapter 10) to represent the stabilizing effects of spatial organization on trophic interactions.

Unfortunately, there are some really important ecosystem-management policy options and issues that we cannot evaluate without explicit spatial analysis and modeling. These include:

1. regulation of particular point-source and diffuse nutrient (and pollution) loadings to coastal ecosystems, where the high economic cost of regulation may lead to requirements for estimating just how far any beneficial effects may extend;

2. siting and determination of required sizes for marine protected areas, in particular whether to implement many small reserves or a few large ones (the "SLOSS" debate), and whether fishing effort displaced from the reserves may end up doing more harm elsewhere;

3. efficacy of measures intended to protect particular life stages (e.g., closing nursery areas to fishing) and source areas where valued populations are thought to display a source-sink metapopulation structure;

4. assessing of possible impacts of climate change on production processes, when it is thought that changes will affect the spatial organization of nutrient-cycling processes (thermal barriers to vertical mixing, presence/absence of currents that drive upwelling and convergence patterns);

5. release-site selection and possible ecological impacts of marine enhancement programs aimed at restoring historically depleted stocks;

6. evaluation of risk to interpretation of results from spatially organized adaptive management experiments from "cross-over" effects caused by the dispersal/migration of organisms among experimental units with different management treatments;

7. design of multispecies and ecosystem-management policies for multiple management jurisdictions that share migratory species and production dynamics that involve the large-scale advection of pelagic organisms.

This list is not meant to be exhaustive or complete; it is a collection of examples that have arisen just in the authors' experience.

Two basic approaches have been used for modeling spatial-policy issues: brute force dynamic analysis and equilibrium analysis. In both approaches, we start with a set of dynamic equations (differential or difference) representing the "site" behavior of key variables at one spatial point or within one relatively homogeneous, representative spatial cell or area. Then we add advection-diffusion terms to the rate equations to represent physical transport, migration, and the random diffusion of organisms in space.

Physical transport rates can generally be obtained from the relatively good hydrodynamic models that are available today, while migration-pattern and biological-dispersal rate terms generally have to be guessed using relatively fragmentary data from tagging programs, spatial-density sampling surveys, and information on the movement of fishers (they often know a lot more about how creatures move than we do, and they use this knowledge effectively).

In the brute force approach, we then attempt to solve the set of partial differential or difference equations explicitly over time (see box 11.2). This generally requires a massive computational effort, especially when we need to predict relatively fine-scale patterns so that the equation system is very "stiff" due to large mixing-rate terms. The main applications of this approach so far in fisheries have been to analyze coastal ecosystem management (mainly nutrient loading) issues, e.g., Fulton (2001). There have been various attempts to develop useful predictive models of spatial organization in large pelagic ecosystems like the eastern tropical Pacific tuna fishing areas, but these models have not yet seen any practical application (Deriso, R.B., IATTC, pers. comm.). Detailed, discrete-time spatial population dynamics models with relatively simple movement rules have been used to compare marine protected area designs for sets of key fish species (Meester et al. 2001), but these have not yet incorporated a full analysis of the trophic interaction issues. Individual-based models have been developed to predict movements of large pelagic fish in relation to environmental (thermal) preferences, that could be linked with the individual-based ecosystem-modeling approach discussed in the next section (Humston et al. 1999).

In the equilibrium approach, we basically attempt to do a multispecies, spatial analog of classical equilibrium yield analysis in single-species assessment. We begin by writing dynamic equations that include trophic interactions and mixing rates among spatial cells, but we then set the time derivatives of these equations to zero and attempt to solve the resulting (very large) nonlinear-equation system for equilibrium abundances and rates. The main implementation of this approach has been "Ecospace" (Walters et al. 1999), included as part of the Ecopath/Ecosim modeling software. Ecospace sets up an Ecosim biomass/numbers equation system for each cell in a map like figure 11.9, and links the cells through dispersal, advection, and, possibly, migration rates. The nonlinear-equation solution method used in Ecospace to find the spatial ecosystem equilibrium is a stepwise procedure that we call "pseudo-time stepping" because the stepwise results are usually close to values obtained by integrating over time. But Ecospace can do this stepping and find a long-term equilibrium in less than 1/1000 the computing time needed for full integration, making it quite practical to use for rapid policy exploration and gaming (a key requirement for effective policy design). Ecospace has been used so far mostly to evaluate MPA options, either from a general perspective (Walters 2000) or for specific ecosystems like the coastal seas around Hong Kong (Pitcher et al. 2000).

BOX 11.2

EFFICIENT SOLUTION OF DIFFERENTIAL EQUATION SYSTEMS THAT INCLUDE
SPATIAL MIXING (ADVECTION, DIFFUSION, MIGRATION) EFFECTS

Models like Ecospace can represent the spatial distributions of variables like
biomass and nutrient concentrations by grids of values at discrete points, where
the concentration at each point is assumed to be representative of a small cell
or box of space around the point. If $X_{ij}(t)$ is the value of some variable for the
row i, column j cell of the grid, to generate time predictions of this value we
typically need to solve a system of rate equations of the form

$$\frac{dX_{ij}(t)}{dt} = f_{ij}(t) - \left[Z_{ij}(t) + \sum_k m_{ij,k} \right] X_{ij}(t) + \sum_k m_{k,ij} X_k(t) \qquad (11.1)$$

Here, f represents the growth of or addition to X at location ij due to processes
like recruitment and nutrient loading, Z_{ij} represents the local (within-cell) instantaneous loss of X_{ij} due to processes like predation (a complex sum of
rate components in models like Ecospace), $m_{ij,k}$ represents the instantaneous
spatial "flow" rates from cell ij across its four neighboring cells k (due to advection, diffusion, migration; here k is shorthand for the four index combinations
$i-1,j; i+1,j; j-1,i$ and $j+1,i$), and $m_{k,ij}$ represents flow rates from the cells
k surrounding cell ij. Note that equation 11.1 is typically "stiff" for ecosystem
models; the $f - Z$ terms are commonly on the order of 0.1–1.0 yr^{-1}, while the
mixing rate m terms are commonly much, much larger.

   If we attempt to solve equation 11.1 by simple forward differencing methods of the form $X(t + \Delta t) = X(t) + \Delta t[dX(t)/dt]$, we typically find that Δt
must be very small indeed (often on the order of minutes) to prevent numerical
instability. The simplest practical way to avoid this instability in models for
which we need to examine long-term behavior is to use an "implicit" solution
method. In implicit methods, we assume that the average dX/dt over the interval Δt is approximately proportional to $X(t + \Delta t)$, rather than to $X(t)$. This
results in the difference equation

$$X_{ij}(t + \Delta t) = X_{ij}(t) + \Delta t \left\{ f_{ij}(t) - \left[Z_{ij}(t) + \sum_k m_{ij,k} \right] X_{ij}(t + \Delta t) \right.$$

$$\left. + \sum_k m_{k,ij} X_k(t + \Delta t) \right\}$$

(*Continued*)

(BOX 11.2 continued)

which can be re-arranged as

$$X_{ij}(t + \Delta t) \left[1/\Delta t + Z_{ij}(t) + \sum_k m_{ij,k} \right] = \left[X_{ij}(t)/\Delta t + f_{ij}(t) \right] \quad (11.2)$$

$$+ \sum_k m_{k,ij} X_k(t + \Delta t)$$

Then, if the $f_{ij}(t)$, $Z_{ij}(t)$, and m's are treated as approximately constant over the time interval Δt, equations 11.2 represent a sparse system of linear equations that can be solved quite efficiently for the $X_{ij}(t + \Delta t)$. In fact, letting Δt become very large represents a way to approximate the equilibrium solution for the spatial X pattern (but generally requiring a further iterative solution to account for the effects of changes in the X's on the growth and mortality rate terms f, and Z).

The most efficient way to solve sparse linear-equation systems like equation 11.2 is by iterative methods, in which a trial estimate of the $X_{ij}(t + \Delta t)$ (e.g., $X_{ij}(t)$) is first provided, then updated iteratively until no change is seen. The crudest such iterative method is the Gauss-Seidel iteration (Mathews 1992; Press et al. 1996), in which the right side of equation 11.2 is calculated and divided by the total loss measure $\left[1/\Delta t + Z_{ij}(t) + \sum_k m_{ij,k} \right]$ to provide a new estimate of $X_{ij}(t + \Delta t)$, and this is repeated (over all ij) until the X's stop changing. Much more efficient methods can be developed by using "successive over-relaxation" (SOR; see Press et al. 1996) while solving for whole rows or columns of X_{ij} at a time (each row or column can be solved exactly given the most recent X estimates for the other rows and columns, as a tridiagonal equation system).

Spatially discretized, continuous-rate models of the equation 11.1 form, along with implicit integration methods for the efficient approximation of the model time solutions, have dramatically improved our ability to examine how local dynamics ($f - Z$ rates) interact with larger-scale spatial processes to determine ecosystem structure and response to factors such as nutrient gradients and protected-area policies. As computing power improves over the next few years, they will likely become one of the central tools of ecosystem modeling and management-policy design. ∎

The most important general prediction obtained so far with Ecospace is shown in figure 11.10. This prediction is that marine protected areas should not work nearly so well as might be expected from simple, single-species arguments about eliminating fishing mortality. Virtually all of the more "realistic" effects included in Ecospace calculations predict reduced impact of protection: (1) trophic cascade effects (reductions in prey density within the MPA) should serve both to limit abundance increase and to make areas

Intermediate predators	Large predators	Demersal zoobenthic feeders	Large zoobenthic feeders
Demersal zooplankton feeders	Small pelagics	Large crustaceans	Molluscs & worms
Heterotrophic benthos	Small crustaceans	Zooplankton	Phytoplankton

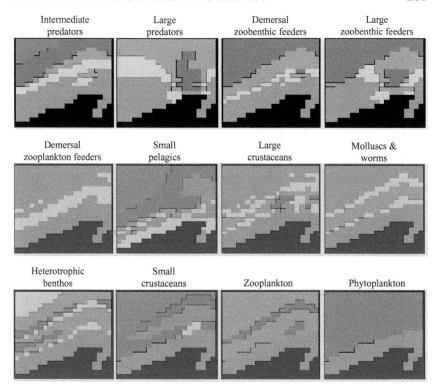

FIGURE 11.9: A typical spatial-grid prediction of an equilibrium ecosystem state using Ecospace. This grid represents the coastal ocean off Brunei, Darussalam, Southeast Asia (from Walters et al. 1999).

outside the MPA look more attractive for foraging fish (e.g., Ley et al. 2002); (2) any initial success at increasing abundance within the MPA should cause an increase in the number of fish dispersing out of it, which should attract increased fishing pressure near the boundary; and (3) high mortality rates just outside the MPA boundary mean that there will be a "dispersal imbalance" for sites just inside the boundary, where movements out of the site will not be balanced by inward movements from outside the MPA, so that abundances at such sites will be depressed despite their being protected. For species with high dispersal rates, the dispersal imbalance effect is predicted by Ecospace to end up causing reduced abundances for considerable distances into the MPA and would effectively void any abundance response at all for very small MPAs.

Two main problems have slowed progress in the development of models like Ecospace as practical policy tools. The first is the lack of detailed spatial data collected under contrasting management treatments (e.g., mapped across and around MPAs), to test whether the model predictions are at least

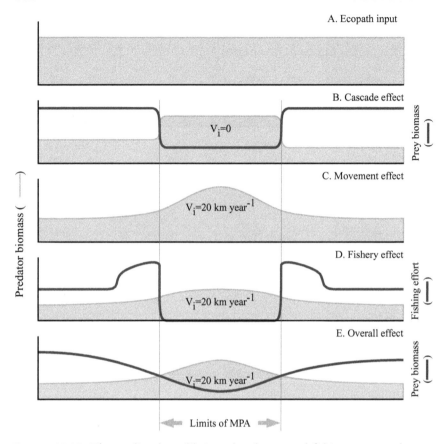

FIGURE 11.10: The predicted equilibrium abundances and fishing patterns along a transect through a hypothetical marine protected area (redrawn from Walters et al. 1999).

qualitatively correct. There is plenty of spatial data on the abundances of marine organisms, but such data almost always have a strong confounding of effects of other factors besides fisheries management (making the data even harder to interpret than time-series data in which the fishing and environmental effects are also typically confounded). The second is a conceptual difficulty in modeling spatial movements associated with the ontogenetic and seasonal migration of mobile organisms. We really do not understand why fish follow the dispersal and migration patterns that most of them (at least in temperate waters) exhibit, and it could be perilous to "hardwire" observed past movement patterns into models like Ecospace. Very likely these patterns are responsive to changes in fishing impacts and associated trophic (feeding and predation risk) conditions and are almost certainly volatile under environmental changes. Worse, it is not clear how to approximate the impacts of seasonal movement patterns (and associated changes in fishing and predation mortality risk) in equilibrium methods like Ecospace, especially when the mortality risk may be concentrated at particular locations along migration routes (e.g., salmon harvest in coastal passes through which the fish migrate to spawn).

11.5 INDIVIDUAL-BASED SIZE- AND SPACE-STRUCTURED MODELS

There are basically two ways that nature can surprise us with behaviors the opposite of what we would predict with a given model: (1) some biological detail that we overlooked turns out to have an important large-scale effect (e.g., the temperature-growth-survival link in the Carnation Creek example, cannibalism by *Mysis* in the Kootenay Lake example); and/or (2) the change is driven by some factor "outside" the breadth of factors considered in the model (e.g., accumulation of nutrients on coastal lands driving destructive algal blooms in coastal bays following droughts). Most biologists assume that errors of the first kind are the most common and important. But in fisheries policy design, we have most often been plagued by errors of the second kind, particularly from unregulated fisher behavior in models that look only at fish dynamics and treat fishing incorrectly as fully controlled. An explicit aim in ecosystem modeling and policy design is to reduce errors of the second kind by defining problems more broadly and recognizing more kinds of dynamic variables. But no matter how good we are at this, we will still be plagued by errors of detail.

Early ecosystem models (and some existing mass-balance approaches) were particularly prone to errors of missing detail because of trying to copy physical-chemical modeling through the use of some single "currency" (like biomass or nutrients) for state representation. Almost always now we admit the need for multiple currencies so that we can "drill in" to look at some dynamic relationships more closely while representing others in simpler, aggregate ways. For example, we commonly mix simple biomass-dynamics calculations for some species with more detailed size-age structured calculations for others, i.e., we mix biomass currency with numbers–body size currency so that we look further down into the state-articulation hierarchy for species interactions of particular interest.

Recently, there has been at least one demonstration that it is possible to dramatically increase the detail included in particular interaction calculations, by moving beyond a numbers-size population-state articulation to an individual-based simulation of interacting species. Jerry Ault and his colleagues (Ault et al. 1999; Ault et al. 2003) examined the interactions of spotted seatrout (*Cynoscion nebulosus*) and shrimp (*Penaeus duorarum*) in Biscayne Bay, Florida. They developed a highly detailed spatial model that included hydrodynamic processes (wind and tidal currents, freshwater inputs, salinity, temperature) and habitat structure (depth, bottom type). They then "seeded" this model with small temporal cohorts (small groups of

individuals) of larval trout and shrimp and simulated their physical transport, movements, foraging, bioenergetics (growth, maturation), and mortality (interaction/predation, fishing, and other causes) on very fine time scales
of a few minutes per simulation time step. High temporal resolution was
used to capture spatial movement and factors like the changing vulnerability of shrimp to trout predation caused by the shrimp's "tide-hopping"
behavior—the periodic use of tidal currents to aid in inshore-offshore ontogenetic migrations. This means that they came very close to explicitly
modeling the fine-scale encounter dynamics and behaviors of foraging arena
theory (see also Beauchamp et al. 1999). Ault et al. examined the predicted
spatial patterns and growth/abundance changes on time scales of up to a
few years, a massive computational exercise that required on the order of
one-half day of supercomputer time per simulated year.

In the end, Ault et al. had to admit that incomplete quantitative data for
many parameters needed in the calculations meant "as a result, the predicted
outputs presented here are not intended to necessarily represent realistic spot-
ted seatrout-pink shrimp community dynamics per se, but are representative
of the kinds of dynamics that a predator-prey community with these demo-
graphic characteristics would be expected to display in a coupled physical
and biological environment." But they then went on to outline research priorities for filling the main gaps. They said, of course, that we need much
more data on just about everything. But then they made some really important points about the functional relationships that troubled them most: rules
that animals follow in movement (in relation to habitat and foraging needs)
and predator-prey interaction rates. Roughly, their very large model parameter set divides into two parts: (1) physiological-bioenergetics parameters
that can be measured through fine-scale (e.g., laboratory) experimentation;
and (2) statistical-ecological parameters like net movements and predator
search-and-encounter rates that represent the combined effects of many very
fine scale behavioral actions that become "visible" to us (measurable as net
rates) only at large enough space-time scales to be very difficult to study experimentally. Thus it is no accident that there were some particularly troublesome data gaps; these gaps involve processes that have been recognized
as critical by ecologists for a very long time but have so far defied practical
experimental study. They come up as "research needs" with monotonous
regularity in ecological modeling studies, and they will continue to do so
until we discover novel ways to deal with them experimentally. Ault et al.
(1999) have shown us that we can meet the modeling challenge of highly
detailed ecological representation if necessary, and that the ball is now in
the court of field and experimental biologists to do a better job with difficult
processes.

Over the next decade or so, computer speeds will doubtless increase to
the point at which we can do the kind of extensive simulation testing and
comparison to data needed to properly evaluate highly detailed models like

Ault et al. (1999, 2003). Even more importantly, we will hopefully have at least a few more examples for which such models have been used to predict the effects of specific policy (or habitat/environmental) change, and for which the change has actually come to pass. Only then will we see if our models have solved the problem of missing details or, instead, fail because of missing details whose importance was not apparent at the time of model development.

CHAPTER 12

Parameterization of Ecosystem Models

WHEN FISHERIES SCIENTISTS TRY to move beyond simple qualitative predictions about ecosystem response to fisheries policy, to deal with quantitative policy issues and the possibility of complex indirect effects, they must face some daunting problems in parameter estimation. Indeed, Hilborn and Walters (1992, p. 448) concluded, "We believe the food web modelling approach is hopeless as an aid to formulating management advice; the number of parameters and assumptions required are enormous." We were not thinking carefully when we said this, and the statement is incorrect in at least three respects. First, we were basically thinking only about using food-web information to improve single-species predictions, failing to anticipate the recent demand for much broader predictions about ecosystem impacts of fishing and ecosystem-scale management options. In the face of this demand, there is not really any choice but to try and deal with complex parameterization problems, unless scientists want to base ecosystem policy predictions on pure intuition and supposed "principles" of ecosystem dynamics. Second, we were thinking in terms of parameter estimation based mainly on fitting models to time-series data, as is done in single-species assessment; that is indeed a hopeless task (with many severely confounded parameter effects). There are other ways to marshall much of the information needed for ecosystem modeling, and those will be the focus of most of this chapter. Third, we invited misunderstanding by using the word "assumptions." "Enormous" numbers of assumptions are, in fact, needed whenever one derives a proposed scheme for making predictions about any natural dynamic system, whether or not one bothers to (or even can) make these explicit. As literally stated, these assumptions are almost always wrong. What they are used for is to define parameters and forms of functional relations that, it is hoped, will provide reasonable *approximations* to the dynamics. Thus, the real issue is not the quality of assumptions but, rather, the quality of approximations. For example, in single-species models the natural mortality rate M is usually assumed to be a single constant, while full well recognizing that this blatantly incorrect assumption may still be a good approximation for many predictive purposes. One of the main things that can be done in ecosystem modeling is to "decompose" M into a set of predation-competition components, as $M = M_a + M_b + M_c + \ldots$. This more detailed approximation may be better or worse than just using a single M, depending on whether scientists are good at approximating the components and whether they are careful to constrain these components to avoid error propagation when the components are added (and integrated to give the net predicted effects over time).

It is important to recognize that the functional relationships in ecosystem models generally require three more or less distinct types of parameters, in terms of the scales at which the parameters are "visible" to experimental observation. First, some physiological and behavioral parameters, like the metabolic and efficiency parameters of growth bioenergetics, or nutrient uptake kinetics of phytoplankton, are measurable at the scale of individual organisms in laboratories and field microcosms. Understandably, there is a very rich literature on such parameters. Second, there are essentially ecological parameters, like the vulnerability exchange rates of foraging arena theory, that are defined by multiple behavioral events of interacting organisms. These parameters are defined not by what any one individual typically does but by the statistical dynamics that appear when individuals interact in space and time, and these parameters become visible only in experiments at larger scales. For obvious logistic reasons, estimates of such parameters are much less common in the literature. Third, there are (in some models) parameters representing events and rates that are so rare or slow that they cannot be practically observed at any scale short of the whole ecosystem, such as colonization rates for habitats that have been historically denuded.

Likewise there are three broad approaches to parameter estimation: (1) micro- and meso-scale process experiments; (2) state-rate estimations using mass balance data; and (3) fitting to temporal and spatial data. In the first approach, we pretend to be able to extrapolate directly from experimental results to larger-scale field rates. In the second approach, we attempt to measure states and rates for some baseline situation or condition, and to infer as many process parameters as possible from the baseline results. In the third, we look for "signals" representing particular parameter effects in time-series and spatial data, by seeking parameter values that improve the fit of a model to historical data when we ask that model to "replay" the dynamic history or measured spatial pattern of a system. The following sections examine each of these approaches, with a particular view to warning prospective ecosystem model users about the pitfalls of each.

As an important tactical point, the following discussion presumes that ecosystem modeling will rarely proceed by assigning a single scientist to sit in an office and do the modeling and estimation alone (as is usually the case in single-species work). Rather, we assume that any parameter estimation will generally involve team efforts that make use of the broadest possible range of knowledge and experience.

12.1 PARAMETERIZING MODELS

Textbooks and papers about dynamic modeling usually invite us to think about the generation of predictions as involving two stages, as shown in figure 12.1. This way of thinking invites us to go off in all sorts of directions in

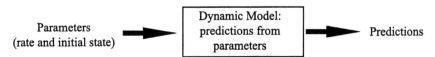

FIGURE 12.1: The traditional way of thinking about predictive models: enter the parameter estimates, make predictions.

search of parameter estimates, which in turn invites severe error propagation when the resulting parameter set is "plugged into" the prediction equations.

A much more powerful way of organizing information to make predictions is as a three-stage process as shown in figure 12.2. In this process, we impose a collection of parameter setup relationships between the input data and the dynamic model equations, such that the setup relationships involve deriving as many of the dynamic-model parameters as possible from a relatively simple set of critical leading parameters.

This approach is used extensively in software packages like Ecosim and drives much of the discussion in the rest of this chapter. We can do several important things in a formal parameter setup procedure (or block of spreadsheet cells), like:

1. calculate some parameter values for functional relations so as to force the functional relations to predict particular "known" or assumed rates at particular known or assumed system states, thereby reducing the number of parameters that have to be entered;
2. constrain the parameter values so as to meet known or independently estimated cumulative values to avoid some kinds of error propagation;
3. run more detailed "submodels" for processes that are difficult to parameterize directly, using the results of these submodels to calculate the aggregate/average parameter values for a simpler ecosystem model.
4. remove "irrelevant" parameters by scaling input information to get rid of arbitrary units of measurement.

We have already discussed this concept in relation to single-species models (chapters 3 and 5), for which we have encouraged the use of important "leading parameters" like unfished biomass B_o as the input unknowns to be estimated by fitting the models to the data (see box 3.1 on page 56). When B_o is used as an input to a spreadsheet age-structured model, the block of spreadsheet cells representing the parameter setup procedure in figure 12.2 uses B_o along with base survivorship and weight at age schedules to calculate a base recruitment R_o needed for B_o to be a naturally sustainable level, initial numbers at age for simulation, and one of the minimum of two parameters needed for a mean stock-recruitment relationship. Then whenever we vary B_o, e.g., by using a minimization/maximization procedure like Solver to search for the best-fitting parameter estimates, the change in B_o is "passed through" the setup calculations to affect a variety of quantities used in the dynamic-model calculations.

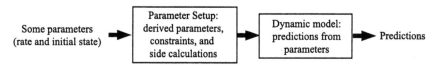

FIGURE 12.2: A more powerful way of thinking about predictive models: some parameter values and baseline information/knowledge is entered into a collection of "parameter setup" relationships, and parameter values from these relationships are then passed to the prediction equations.

So a formal parameter-setup interface can help us move in two quite different directions. We can use it to "elaborate" simple information (e.g., B_o) into a more detailed set of parameters for more complex (e.g., age-structured) simulations. Or we can use it to "collapse" more detailed entry information into simpler aggregate parameters, e.g., by running a detailed, individual-based model for how predators search and how prey behave, so as to estimate average predator functional response parameters. Some day, such "hierarchic" parameter-setup procedures might allow us to put the whole issue of simple versus complex models (see chapter 11) to rest, by having the procedures systematically and deliberately examine (and explicitly link) the key predictions from various models to one another and to particular, critical field measurements (such as total mortality rate, Z).

12.2 PARAMETER ESTIMATES FROM EXPERIMENTAL DATA

There is a growing literature that summarizes available parameter estimates for ecosystem models; for general reviews, see Jorgensen (1995, 2000). Microscale parameters that can be measured on individual organisms, particularly bioenergetics parameters, are available for a wide variety of organisms. Much information for fish is summarized in the general database system Fishbase (*www.fishbase.org*). For other reviews and summaries related to bioenergetics (growth and metabolism), see Hartman and Brandt (1995), McCann and Schuter (1997), Schindler and Eby (1997), and Bartell et al. (1999). A particularly interesting pattern is noted by Schindler and Eby (1997), who review growth, bioenergetics, and nutrient-cycling parameters for fish. They show that feeding rates calculated using bioenergetics parameters and field-growth data have averaged only 26% of the maximum rates expected from laboratory feeding conditions. This is precisely the sort of divergence in field versus laboratory rates expected from foraging arena theory (fish feeding at much lower rates than possible due to the effects of predator-avoidance behaviors). Another source of microscale (individual organism) parameter values is comparative studies of allometric patterns (e.g., Moloney and Field 1989), though the information must be used with great care (Cyr and Pace 1993).

It is debatable whether we need to include most of the readily available physiological parameters in overall ecosystem models. As noted in the previous chapter, Fulton (2001, 2002a, 2002b) has shown that replacement of detailed nutrient pool and ratio parameters with simple constant-ratio assumptions does not appear to degrade the performance of coastal-ecosystem model predictions for phytoplankton and, in fact, may even improve performance. Likewise, we can usually see little effect from replacing complicated bioenergetics calculations with simple efficiency ratios of growth to food consumption. What is happening in cases like these is that strong individual variation in parameter values, particularly in relation to body size, ends up being averaged out or hidden when aggregated across individuals and over time to result in population-scale rates. This effect can be easily checked (to avoid silly arguments about whether some detail parameter is needed) by using "mini-models" and side calculations in the parameter-setup component of figure 12.2. For example, it is easy to construct a bioenergetics model for individual growth of the form

$$dw/dt = (1 - A)(1 - D)q(w) - mw^{0.8} \qquad (12.1)$$

where w = body weight, A = assimilation efficiency, D = proportion of assimilated intake lost to specific dynamic action and activity proportional to food consumption, $q(w)$ = size-dependent feeding rate (most likely varying approximately as $q(w) = kw^{2/3}$), and $mw^{0.8}$ = routine metabolic rate as a function of body weight. For readers who like to see lots of parameters, include the effects of temperature on the rates and modify equation 12.1 to include the temperature effects on the rate components. Models of this form obviously predict a strong variation in growth efficiency $g(w) = (dw/dt)/q(w)$ with w. But if we are developing an ecosystem model that aggregates individuals into biomass "pools," the growth component of $dB/dt = N(w)dw/dt + wdN/dt$ is the integral over some range of w of $N(w)dw/dt$, and the average g_B = (total growth)/(total biomass) is given by the integral over w of $N(w)dw/dt$ divided by the integral over w of $N(w)w$. Unless there is a wild and systematic variation in the composition proportions $N(w)/N$ over time, g_B will likely be quite stable. This damping effect is illustrated with an age-structured model in figure 12.3.

Even in this example, when we set the parameters to cause the age-specific growth efficiencies to vary from as high as 0.37 for age-1 fish to near 0.0 for age-20 fish, the population-scale efficiency could be reasonably approximated by the mean value 0.1, or somewhat better by using the overall mortality rate (and/or simulated mean age) to correct for the systematic increase (to around 0.15) under severe overfishing.

The literature suddenly becomes very sparse when we move from "physiological" to ecological scales of observation and experimentation, i.e., to dealing with parameters that do not become visible except at field scales in time and space. Ecologists have taken three basic approaches to estimating such parameters, particularly those related to trophic interactions (the functional responses of predators to prey and predator density):

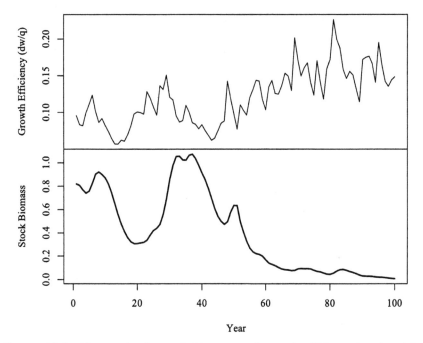

FIGURE 12.3: Changes in the stock biomass and growth efficiency (total weight growth/total food consumption) for a simulated population that has a strong natural variation in recruitment and is subject to overfishing after year 50. Higher growth efficiencies correspond to years when the stock has higher proportions of young fish, i.e., early periods of strong recruitment and after overfishing.

1. expensive meso-scale experiments;
2. experimental components (Holling 1959) measurement of the finer-scale activities and events that make up the process, and the integration of these measurements by means of individual-based models;
3. analysis of spatial data (e.g., measuring predator diets in areas of different prey density and composition) to take advantage of natural contrasts.

For example, there are very few field experimental measurements of the critical rate of effective search "a" parameter for predator functional responses (see box 10.1; equations 10.3–10.6), except for plankton organisms whose feeding rates can be measured in microcosms. We can use simple field and laboratory observations to measure experimental components of the rate of effective search. Since "a" is an area or volume swept per unit of time searching, it can be represented as a product of three components: speed times reactive distance (or area) times probability of detection and capture of prey. Unfortunately, such component calculations do not have much meaning at field scales when the search process (and resulting prey capture rate) is greatly complicated by prey behaviors, diurnal changes in search conditions (e.g., light), and time-allocation decisions by individual predators (e.g., short feeding bouts to avoid predation risk). These same complexities are what

make it extremely difficult to design mesocosm-scale experiments for measuring the predator's functional-response parameters directly, since almost
anything we do to contain the process for measurement will also alter the
behavior of both the predator and its prey (e.g., predator's using enclosure
walls to trap prey). So we could reduce or "decompose" the estimation
of the "*a*" parameters for trophic interactions by including experimental
components calculations and/or detailed individual-based simulations in the
parameter-setup interface of figure 12.2, but the results of those calculations
would be suspect because it would be easy to omit or miscalculate some
critical component (like the proportion of time spent actually foraging).

Ecologists have not exploited opportunities for estimating functional response parameters from spatial and temporal comparisons of feeding patterns over variable prey abundances. There is a very large literature on
comparing prey composition in diets to prey composition in the environment to determine prey selection and calculate so-called "electivities" (the
ratios of rates of effective search for alternate prey), but such studies do
not generally even attempt to estimate and report overall rates of effective
search. Fine-scale observations of predator feeding rates in relation to prey
densities are difficult to conduct and interpret; in fact, we often see negative
correlations between prey occurrence in diets and local prey density, because
predator feeding has already depleted a local prey concentration by the time
we can observe it.

12.3   ESTIMATING PARAMETERS FROM MASS BALANCE SNAPSHOTS

Suppose that we can somehow assemble a set of baseline (or leading parameter) values for abundances, rates of population or biomass change, and
trophic flows (consumption rates of prey) for some particular reference period, for a whole ecosystem from field-scale measurements on that ecosystem.
Suppose further that we want to ensure that the ecosystem model parameters are chosen to ensure that the model predicts exactly the baseline flows
and rates of change whenever the model system is in exactly the baseline
abundance (and physical/chemical forcing) state. Then we can (and, in fact,
must) use the baseline input values (in the parameter setup procedure, figure 12.2) to calculate at least some of the aggregate parameters that may be
difficult or impossible to estimate by the approaches outlined in the previous
section.

Consider, e.g., the basic model from foraging arena theory that is used
to predict consumption flows along food-web links in Ecosim (eq. 10.3 on
page 237):

$$Q_{ij} = \frac{a_{ij}v_{ij}B_iB_j}{(v_{ij} + \acute{v}_{ij} + a_{ij}B_j)}.$$

Suppose we have used one of the state-rate reconstruction tools (Ecopath
or MSVPA) described later in this section to obtain baseline rate and state

estimates $Q_{ij}^{(o)}, B_i^{(o)}$, and $B_j^{(o)}$, and we want to be sure that a_{ij}, v_{ij}, and $\acute{v}_{ij}$ are chosen so as to predict the baseline predation rate $Q_{ij}^{(o)}$ whenever the modeled prey biomass is $B_i^{(o)}$ and the effective predator abundance is $B_j^{(o)}$ (if handling-time effects are to be included in the calculation, we need to extend the "whenever" to say "whenever all modeled prey biomasses are at $B_i^{(o)}$"). To achieve this, the parameter values must satisfy the constraint

$$Q_{ij}^{(o)} = \frac{a_{ij} v_{ij} B_i^{(o)} B_j^{(o)}}{(v_{ij} + \acute{v}_{ij} + a_{ij} B_j^{(o)})} \tag{12.2}$$

We can use this constraint to calculate either v_{ij} or $\acute{v}_{ij}$ given a separate estimate of a_{ij}, or calculate a_{ij} given a separate estimate of v_{ij} and $\acute{v}_{ij}$ (just solve eq. 12.2 for one or the other parameter). In Ecosim, users are required to enter estimates of v_{ij} (and $\acute{v}_{ij}$ is assumed to be a constant multiple of v_{ij}), to represent alternative hypotheses about "top-down" (high v_{ij}) versus "bottom-up" (low v_{ij}) control of the interaction rates. Solving equation 12.2 for a_{ij} we obtain:

$$a_{ij} = \frac{v_{ij} + \acute{v}_{ij}}{B_j^{(o)} (v_{ij} B_i^{(o)} / Q_{ij}^{(o)} - 1)} = \frac{v_{ij} + \acute{v}_{ij}}{B_j^{(o)} (v_{ij} / M_{ij}^{(o)} - 1)} \tag{12.3}$$

Our preference is to solve equation 12.2 for a_{ij} and specify the vulnerability exchange parameters because it is very difficult to estimate or infer rates of effective search (a_{ij}) from field observations. There are, however, alternative options for estimating the vulnerability exchange parameters (see box 12.1). For example, Cox et al. (2002) provide an alternative explanation for estimating a_{ij}, where v_{ij} represents the maximum predation rate on prey i in the presence of infinite predator biomass (i.e., the asymptote of Q_{ij}/B_i versus B_j). Under this definition, v_{ij} is a multiple of the base predation mortality rate, and Ecosim users specify the ratio of maximum predation mortality relative to base predation mortality. There is at least some hope of measuring these ratios from contrasting observations on predator-prey abundances and diet data (e.g., Lilly et al. 2000).

Calculations like this obviously do not fully solve the parameter estimation problem. But they do let us focus our attention on fewer, particular parameters (like the vulnerability exchange rates v_{ij}; see box 12.1) while being sure that others are assigned values that make sense given the baseline field information. This is exactly like using a B_o parameter in a single-species assessment to calculate a base recruitment rate R_o, noting that the equation for mean recruitment as a function of stock size must pass through (B_o, R_o), and turning our attention to the other, really critical stock-recruitment parameter, which is the slope of the recruitment curve at low stock sizes. Just as we need to observe recruitment at low stock sizes to estimate that slope, so we must somehow obtain observations of Q_{ij} (or M_{ij} or the effects of varying M_{ij}) over a range of B_j values in order to estimate v_{ij} (or a_{ij} given v_{ij}).

BOX 12.1
OPTIONS FOR ESTIMATING TROPHIC VULNERABILITY EXCHANGE PARAMETERS

Predictions about stability, biodiversity, and compensatory response to fishing in models that include trophic interactions are critically sensitive to any assumed values of the vulnerability exchange parameters v_{ij} in trophic-flow/functional-response equations like equation 10.3. These parameters represent the complex effect of prey and predator behaviors on the instantaneous abundance of prey actually available to predators at any moment, and of behavioral and physical factors that limit the flow of prey between safe and vulnerable states. Low vulnerabilities imply a global stability in abundances, a lack of predator-prey cycles, a potentially high biodiversity, and strong compensatory responses in per-capita predator performance when predator abundance is reduced by factors like fishing. One of the biggest challenges to ecosystem analysts is to find ways of estimating these parameters.

Since the exchange rates first become "visible" (in terms of their effects on interaction rates) at relatively large space-time scales (at least hundreds of meters or days), they are very difficult to measure through direct experimental manipulations of prey and predator abundances. There are at least four options for getting at least crude estimates of them from sources besides direct field experiments:

1. **Calculations based on movement/exchange-rate data.** Some vulnerability exchange rates mainly represent relatively simple advection/diffusion processes, e.g., the delivery of zooplankton organisms to reef planktivores or juvenile fishes that use littoral areas as refuges from predation. In such cases, we may be able to estimate v_{ij}'s just as the rates of physical water exchange (and/or diffusive movements of the prey) between foraging arenas (local feeding areas) and the larger areas over which the prey are distributed.

2. **Analysis of changes in prey mortality rate Z with changes in predator abundance.** Sometimes we have direct measures of the total prey mortality rates (usually from size/age composition data) at different predator abundances (either as time-series data or as mass-balance "snapshots" for a few periods); plots of Z against predator abundance for such data should be relatively flat if the predator is estimated to cause significant total mortality but if v_{ij} is low.

3. **Examination of long-term changes in predator abundance.** If baseline trophic interaction-rate estimates are based on data from a severely depleted predator population, and if we know roughly how much more abundant the predator used to be (and that it ate the same prey back then), we can set lower bounds on the v_{ij}, at values large enough to

(Continued)

(*BOX 12.1 continued*)

have at least "supplied" enough prey to support that larger predator population.

4. **Fitting of ecosystem models to time-series data.** The v_{ij} play much the same role in ecosystem models as the slopes of stock-recruitment curves or population "r" values in a single-species assessment. For systems that have been highly disturbed by fishing, and when we can see strong transient responses in the historical data, we may be able to at least set lower bounds on the v_{ij}, by using the fact that low v_{ij} greatly "dampen" simulated responses to disturbances like fishing.

None of these approaches is very precise or robust to possible confounding factors (other things that may have created spurious patterns) in the data that are typically available. This is a very important research area for both field ecologists and theoreticians. ■

Two main methods are now being used widely to provide baseline state $B_i^{(o)}$ and trophic flow rate $Q_{ij}^{(o)}$ mass-balance estimates that can be used in parameter-setup calculations like equation 12.3. Ecopath (Christensen and Pauly 1992) is used to establish a single "snapshot" of states-rates for one reference time or reference year, using relatively simple aggregate data on biomasses, mortality rates, food-consumption rates, and diet compositions. MSVPA (see review in Magnusson 1995) is used to estimate time series of more detailed states-rates for cases in which time series of age-size structured data are available. These methods are not mutually exclusive; for example, V. Christensen (pers. comm.) has used MSVPA results for the North Sea fish community to calculate overall biomasses and trophic flows for some components of that ecosystem, and has combined these in Ecopath with information on creatures (mainly lower trophic levels) for which the MSVPA-type calculation is inappropriate (or for which detailed data are lacking).

The basic idea of Ecopath is to estimate the missing components in a mass-balance relationship, which states that every biomass pool i must satisfy:

Production = predation + catch + net biomass change + net migration

$$PB_iEE_iB_i = \Sigma_j QB_j DC_{ij} B_j + C_i + BA_i + NM_i \qquad (12.4)$$

Where the standard notation used by Ecopath users is:

PB_i is the net production per biomass (food consumption-respiration) of i

B_i is the biomass of i (so net production of $i = PB_i B_i$)

QB_j is the total food consumption per biomass of biomass pool j

DC_{ij} is the proportion of QB_j that pool j species obtain from pool i

C_i is the total mortality on i caused by fishing (landings + discards)

BA_i is the biomass change of i over the Ecopath base-reference-unit time step (usually one year)

NM_i is the net biomass migration (immigration-emigration) for i

EE_i is the proportion of pool i production that is assumed or estimated to be accounted for by the predation loss, catch, and net-biomass-change terms on the right side of equation 12.4.

Notice that there is no assumption in equation 12.4 that the ecosystem is in any sort of dynamic equilibrium or stasis; the term "balance" in mass-balance analysis refers just to the idea that the gain term (production) has to be balanced with, i.e., must be accounted for by biomass change and the set of loss terms on the right side of equation 12.4. The parameter EE_i is called "ecotrophic efficiency" by Ecopath users; this is an unfortunate term that connotes something like an ecological conversion efficiency, when, in fact, all that is actually meant is something about the efficiency of the model developer at accounting for components of loss. In most published Ecopath assessments, the BA and NM terms are, in fact, set to zero, so the initial input data for $B, PB, QB,$ and DC are assumed to be for a situation in which the system is near or approximately at a dynamic equilibrium.

The entry data for Ecopath consist of the diet composition matrix DC_{ij} (for all i, j trophic-linkage combinations), along with three of the four pool-specific parameters $B_i, PB_i, QB_i,$ and EE_i for every pool i. The mass-balance equation system in equation 12.4 is then solved for the remaining parameters; this solution involves a system of linear equations for any B_i's that are not directly entered, along with simple solutions of equation 12.4 for whatever other parameter is missing for each pool for which B_i is provided. DC estimates, which define the baseline food-web structure, usually come from field stomach-contents sampling. B_i estimates are commonly obtained from direct biomass surveys, from the results of single-species stock assessments, or even from MSVPA biomass reconstructions. PB estimates are most often taken from estimates of the total mortality rate Z ($PB = Z$ at equilibrium for populations closed to migration, Allen 1973), with the Z obtained in turn from information on size-age composition and longevity (e.g., catch curves). QB estimates can sometimes be obtained from experimental studies of the daily ration but more often are back calculated from growth data using bioenergetics models (see $www.Fishbase.org$ for alternative methods). We must basically assemble all these sorts of information for any ecosystem model that involves trophic-linkage predictions; Ecopath can be viewed just as a relatively simple and convenient framework that facilitates such information synthesis, with the added bonus of using mass-balance assumptions to fill in some information gaps.

MSVPA is basically a generalization of the idea of virtual population analysis (chapter 5) as a method for reconstructing historical abundances. The idea is to approximate the time dynamics of numbers of creatures $N_{i,a,t}$ in a population i by age a over time t with an equation of the form

$$N_{i,a+1,t+1} = (N_{i,a,t} - C_{i,a,t} - \Sigma_j QB_j DC_{i,a,j,t} B_{jt})s_{ia} \qquad (12.5)$$

Where

$C_{i,a,t}$ is the estimated number of type-i, age-a animals killed by fishing (harvested + discarded)

$QB_j DC_{i,a,j,t} B_{jt}$ is the estimated consumption of i,a type animals by predators of type j (j index would usually run over ages-sizes of predator species)

s_{ia} is the survival rate of animals not harvested or eaten by predators (so $1 - s_{ia}$ represents the mortality not accounted for, like $(1 - EE_i)PB_i$ in Ecopath).

The concept in MSVPA is to solve equation 12.5 backward in time, as

$$N_{i,a,t} = N_{i,a+1,t+1}/s_{ia} + C_{i,a,t} + \Sigma_j QB_j DC_{i,a,j,t} B_{jt} \qquad (12.6)$$

The same approaches can be used for setting the terminal numbers $N_{i,a,T}$ for this recursion as are used in single-species VPA. Since the estimates of various predator abundances $B_{jt} = N_{a,j,t} w_{a,j,t}$ are reconstructed simultaneously by the same recursive method, it is usually necessary to apply equation 12.6 at least twice: on the first pass, predator biomasses B_{jt} can approximated by the values $w_{j,t}(N_{j,a+1,t+1}/s_{ja+1} + C_{j,a,t})$ that do not include predation loss, then resulting estimates that do include predation losses are used to provide improved estimates of B_{jt}.

Notice that while the MSVPA equation 12.6 looks superficially very different from the Ecopath mass-balance relationship equation 12.5, it is driven by essentially the same basic inputs as Ecopath: (1) estimates of nonpredation mortality via s or EE; (2) human removal rates C; and (3) predation rates estimated as products of per-capita consumption rates QB, diet compositions DC, and abundances B. There is the potential with MSVPA to provide more precise pictures of age-size–structured predation impacts and to examine how these have varied over time (so as to parameterize or check alternative models for predicting future diet-composition changes), but this potential is not free. MSVPA has two main drawbacks: (1) the calculation is mainly useful for age-structured populations and does not encourage the use of "bottom up" information on the dynamics of short-lived creatures (lower trophic levels); and (2) the species/age/time disaggregation of diet compositions DC creates a requirement for truly massive stomach-contents sampling (or very good models for predicting diet composition from prey abundance and size/age characteristics). Sample size requirements for diet composition are further increased by the need to track seasonal changes in diet associated with movement and "pulses" of availability of particular prey, and by the annoying fact that high proportions of stomachs are empty in typical field sampling (see earlier discussions about why fish typically feed at much lower rates than we might expect based on laboratory studies of maximum feeding rates). But now that this massive diet sampling has been done in a few places, particularly the North Sea, it does appear likely that good models of prey-selection patterns can be developed and used to interpolate or approximate DC patterns for at least some fishes, so that direct stomach-contents sampling can be reduced to fairly infrequent checks for "surprising" changes in feeding patterns (e.g., due to unexpected changes in the availability of particular prey species).

The collection and synthesis of historical diet-composition data is easily the most difficult, frustrating, and potentially misleading step in the development of both Ecopath and MSVPA state/rate reconstructions. But on the positive side, one easily overlooked benefit of estimating the parameters for trophic interaction models from field diet data is that the resulting models implicitly capture some very important spatial effects. That is, observed DC patterns depend on some obvious local biology details like prey/predator size ratios in relation to predator mouth sizes, on fine-scale search and escape behaviors, and on diurnal/seasonal activity patterns. But observed DC also depends on larger-scale spatial "overlap" patterns, defined by a host of life-history factors ranging from resting habitat choices and feeding movements to ontogenetic and seasonal habitat shifts and migrations. Hence the parameters estimated directly from DC "contain" these larger-scale spatial effects; indeed, such effects are probably the main reason we commonly see predators failing to eat particular prey species that appear entirely suitable and would be eaten with relish under laboratory conditions. This containment or hiding of spatial relationships in DC data should make us question the need for very complex, explicit models of trophic interactions, like the Ault et al. (1999, 2003) model discussed in chapter 11.

There is one difficult problem that users of diet-composition data must face. Most diet-composition studies have been aimed at assessing trophic support relationships, i.e., what prey are important to each predator. There has been a tendency to overlook or not bother reporting "rare" prey items, especially small fishes that are difficult to identify and are digested rapidly. But think for a moment about how diet composition enters calculations of prey impact as measured by the prey mortality-rate components $M_{ij} = Q_{ij}/B_i$ (i = prey, j = predator). Articulating this calculation explicitly in terms of diet composition, we see that M_{ij} will vary as $M_{ij} = DC_{ij}QB_jB_j/B_i$. What if j is a small body-size, abundant species that has a very high total consumption QB_jB_j, and if this species "incidentally" takes a prey fish i that is the juvenile of some larger, rarer species (so that B_i is very small)? In such cases, M_{ij} can be very large, i.e., the impact on the juvenile of the large species can be very large, even if DC_{ij} is "trivially" small. Some really important relationships that can create complex dynamics (e.g., cycles and multiple equilibria), such as cannibalism and the predator-prey role reversals associated with trophic triangles or cultivation/depensation effects, are exactly of this kind. "Incidental" predation may also be one of the reasons we have had so much difficulty explaining recruitment variation. For example, in a recent Ecopath assessment for the Georgia Strait, British Columbia, we estimated Pacific hake biomass to have averaged around 10 t/km^2 in recent years and to have had a QB of around 5.0. We estimated the juvenile chinook salmon biomass to have averaged around 0.2 t/km^2. Juvenile chinook suffer a total instantaneous mortality rate Z of around 2 (yr^{-1}) during their first year of ocean life. Hake could cause half of this mortality, if only 0.4% of their diet were juvenile chinook salmon. We do not even know if hake eat juvenile salmon (but they are certainly capable of doing so), but we do know that the stomach-contents sampling of hake would likely have missed such a "minor" component of their diet.

In the long run, our best hope for recognizing when small diet contributions are important to the prey but not the predator is to develop technologies for directly measuring the partitioning of mortality rates M_{ij}. Existing assessment techniques allow fairly good estimates of total mortality rates Z, and sometimes the fishing rate components F of these Z. What we lack today are methods to directly apportion $M = Z - F$ into components M_{ij}; to do that, we will need some sort of tagging or long-term direct-observation systems that give counts of animals lost to particular mortality agents, including predators j. For example, we may someday be able to put small transmitting tags in fish, that provide just enough of a signal for an autonomous, camera-equipped vehicle to follow the fish and record what happens to it. But until such technologies do become available, we strongly recommend "driving" trophic interaction models with multiple alternative hypotheses about M_{ij} partitioning, rather than including just the M_{ij} estimates that can be well supported with DC_{ij} measurements. That is, enter Ecopath or MSVPA assessments not just with DC_{ij} data but also with at least some M_{ij} values representing alternative partitioning of Z among probable or possible mortality agents. Wherever such an M_{ij} is to enter the analysis but the software expects the entry of DC_{ij}, back calculate the DC_{ij} associated with the hypothesized M_{ij} value, as $DC_{ij} = M_{ij}B_i/(QB_jB_j)$. Mass-balance relationships (eq. 12.4) can, in fact, be restated entirely in terms of such "mortality partitioning" inputs (as $EE_iPB_i = \Sigma_jM_{ij} + F_i + (BA_i + NM_i)/B_i$), but such restatements will not be particularly helpful (except as a way to define alternative impact hypotheses) until we have a practical way to measure ecosystem-scale M_{ij} patterns directly.

In our experience, the two most important general warning signs of important (for policy) effects of possible errors in state/rate initialization are suspicious values for the total mortality rate Z and ecotrophic efficiency EE. Biologists often worry about estimating QB (a relatively simple bioenergetics problem), but when diagnosing model weakness (both Ecopath-based and more complex), they should worry much more about Z and EE. EE is defined in Ecopath as $EE = $ (predation $+$ catch)$/Z$. High EE means that a model "recognizes" most of production to already be "appropriated," implying the possibility of strong fishing and also biological competition effects among creatures/fisheries causing the high EE, should any of them change. Low EE can mean: (1) inadequate accounting (the model omits important natural mortality agents); (2) a predation mortality rate limited by predation impacts on the predators from higher trophic levels (if there is a strong "trophic cascade" effect, we expect to see low EE at the top trophic level, high at the next down, low at the next down from that, etc); and/or (3) the next trophic level up representing an "opportunity for invasion," i.e., vulnerability to colonization by some creature that can capture the production without being subject to whatever agents (predators, fishing) are preventing predators from capturing it already.

12.4 CHALLENGING ECOSYSTEM MODELS WITH DATA

The main aim of developing ecosystem models in fisheries contexts is to provide tools for comparing and screening broader policy options, under broader management objectives, than have been considered in traditional single-species management. If policy comparisons based on ecosystem models are to be considered, we need at least to demonstrate that the models can broadly reproduce (would have predicted) historical and spatial patterns of response to known disturbance regimes. Such demonstrations are obviously not a guarantee of correct predictions about future policy impacts or even the ordering of future performance under different options, since the models may incorrectly explain what has happened in the past and/or may incorrectly extrapolate state/rate patterns for "novel" system states (and policies) for which we have no historical experience.

Fisheries scientists are beginning to fit ecosystem models to time-series and spatial data using essentially the same nonlinear statistical search procedures as have long been used to estimate parameters for single-species stock-assessment models. As in the single-species case, the basic idea is to "drive" a model with a set of known or hypothesized historical inputs or disturbances, then vary selected model parameters so as to improve its ability to hindcast or fit a set of measured historical response measurements. For example, the Ecopath/Ecosim (EwE) software permits users to enter both historical forcing inputs and response data via a simple spreadsheet that defines a historical reference period and the data available for each year over that period. EwE forcing inputs can (or must in the case of fishing-effort or fishing-rate inputs) include:

1. annual fishing efforts by gear type and/or estimates of historical fishing-mortality rates by year and species or biomass pool;

2. "forced" biomasses over time for particular species that have likely impacted species included in the model, but are not themselves modeled for various reasons;
3. predictors or indices of historical changes in basic productivity, such as relative upwelling rates;
4. predictors or indices of relative larval-settlement rates (early recruitment success) for species represented by age-structured calculations.

EwE performance or response data can include time series of:

1. relative (e.g., survey CPUE) or absolute abundances by biomass pool type;
2. direct estimates of total mortality rate Z, from the analysis of tagging or age/size composition data;
3. mean body weight for split pool groups;
4. historical catches.

It would be possible to treat catches (rather than efforts or calculated historical fishing-mortality rates) as inputs (so-called "conditioning on catch" approach) as is popular in single-species assessments, but that approach commonly causes severe dynamic instability in multispecies simulations (depensatory effect of catches when simulated abundances decline for any reason) and does not permit an analysis of the multispecies effects of nonselective fishing gears.

When we fit surplus production or age-structured models to single-species historical data, our main aim is usually to obtain estimates of four things: (1) historical population size or scale (B_o) by examining how much relative abundance and population size/age composition has been altered by "known" total removals (catches); (2) historical changes in mortality rates (and, hopefully, how much of this change has been due to fishing); (3) strength of compensatory responses, particularly in recruitment rates, measured by the slope of the mean stock-recruitment relationship at low stock sizes; and (4) temporal patterns of "process errors" or "recruitment anomalies" representing the effects of past changes in factors unrelated to (not predictable from) changes in stock size. We can attempt to achieve these same aims in ecosystem-model fitting, plus hopefully explaining at least some of the single-species process errors as likely having been due to trophic-interaction effects. There is a larger set of scale (biomass) parameters to be varied, but this variation can usually be tightly constrained by mass-balance relationships (e.g., biomass combinations that would imply any $EE > 1.0$ in Ecopath represent the proposition that more was eaten than could be accounted for by production and biomass change, implying spontaneous creation). The strength of compensatory responses is represented in ecosystem models largely by predation functional-response parameters, particularly the vulnerability exchange parameters v_{ij} that represent how predator and prey habitat use and behavior limit interaction rates. These are particularly good candidate parameters to include in fitting exercises (i.e., that we might hope to estimate from the historical patterns of change in abundances), since they represent the effects of mesoscale biological processes that are extremely difficult to measure directly in the field. Further, they are critical in predictions about future ecosystem stability and biodiversity.

One might think that with so many historical forcing inputs and model parameters to vary, it would be very easy to fit ecosystem models to data and to obtain spuriously good fits (the overparameterization problem: many parameter combinations would give equally good fits). In fact, our experience to date with EwE has been quite the contrary; it can be difficult indeed to get a complex model to fit multiple time series of relative-abundance and mortality-rate data, that often vary in complex and apparently unrelated ways not easily explained by known historical fishing or environmental factors or by "known" trophic interactions.

Generally, we learn a great deal just by the initial qualitative comparisons of observed versus modeled trends, before any formal fitting exercises are attempted. We always see at least a few gross differences between these trends. In such cases, we can usually identify some important (often obvious) component missing from the model (such as forgetting to include some major historical fishery development) or some grossly incorrect initial parameter setting. One very common initial parameter-setting problem is to have foraging arena vulnerability-exchange-rate values (v_{ij}) that are too low to allow the model to predict strong change at all, i.e., observed historical patterns of stock decline or recovery (low v_{ij} cause stronger compensatory responses to increased mortality, and a reduced capacity to increase following reductions in mortality). We may also identify likely temporal changes caused by "regime shifts." For example, figure 12.4 shows two fits of Ecosim to abundance time-series data for monk seals and rock lobsters in the Northwest Hawaiian Islands, obtained by Jeffrey Polovina using a slightly expanded version of his original model of one reef in the region (French Frigate Shoal; Polovina's 1984 analysis of this system was the initial formulation of Ecopath). Assuming only trophic interaction effects (fishery-induced reduction in the abundance of lobsters as a food of monk seals), we were unable to match either the latter years of the monk seal decline or the persistent low abundance of lobsters following recent fishery reductions. But when we arbitrarily reduced the average phytoplankton primary productivity by 40% after 1989, to represent roughly the decline in chlorophyll concentrations that has been observed in the region following an apparent oceanographic regime shift, the model "suddenly" explained both data series much better, without any formal fitting or "tuning."

Spatial comparisons can also be very helpful in identifying missing model components even when time-series data are not available for elaborate fitting exercises. For example, Rodriguez-Ruiz (2001) developed Ecopath models to compare fish communities in trawled versus untrawled *Posidonia* beds off the coast of Spain. She then used these to initialize Ecosim simulations of the effect of trawling on unfished beds and the removal of trawling from fished beds. Her initial biomass predictions contained only the direct effects

Northwest Hawaiian Islands
(French Frigate Shoals)

Fishing effort:

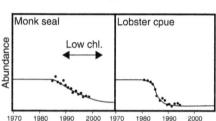

Initial ecosim runs: fishing + trophic interactions only did not explain the monk seal decline, predicted lobster recovery.

Satellite chlorophyll data indicate persistent 40-50% decline in primary production around 1990; this "explains" both continued monk seal decline and persistent low lobster abundance.

FIGURE 12.4: The effect of including the effects of the 1989 oceanographic regime shift on the ability of Ecosim to explain changes in monk seals following the development of a fishery on one of their prey, rock lobsters, in the Northwest Hawaiian Islands. Results courtesy of Jeffrey Polovina, NMFS Honolulu.

of fishing and did not match the observed biomass differences between her trawled and untrawled areas for most fish species (figure 12.5). She then included trophic mediation effects in the model by making vulnerabilities (v_{ij}) of some of the fish species depend on the simulated biomass of *Posidonia*, with a simple curve that assigned lower vulnerabilities at high *Posidonia* biomass (more cover from predators). She also adjusted some of the v_{ij} to further strengthen the predator-prey interaction effects for the study site with reduced *Posidonia*. As shown in the right panel of figure 12.5, these quite credible qualitative changes were enough to considerably improve the model/data comparison, at least to give the right qualitative response of simulated community structure to trawling.

When we attempt to move beyond qualitative evaluations of possible missing model components or driving variables, we have to face two difficult choices: (1) what to use as a statistical criterion or measure of agreement between the model and the data, noting that this measure must aggregate across multiple data types; and (2) which parameters to vary in searches for better parameter estimates. Most analysts today would agree that the "best" answer to question (1) is a likelihood function. For example, Ecosim uses

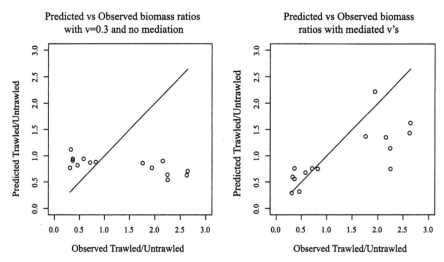

FIGURE 12.5: The ratios of the biomasses of various fishes in trawled vs. untrawled *Posidonia* beds off the coast of Spain, compared to Ecosim predictions of these ratios. The predictions in the left panel include only direct fishing effects, while the predictions in the right panel include "trophic mediation" effects of *Posidonia* abundance on the vulnerability of some of the fish species to their predators. From Rodriguez-Ruiz 2001.

a likelihood function based on the assumption that time-series data in fisheries generally involve measurements with log-normal error, and we assume independence of observations among data types and over time within each data type. Under these assumptions, the log-likelihood for the data, evaluated at the conditional maximum likelihood estimates for all variance parameters (see box 5.1), can be written as:

$$LnL = -\Sigma_d(n_d/2)\,ln(SS_d) \qquad (12.7)$$

where the index d refers to data type (e.g., relative abundance, Z, catch), n_d is the number of observations of data type d, and SS_d is a sum of squared deviations that is calculated in two different ways depending on whether data type d is treated as an absolute or relative state/rate measure. (Note that we have stripped the normal likelihood function of all terms that depend only on the data or reduce to additive constants under evaluation at the conditional maximum-likelihood estimate of variance). Relative measures are assumed to be observations of the functional form $y_{dt} = q_d X_{dt}$, where y_{dt} is the observed value (e.g., a survey CPUE), q_d is an unknown scaling (commonly catchability) parameter, and X_{dt} is the corresponding system-state variable value (e.g., a total biomass B_t). For relative measures, SS_d is calculated in two steps: (1) compute a "z-statistic" $z_{dt} = ln(y_{dt}/X_{dt})$ for each t, and the arithmetic mean $\bar{z}_d$ of these statistics ($\bar{z}_d$ is the conditional-maximum-likelihood estimate of $ln(q_d)$); and (2) compute $SS_d = \Sigma_t(z_{dt} - \bar{z}_d)^2$. For absolute state/rate measures like catches, SS_d is computed as $SS_d = \Sigma_t(y_{dt} - X_{dt})^2$

(i.e., no need to "correct" for the observation-scaling parameter q_d). A variation on equation 12.7 would be to integrate over the possible values of the variance parameters, using a weakly informative prior like 1/variance; this gives the same form of likelihood, but with the $n_d/2$ terms replaced by $(n_d - 1)/2$ (see Walters and Ludwig 1994). Note that in all these SS formulations, the dependence of the likelihood function on the unknown dynamic parameters enters the function via the effect of the parameters on the predicted state/rate values X_{dt}. Though we have encountered few obvious problems with the "automatic" data weighting in equation 12.7 in over a dozen case examples of Ecosim fitting, we do allow users to unweight suspect data by reducing the "effective n_d" for each d by applying an arbitrary multiplier weight to each $(n_d/2) \, ln(SS_d)$.

There is much debate about which parameters to treat as variable or unknown when maximizing criteria like equation 12.7. The safest strategy to avoid blatant overfitting of the data is to restrict fitting searches to those parameters that cannot be estimated by any analysis of independent data. In the Ecopath/Ecosim system, this means treating the "physiological" parameters QB_i, the "easily" estimated growth/survival and diet parameters PB_i and DC_{ij}, and at least some scaling parameters B_i, as well-known. Then at least initially, parameter-estimation searches can be restricted to a few uncertain B_i and EE_i for biomass pools for which B_i cannot be estimated independently and to whichever of the critical vulnerability-exchange parameters v_{ij} cannot be estimated by one of the alternative methods suggested in box 12.1.

Unfortunately, just being careful about which time-independent parameters to estimate is not the whole answer to the question about what should be treated as unknown in ecosystem model–fitting exercises: a much more difficult issue is whether or not to allow fitting procedures to help us "reconstruct" the impacts of historical forcing factors that have likely affected variables like primary production rates but are not well indexed or predicted by particular environmental indices for which we have historical data (e.g., upwelling rates). In single-species stock assessments, we routinely attempt to estimate such time-series patterns, usually as recruitment anomalies, by including a collection of individual anomaly time values in the fitting parameter set (and penalizing large anomalies that we consider unlikely a priori). In this terminology, the question becomes: should we attempt to reconstruct ecosystem-scale production/interaction anomaly patterns? One very good reason for at least trying to do so is that we sometimes see positive covariation (in abundance patterns) over time among species even across two or more trophic levels, indicative of strong changes in "bottom-up" forcing of abundance changes; see, e.g., Gaard et al. (2002). If we are fitting an ecosystem model to multiple abundance series that have not, in fact, been impacted by such ecosystem-scale productivity changes, variations in these series (due presumably to species-scale environmental effects) should not be correlated, and any imposition of correlated variation in the predicted values (via simulated production anomalies) should degrade rather than improve the fit to the data. Based on this reasoning, Ecosim allows users to attempt to estimate annual primary production anomalies (multipliers on selected primary-producer production-rate calculations) along with parameters like v_{ij}. Using this feature, we have found that the shared effects attributed to primary production anomalies can explain up to 50% of the total variance in multiple time series of relative abundance and Z data.

Figure 12.6 shows a smorgasbord of results from early attempts to compare and fit Ecosim to time-series data for a variety of aquatic ecosystems. For these examples, model fitting was restricted to the estimation of a few overall or selected vulnerability-exchange coefficients (v_{ij}) and to primary production anomaly sequences, along with a variety of qualitative model changes based on obvious initial errors in model formulation. In several cases, there was at least some "tuning" or exploration of alternative values for Ecopath diet composition DC_{ij} values, mainly in relation to suspected occurrences of important predation effects due to the occasional consumption of rare prey by an abundant predator (see section above).

These models are all under development, and details will be published by at least the following authors: Kerim Aydin, Chris Boggs, Villy Christensen, Sean Cox, Daniel Pauly, James Kitchell, Jeffrey Polovina, and the authors of this book. Most are available for download at *www.Ecopath.org*.

What can we learn from these examples, besides the obvious fact that there is much variability in harvested ecosystems? Here is a short list of the main lessons that we have taken from fitting exercises with Ecosim:

1. **It is not true that "everything depends on everything else."** An ecosystem model can explain some patterns quite well, while completely missing others. Trophic and mediation linkages do not create such tight coupling that every linkage must be well described in order to make useful policy predictions about some policy issues such as overfishing (we knew this already from experience with single-species models).

2. **Major and persistent declines are most often caused by the direct effects of fishing.** Most of the declining species for which we have direct data in figure 12.6 are ones for which data were collected because of fisheries management interest, and the declines are well explained just by changes in fishing mortality. If there have been declines in many other species due to indirect effects such as competition with fisheries for food, most of these declines have not been well documented.

3. **Strong "top down" control of abundance by predators does occur but appears to be relatively uncommon.** Examples in which reductions in predator abundance lead to increases in the abundance/production of prey are visible mainly in a few well-recognized patterns, such as increases in shrimp and cephalopods following fishery-induced reductions in large fish (Gulf of Thailand; also well-documented for shrimp-cod interactions, Worm and Myers 2003; Berenboim et al. 2000; Lilly et al. 2000) and increases in small pelagic fishes following

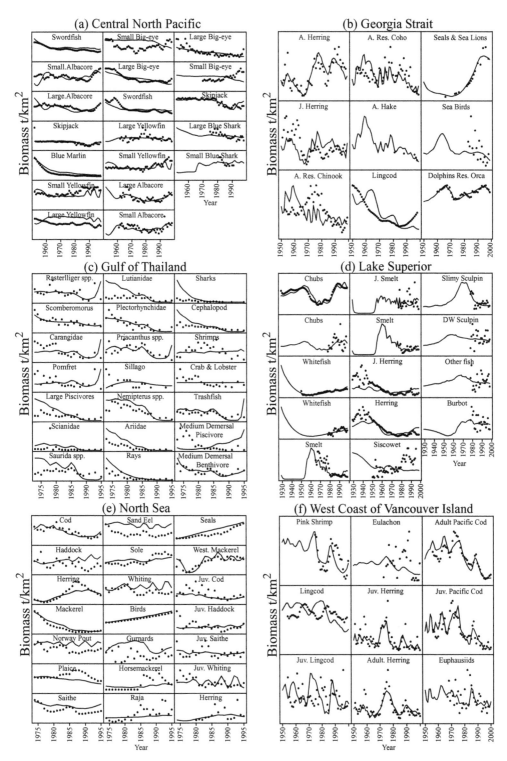

FIGURE 12.6: Sample Ecosim fits to time-series data for aquatic ecosystems. Each block of panels represents one modeled system, and each panel shows observed (dots) and simulated (lines) changes in biomass.

declines in piscivores (Carscadden et al. 2001). We have so far found relatively strong apparent effects of changes in predator abundances on recruitment (i.e., the ecosystem model is helpful in explaining recruitment variations) for only a few cases, such as skipjack tuna in the Central North Pacific model (Cox et al. 2002).

4. **Strong "bottom up" environmental forcing of primary production change on multiple-year time scales is visible in some ecosystems but is by no means a dominant cause of variation in all systems.** In a way, this is a modeler's paraphrase of one main finding in Jackson et al. (2001) from qualitative comparisons of long-term change in a variety of ecosystems. The inclusion of primary production anomaly patterns in Ecosim model fitting has ranged from being crucial to explain positive covariation in abundances across several trophic levels (West Coast Vancouver Island, Georgia Strait), to being not worth including based on the lack of improvement in model fit (Central North Pacific, Gulf of Thailand, and, surprisingly, the North Sea).

5. **Complex dynamical behavior (cycles, multiple equilibria, chaotic changes) caused by trophic interactions is not commonly visible at the spatial scale of regional coastal ecosystems.** The time-series data in figure 12.6 are probably representative of most fairly large aquatic ecosystems. If complex behaviors caused by trophic interactions are indeed important, these appear to be averaged out or masked by using aggregate data for larger spatial scales. Only the Eastern Bering Sea (not shown in fig. 12.6) shows evidence of a "flip" in structure possibly representing a multiple-equilibrium dynamic.

6. **Time-series estimates of total mortality rates Z can be extremely helpful in identifying possible missing pieces in models.** For example, in the Georgia Strait model, we had direct estimates of early ocean (juvenile) mortality rates for coho and chinook salmon, from coded-wire tagging that is used to monitor hatchery-stock performance and contributions to mixed-stock salmon fisheries. We were unable to explain some historical increases in Z until we included time-forcing patterns of hatchery-stocking rates as recruitment "multipliers." The Ecosim model then "explained" the "missing" component of increase in Z as being due to intraspecific competition (the foraging arena effect, driving simulated juveniles to spend more time foraging and hence at risk to predation as their total abundance has increased under hatchery stocking).

7. **It is just as important to be suspicious of "real" time-series data as it is to be suspicious of model predictions/hindcasts.** In several cases, we have spent considerable time agonizing about why Ecosim failed to match some pattern in time-series data, only to discover that the data were flawed because of factors like changes in monitoring methods and inappropriate assumptions about the constancy of the natural mortality rate in single-species assessments.

Ecopath/Ecosim and related ecosystem-modeling tools are now being applied to a wide variety of aquatic ecosystems, and it is likely that comparative analyses will reveal some fascinating patterns, particularly in relation to points (3) and (4), over the next few years.

An important challenge for ecosystem models, and one that has not yet been widely used, would be to see how well they can do at hindcasting large-scale spatial patterns and gradients. Some work has been done with Ecospace to see if patterns in fish production and yield across the Atlantic Ocean can be "mapped" from spatial patterns in primary production and habitat factors like depth, but that work has been frustrated by difficulties in obtaining spatial information on key disturbance factors like fishing effort; it is much, much easier to assemble aggregate time-series data than to gather detailed spatial data on fishing activity (V. Christensen, pers. comm.).

PART SIX

Strategies for Ecosystem Management

Marine Enhancement Programs

ARTIFICIAL PROPAGATION used to be considered an option mainly for freshwater and anadromous fishes; we have known how to grow things like carp and eels for many centuries, we have had large-scale hatchery rearing of salmonids for over a century, and we've even had hatcheries for Atlantic cod in the 1890s (Nielsen 1894 cited in Hutchings et al. 1993). But this situation is changing rapidly, with diligent work mainly by marine aquarists and aquaculturists showing how to rear practically any fish or invertebrate in captivity, most often with spectacular success in terms of the large numbers of juveniles that can be grown and potentially released into nature. So it is understandable that we are now seeing loud demands to use marine enhancement as a central tool to deal with practically every situation for which there aren't as many fish as there used to be (Hilborn and Winton 1993; Leber 2002; Stickney and McVey 2002).

Artificial propagation has long had an almost irresistible attraction for fisheries managers and fishers. Most fish are highly fecund and have correspondingly high egg and early juvenile mortality rates in nature. So when technologies can somehow bypass these early mortalities, there is the potential to use relatively few spawning fish to meet a wide range of possible objectives. Fish can be provided for harvest in places where fish can no longer reproduce naturally due to rearing habitat loss (e.g., above large dams that block passage for anadramous fish). Stocking can be used to try and keep up with the demand for fish when the fishing effort is high and cannot be controlled for political or social reasons. It can also be used to accelerate the rebuilding of wild populations that were historically overfished, so as to restore harvesting benefits sooner. As a last resort, "conservation hatcheries" (Brown and Day 2002) can be used to provide at least some recruitment for genetically unique populations that are faced with the risk of accidental extinction or incidental fishing mortality (bycatch impact) that cannot be prevented without high economic loss to other fisheries. So marine enhancement programs have the apparent potential to become a "quick fix" or relatively painless technological solution to a wide variety of the problems of marine ecosystem management. At best, they may be an important new component of marine ecosystem management; at worst, they may lull fishery managers into false confidence and, hence, act as yet another excuse for inaction and delay in the development of effective management/restoration programs.

But where they have been proposed or implemented, enhancement programs have engendered bitter scientific debate about whether they work at all and whether they do more harm than good. So if there is a widespread response to the demand for them, marine enhancement programs will soon become one of the most important issues in marine ecosystem management, just behind overfishing and climate change and possibly well ahead of coastal habitat loss. This chapter reviews the main arguments from that debate and suggests a framework for monitoring and evaluation so that the debates can ultimately be settled by sound scientific evidence.

There is a crucial thing to understand about the human side of marine enhancement before the reader even begins to examine the fisheries science of the issue. *People who actually do enhancement can rarely be trusted to evaluate objectively and honestly whether they are doing more good than harm, and can be counted on to defend their work no matter what the data might say.* Today, Walters is often called a vicious critic of hatchery programs. But he grew up around trout hatcheries in eastern California where his grandfather had been a zealous early promoter and sometimes manager of hatchery propagation. His first gainful employment was cleaning hatchery raceways. It was brutal, filthy, sometimes even frightening work (with the constant fear of disease outbreaks and water system failures). Yet the products of this labor were highly visible: raceways full of trout fry, thousands of fish dropping into mountain lakes by aerial planting, and smiling fishers with stringers of "catchable size" hatchery trout. There is great satisfaction in seeing the immediate rewards of hard work. And he can still vividly remember screaming in outrage at "city biologists" who would occasionally turn up in his management region and question whether the work was worth doing. It was many years after leaving the region before he could look back and ask himself honestly whether those biologists had valid points, and whether he should have been doing some very different things to at least find out whether his labors were bearing poison fruit.

A simple spreadsheet model (box 13.1) can be used to illustrate some of the potentials and pitfalls of marine enhancement discussed in this chapter. This spreadsheet accounts for wild population stock-recruitment and age-structure dynamics, hatchery contribution to recruitment and competition between wild and hatchery recruits, possible reproductive impairment of hatchery fish that spawn in the wild, and dynamic responses of fishing effort to changes in the total abundance of fish.

BOX 13.1

MODEL FOR EXAMINING INTERACTIONS OF WILD AND HATCHERY FISH

1. Assume two types of fish, one resulting from wild reproduction and one from hatchery releases. Assume the age at full vulnerability to fishing and the age at maturity roughly coincide, and model age-structured biomass B_t and numbers N_t dynamics for wild (W) and hatchery (H) fish after recruitment with delay-difference relationships:

Wild fish

$$BW_{t+1} = e^{-F_t - M}[\alpha NW_t + \rho BW_t] + w_k RW_{t+1}$$

$$NW_{t+1} = e^{-F_t - M}NW_t + RW_{t+1}$$

Hatchery fish

$$BH_{t+1} = e^{-F_t - M}[\alpha NH_t + \rho BH_t] + w_k RH_{t+1}$$

$$NH_{t+1} = e^{-F_t - M}NH_t + RH_{t+1}$$

where α and ρ are growth parameters, w_k is the body weight at recruitment, M is the natural mortality, and R_{t+1} is the number of fish reaching recruitment size w_k in year $t + 1$.

2. Predict early juvenile production JW_t (wild) and JH_t (hatchery) for the two types each year by partitioning egg production between wild and hatchery spawning, assuming a proportion PH_t of total eggs is taken into hatcheries. Account for the possible reduced reproductive success for hatchery × wild and hatchery × hatchery matings in the wild:

$$JH_t = g\phi PH_t(BH_t + BW_t)$$

$$JW_t = \phi(1 - PH_t)(BH_t + BW_t)[W_t^2 + r_{HW}2W_t(1 - W_t) + r_{HH}(1 - W_t)^2]$$

where $W_t = BW_t/(BH_t + BW_t)$ is the proportion of total eggs produced by fish that had been spawned in the wild, ϕ is the average early juvenile production per wild spawning biomass, g is the relative fry production per spawner in the hatchery, r_{HW} is the relative reproductive success of $H \times W$ spawnings, and r_{HH} is the relative reproductive success of $H \times H$ spawnings.

3. Predict the survival rate SJ_t from the early juvenile stage to recruitment, assuming a Beverton-Holt form for density-dependence in the juvenile

(*Continued*)

(*BOX 13.1 continued*)

mortality rate, with all juveniles (both hatchery and wild) involved in this competitive interaction:

$$SJ_t = S_{max}/[1 + \kappa(JW_t + JH_t)]$$
$$RW_{t+k-1} = SJ_t JW_t$$
$$RH_{t+k-1} = SJ_t JH_t$$

where S_{max} = maximum survival at low densities, κ = carrying capacity parameter.

4. Predict the fishing-mortality rate F_t each year as $F_t = q_t f_t$, where q_t = density-dependent catchability and f_t = fishing effort. Assume q_t varies with the total adult biomass as

$$q_t = \frac{q_{max}}{1 + (q_{max}/q_o - 1)(BW_t + BH_t)/B_o}$$

where q_{max} is the maximum catchability at low population density and q_o is the catchability when the total biomass is at the unfished level B_o.

5. Assume the fishing effort f_t varies with total abundance and harvest regulation approach, according to one of the following options:

   (a) **No regulation:** $f_t = f_o(BW_t + BH_t)/B_o$, where f_o is the effort that would occur if the stock size were B_o (i.e., effort proportional to the total biomass).
   (b) **MSY regulation:** $f_t = F_{MSY}/q_t$, where F_{MSY} is the fishing rate for MSY.
   (c) **Effort response with quota ceiling** (Hilborn plan; see Walters 1986): $f_t = min\{f_o(BW_t + BH_t)/B_o, -\ln[1 - Q/(BW_t + BH_t)]/q_t\}$, where Q = fixed quota.

6. Suggested parameter settings, using an assumed unfished biomass level B_o for the wild population and assumed growth and natural mortality parameters:

$$\bar{w}_o = [e^{-M}\alpha + w_k(1 - e^{-M})]/[1 - \rho e^{-M}]$$
$$N_o = B_o/\bar{w}_o$$
$$R_o = (1 - e^{-M})N_o$$

let $\phi = 1$ (since it is confounded with S_{max} and κ); use Myers' metanalysis results to set

$$S_{max} = K_{myers}R_o/B_o, \quad \text{with } K_{myers} \text{ usually in the range of 5 to 20}$$
$$\kappa = (K_{myers} - 1)/B_o$$

(*Continued*)

(*BOX 13.1 continued*)

For hatchery production performance and wild viability, set $g = 10$, $r_{HW} = 0.7$, $r_{HH} = 0.3$ (to simulate a relatively good survival of juveniles in the hatchery but a poor reproductive performance of hatchery fish in the wild). Assume strong density dependence in q: set $q_{max} = 5q_o$, and set q_o to give a high F_t when $f_t = f_o$.

7. Initial policy situations:
   —set fish wild stock initially to a healthy level ($F_t \approx F_{MSY}$) before starting hatchery
   —or overfish the stock to near extinction before starting the hatchery, and set the effort response slope f_o to ensure F_t remains large when BW_t is low.

The most important initial lesson to learn from this model is that effort-response dynamics completely dominate the predictions unless effort is closely regulated, leading to continued or even more extreme overfishing as hatchery production is increased. The details of hatchery-wild stock interaction and reproductive performance matter only for scenarios in which fishing is controlled so as to allow at least some chance of wild stock recovery. ∎

13.1 THINGS THAT CAN GO WRONG

Over the years, biologists have identified at least six things that can cause trouble in marine enhancement programs, such that whoever foots the bill for the program (usually the public, or at least license-buying fishers) ends up with net benefits less than the program costs. Only the first of these has to do with the "direct" failure of a program to produce recruitment (fish that survive long enough for someone to catch or to use to preserve a gene pool); most have to do with some "indirect" effects on total abundance and productivity of the target stock(s) and other stocks/species.

Failure to Produce Fish That Successfully Recruit to the Harvested and Spawning Population

It might surprise you to hear that there are hatcheries out there that have been operated for decades, without anyone having taken the trouble to find out whether the fish released from the hatchery have even survived to reach harvestable and spawning sizes. For example, there are "mitigation" hatcheries for salmonids on the Columbia River, constructed to make up for the loss of production above impassible dams, for which we cannot be sure that even enough hatchery fish survived to replace their hatchery parents. Such hatcheries may have "robbed" eggs from the remaining wild populations in order to continue operation. Aerial planting programs continue in mountainous areas of North America, despite evidence that once a wild stock becomes well established, virtually all of the juveniles that rain down across a lake's surface from the air may be cannibalized before they can reach the safety of shoreline areas.

Lots of things can go wrong when naïve fish that have been reared in a protected hatchery environment (the reason they survive well in the first place) suddenly find themselves facing the rigors of natural conditions. They may be incapable of finding food, or look for food in the wrong places. They are fodder for predators. They may be physiologically stressed so as to be made vulnerable to disease transmission from the remaining wild fish. Even after many years of trying to produce healthier and stronger hatchery salmonids in the Pacific Northwest, we still find from Coded Wire Tagging (CWT) experiments that survival-rate estimates for hatchery fish are typically about half that for nearby wild stocks. See, e.g., the survival trend data for coho salmon in Walters and Ward (1998). We think this difference is due mainly to predation shortly after hatchery release, by a wide variety of predators that are attracted to the high densities of naïve fish (birds, seals, sculpins, you name it).

However, even a considerable depression in the post-release survival of hatchery fish compared with that of wild fish is not necessarily a serious issue, provided the net egg-to-recruitment survival rate is still higher than for wild spawning fish. It does not bother us much to lose 50% of hatchery Pacific salmonids right after release when we have already improved egg-to-release survival tenfold; it just means that we get five times as many recruits per egg rather than 10 times as many as wild spawning would produce.

Direct Exploitation of Wild Fish to Provide Hatchery Seed Stock

Often we are loath for various good reasons (disease, genetics) to use eggs from other populations to stock a hatchery. When the hatchery is used to try to rebuild or conserve a severely depressed natural population, fish taken into the hatchery for spawning can represent a high proportion of the total egg-production potential for that wild population. From the standpoint of wild population productivity and recruitment, this hatchery removal is then a serious exploitation rate, as surely as if the eggs (and the fish if the hatchery removal is permanent or fatal) had been harvested for consumption.

This means that when a stock is already severely depressed, we have to depress its potential reproduction even further to even initiate a hatchery-based rebuilding program. That is, we have to take an immediate risk of the sort mentioned above that the hatchery fish will not survive as well as if the eggs had been left in the wild, just to get the stock rebuilding program started. This is obviously a trade-off and risk-management problem, and the worst thing anyone can do is to proclaim confidently that they have some delivered wisdom that the gamble will or will not succeed. Only time, and actual release experience, will resolve the uncertainties about the balance of positive and negative effects of the hatchery exploitation.

Post-Release Competition between Hatchery and Remaining Wild Juvenile Fish

As we have seen (chapter 6), one of the most striking features in the population dynamics of most fish is that average recruitment is nearly independent of egg deposition over a wide range of spawning-stock sizes. We have seen that it is incorrect to interpret this observation as meaning that recruitment is "limited" by habitat or environmental factors; rather, it implies that there is a strong density-related improvement in the egg-recruitment survival rate when the number of eggs is reduced by factors like fishing. That is, there is a strong change in the apparent effects of intraspecific competition among larvae or juveniles, such that survival is much improved when fewer larvae or juveniles enter the prerecruit life stages. We have seen that the most obvious cause of this apparent competition and its linkage to survival is the "foraging arena" effect of restricted habitat use by juveniles, driven by predation risk. Restricted habitat use implies intensified competition in local foraging arenas near sites of reduced predation risk, and increased juvenile abundance means reduced resource availability in these arenas and hence a requirement for juveniles to spend more time feeding and at risk to predation when juvenile abundance is high. We cannot predict the severity of this competition/predation effect just by measuring overall resource availability (e.g., the total food production) at large ecological scales. Absent some special effects such as intense cannibalism, we generally expect the average survival rate from egg to recruitment to vary according to the Beverton-Holt relationship:

$$s_t = \frac{s_o}{(1 + \kappa \epsilon_t)} \tag{13.1}$$

where s_t is the total proportion of wild eggs that survive to recruitment, s_o is the maximum average survival rate of these eggs in the absence of intraspecific competition, κ is a population-size scaling parameter that combines the effects of habitat size and maximum survival rate, and ϵ_t is the total number of eggs beginning the juvenile life stage. Obviously, equation 13.1 implies a declining juvenile survival rate of fish rearing in the wild if enhancement causes the effective egg deposition ϵ_t to increase.

Unfortunately, while equation 13.1 is very well supported by field data for wild populations, it should not be used directly to support claims that enhancement will necessarily cause negative survival effects for wild fish. That is because density dependence in the survival rate can occur at any one or several life-history stages before recruitment to the adult population. If the enhancement method holds fish until they are past such stanzas, then those enhanced fish are not involved in the competitive relationship that leads to equation 13.1, and releasing them into the field will not cause survival problems for wild fish.

Even more unfortunately, we generally do not know precisely where in the juvenile life of fishes the density dependence that we see during overall recruitment has actually occurred. There is some evidence that density-dependent effects usually drop off over time/age as juveniles grow and can avoid more and more predation risk while foraging (Rose et al. 2001; Myers and Cadigan 1993). But in other cases, strong density dependence may occur at particular times, e.g., just after larval settlement. About all we can say in general is that there is a risk that we will see density-dependent effects at any life stage at which the fish still show a predation-driven restriction in habitat use and fairly high natural mortality rates.

Consider the policy implication if there is strong density dependence in survival rates for some life stage(s) after hatchery release. In that case, total recruitment may be limited not by wild or hatchery egg production but, rather, by competitive interactions. This means that releasing fish from the hatchery will not increase total recruitment at all, i.e., hatchery production will produce no net benefit of any sort at the population level but will incur production costs.

The fool's bargain gets worse when fish originally produced in the hatchery may return to some specific area (e.g., the hatchery itself) at some later time to spawn. In that situation, people who are positioned to observe and benefit from the hatchery fish (e.g., harvest them near the hatchery) will see a very real, local positive benefit of the hatchery program (they see lots of fish that weren't there before) and will be blind to the negative effects that these fish have had earlier in life on the survival rates of wild fish.

We have had a few surprises in Pacific salmon enhancement when we assumed that hatchery rearing was bypassing the important competitive life stages. The dogma in Pacific salmon biology, at least for species whose early rearing is in lakes and streams, has been that density-dependent effects occur mainly in the obviously restricted freshwater habitats but not in the ocean. If this is true, we can bypass the fool's bargain just by rearing hatchery fish to the smolt (ocean migrant) life stage. This is what we assumed for coho salmon in the Georgia Strait, British Columbia, and over the 1970s we developed the hatchery capacity to double the number of smolts entering the ocean environment (fig. 13.1, bottom panel). It was fully expected that this would result in a doubling of the Georgia Strait catch. But in the end, the catch did not increase at all (fig. 13.1, top panel) and has even dropped off dramatically in the last few years. That is, the Georgia Strait acted as though it had a limited carrying capacity for coho salmon, implying an inverse marine-survival relationship like equation 13.1. As hatchery production increased, and the hatchery contribution to total catch increased correspondingly, the wild catch and spawning stock decreased so as to just about match the hatchery gain. Such apparent density dependence in ocean survival rates has been

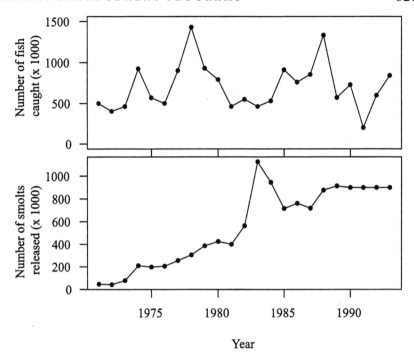

FIGURE 13.1: The number of coho smolts released from hatchery programs and the total coho landings in the Georgia Strait, British Columbia. Increased smolt production did not result in the projected increase in coho catches (DFO 1999).

seen in other salmon stocks (Hilborn 2002; Hilborn and Eggers 2000; Levin et al. 2001; Peterman 1991; Noakes et al. 2000).

Such surprises have engendered even nastier scientific and policy debates. While we may suspect that the wild stock has been driven down by enhancement production and a limited carrying capacity like the Georgia Strait example, we can never prove (without experimentally stopping the hatchery production entirely) that competition was, in fact, what caused the wild stock decline. There are many reasons for wild populations to decline (and trigger demands for hatchery production in the first place), like overfishing and loss of natural rearing habitat and changes in oceanographic conditions or ocean productivity. And we can be sure that proponents of hatchery production will argue for all of these and will claim that they are saving the fishery by compensating for the loss of wild production. We should certainly not expect them to admit that they may be causing that loss in the first place.

Increase in Predation and Disease Risk for Remaining Wild Fish

Concentrations of just about any creature in nature are likely to stimulate "numerical responses" by whatever other organisms see that creature as a resource. So birds, seals, and various fishes concentrate downstream of salmon hatcheries and form a predation gauntlet for migrating juveniles, both wild and enhanced. High rearing densities under culture are also likely to cause enhanced disease transmission, and it will be the surviving "Typhoid Marys" from hatchery disease-outbreak events that are later released to mix with wild fish. Some diseases transmitted during such events may show a delayed expression (and hence not be detected and treated) in the hatchery.

A perverse and perhaps unusual variation on this problem can arise when several year-classes of juveniles share the same nursery area and older juveniles either cannibalize or outcompete younger ones. In this case, the natural recruitment dynamics may exhibit a cyclic pattern, with the nursery area dominated at any time by a single rearing cohort and successful recruitment occurring only once per the number of years that juveniles spend in the nursery area. A striking example of this phenomenon is the catadromous and highly piscivorous barramundi discussed in chapter 6 (see fig. 6.6 on page 143). In situations like this, stocking is likely to be either ineffective at least every other year (if juveniles are stocked at age 0) or routinely destructive of wild recruitment (if juveniles are stocked at ages/sizes large enough to bypass the cannibalism/predation bottleneck and suppress wild recruitment every year).

Selection under Hatchery Conditions for Traits That Are Inappropriate in the Wild, so Hatchery Fish That Later Spawn with Wild Fish Can Create a "Sterile Male" Effect

Long-term propagation of fish in hatcheries can obviously lead to selection for phenotypic traits that are appropriate under hatchery conditions but deleterious in the wild. Historically, this problem was exacerbated by poor breeding practices that created small effective population sizes, like fertilizing all hatchery eggs with sperm from one or a few males. But there have been long-term declines in the post-release survival rates of hatchery fish in a few cases in which there has been deliberate and careful attention to sound breeding practices, e.g., salmonid hatcheries in southern British Columbia (Coronado and Hilborn 1998; Walters and Ward 1998). We have tended to explain away these declines as due to bad luck or the coincidence of hatchery development with unfavorable changes in other factors like ocean temperatures. But there remains a nagging suspicion that we are somehow selecting for the wrong kind of fish.

To emphasize the risks of genetic change, hatchery programs have been called "breeding programs for the culls" because they very obviously (and intentionally) keep a large number of juveniles alive that would have died under natural rearing conditions. It is quite likely that high natural mortality rates in nature are to at least some degree selective with respect to heritable traits. In particular, natural mortality probably involves balancing selection with respect to a variety of quantitative traits (like spawn timing, size at maturity, and aggressive behavior), and we find that such traits can be altered quite dramatically under hatchery selection.

What these arguments mean is that we are uncertain if hatchery fish are fully capable of producing normal and wild-viable offspring if/when these fish spawn in the wild or mate with wild-type fish. It has been extremely difficult to demonstrate reproductive impairment of hatchery × hatchery or hatchery × wild cross fish under wild conditions, because we would need to show not only whether such spawnings had been immediately successful but also whether the juveniles produced by such matings had normal survival rates. The main field evidence for the possible bad effects has come mainly from studies of Atlantic salmon in Scandinavia, where fish from highly domesticated farm strains regularly "invade" wild populations (Einum and Fleming 2000; Fleming et al. 2000; Fleming and Einum 1997; Fleming et al. 1996). The Atlantic salmon studies show depression of as much as two-thirds in reproductive success, indicating that the farm fish do represent a major threat to wild populations.

An extreme example of the possible genetic effects of hatchery fish on wild populations would be the "Trojan gene effect" suggested for genetically modified organisms (GMO) that might escape into the wild from aquaculture (Muir and Howard 1999; Hedrick 2001). The notion here is that a GMO might have some phenotypic characteristic that gives it a distinct mating advantage over wild fish (e.g., a much larger body size), such that the characteristic is passed on to offspring of GMO × wild matings, but this characteristic might be fatal to juveniles in the wild (e.g., rapid growth and large body size may be highly deleterious for juveniles attempting to rear in the wild). In this case, the overall wild reproductive success would be decreased in direct proportion to the relative abundance of GMO-type individuals.

Attraction of Fishing Effort by Unregulated Fisheries, Which Then Hits the Remaining Wild Fish Even Harder Than Would Otherwise Happen

Fishers are anything but stupid and show an almost uncanny ability to detect and respond to changes in the spatial distribution and abundance of fish (chapter 9). Even when we pretend to regulate the total fishing effort through methods like license limitation, there is generally considerable "latent effort" waiting to be exerted should high abundances of fish become available. This means that any policy innovation or success that results in more fish in the water can be counted upon also to result in more hooks, nets, or whatever in the water. Except under a few special circumstances, that gear in the water will then mean higher fishing-mortality rates for any fish that happen to be present, hatchery or wild. This problem can reach almost ludicrous extremes when the effort is unregulated and a management agency uses a simple per-fisher performance measure like catch-per-effort to decide when to stock more fish. Then increases in stocking result in CPUE increases, which attract more effort, which drives CPUE back down and triggers more stocking, in a runaway feedback that may end only when the management agency can no longer finance increased stocking. If you do not think management agencies can get trapped into this sort of pathology, try trout fishing in California some time.

There are at least four circumstances or policy options that can help prevent hatchery production from causing increased fishing mortality rates for depressed wild populations:

1. *Gear-saturation/handling-time effects.* When fishing gear can handle only a limited number of fish or when each capture occupies a considerable portion of the fisher's time, it is possible for enhanced fish to "buffer" wild fish. In such (rare) cases, the increased abundance of hatchery fish can essentially fill the gear/time so that the effective effort to which wild fish are at risk is reduced. Generally simple measures like daily bag limits are not effective at causing this effect, because they are most often set so high that few fishers are affected by them in the first place.

2. *Spatial separation of enhanced and wild fish.* Sometimes, enhancement can increase the local abundance of fish in areas not frequented by wild fish, attracting effort to move away from areas where wild fish are concentrated. But this is at best a short-term effect: if it is successful at helping wild abundance to rebuild, that wild abundance will soon begin attracting the effort again.

3. *Marking to permit the selective retention of enhanced fish.* There is a test under way in the Pacific Northwest to mark all hatchery fish so that fishers can retain these while releasing wild fish. The obvious concern with this approach is the mortality of wild fish caused by the capture and release process. Release mortality is difficult to evaluate because (a) it may be delayed (cannot hold fish for just a few hours to see if they survive), (b) it may involve indirect effects such as increased vulnerability to predators after release, and (c) it is hard to separate from tagging effects when estimated by marking the released fish for later survival estimation.

4. *A "Hilborn Plan" in which a quota restriction of the total catch drives the fishing effort down.* During the 1970s Ray Hilborn proposed a clever idea for combining quota management (limitation of total allowable catch) with enhancement to promote wild-stock rebuilding (Walters 1986). Suppose the total catch from a fishery is effectively "capped" with a fixed (not changing from year to year) quota, and large-scale enhancement is used to increase the number of fish available to fill that quota. Then as the enhanced (and later wild) stock grows, the fishing effort or time needed to reach the quota will drop, so wild fish will be subject to a progressively lower fishing effort over time. While this concept is attractive in principle, it relies upon there not being strong competitive interactions between wild and enhanced fish and upon there being a very strong management agency. It would create a situation such that fishers see more fish every year, yet are subject to ever more restrictive

fishing regulation; few management agencies could withstand the uproar that this combination would be almost certain to bring.

It is generally unwise to use effects (1) and (2) as excuses for inaction about limiting the total fishing effort. Recall from chapter 9 that we can think of the fishing mortality F_t as the product of fishing effort (f_t) times the proportion of stock caught per effort (q_t), $F_t = q_t f_t$, where q_t represents the area (a_s) "swept" by a unit of fishing effort divided by the area (A_c) over which fish and fishing are concentrated, so $q_t = a_s/A_c$. Density-related dispersal (and the deliberate spreading of enhanced fish) very often leads to A_c being about proportional to the total stock biomass B_t, i.e., $A_c \approx k_1 B_t$, where $1/k_1$ is the average density of fish per unit habitat area actually used. Combining these relationships results in the prediction that F_t will vary approximately as $F_t \approx f_t a_s/(k_1 B_t)$. If the fishing effort f_t were held constant, this relationship would mean a strong dilution of the fishing mortality rate with increases in stock biomass B_t and area occupied (fishers sweep the same area $f_t a_c$ each year, but this area decreases relative to the total area $k_1 B_t$ occupied by fish). But suppose that the total fishing effort attracted to area A_c is about proportional to the total stock size, a pattern that we often see especially in recreational fisheries. In that case, f_t is not constant but, instead, varies roughly as $f_t \approx k_2 B_t$. Substituting this response relationship into the prediction for F_t results in $F_t \approx a_s k_2/k_1$, i.e., *no variation at all in the fishing rate despite possibly large changes in the fishing effort and area occupied by the stock!* If a combined relationship even roughly like this does result from enhancement, and if F_t is too high to start with, then the stock will remain overfished (with respect to the wild stock) despite any positive effects that enhancement might have on the total recruitment. The moment that this enhancement is stopped (if it is being used as a temporary measure to rebuild the wild stock in the first place), the wild stock will begin declining to the same low abundance that triggered the call for enhancement in the first place.

Effort-response dynamics are particularly dangerous in management when there is confusion about the purpose of enhancement in the first place (rebuilding wild stocks versus providing fishing opportunities), and when people do not understand that increased fishing effort causes increased fishing mortality. Too often people seem to accept the notion that more fish (due to enhancement) may mean that more can be safely caught, without realizing that catching more is not, in fact, safe if it involves allowing more fishing effort. This confusion has not been helped by scientists warning that fishing effort is often not a good predictor of the fishing mortality rate due to changes in fishing technology and increases in catchability (the proportion of stock caught per unit effort) with decreases in stock size. That scientific warning is a valid one, but it is not about stock-enhancement situations; rather, it is about the difficulty of bringing fishing mortality rates under control during periods of stock decline and improving technology.

13.2 CRITICAL STEPS IN ENHANCEMENT PROGRAM DESIGN

Some of the risks and problems identified in the previous section can be avoided or mitigated by approaching enhancement program design and implementation with sensible caution. Below are a few key elements, or required steps, that a "Code of Responsible Conduct" for marine enhancement program development might contain.

Make Certain That Management Priorities and Acceptable Trade-offs Are Absolutely Clear

In most management jurisdictions today, the maintenance and restoration of healthy wild populations is considered a top priority, for legal, ethical, and long-run economic reasons. There are few cases for which we would be willing (or it would be legally acceptable) to knowingly "write off" natural production in favor of the relatively risky option of basing production solely on hatchery recruitment, especially in view of the gross uncertainty about the long-term sustainability of hatchery-based production systems.

If the central objective of management is, in fact, to have healthy, productive wild populations, then enhancement should be viewed solely as a stop-gap tool to accelerate wild-stock rebuilding when natural recruitment has been impaired through habitat damage or overfishing. The top management priority should be to redress the original and potential future causes of stock decline, which in most marine cases would mean reducing fishing mortality rates. Until that cause of decline and low current abundance is removed, enhancement is likely only to exaggerate the problem (mainly via fishing effort responses and competition between wild and hatchery fish).

Even when wild recruitment cannot be restored because of irreversible loss of juvenile-rearing habitat, it is not always true that a hatchery-based replacement of recruitment can be done without trade-offs and loss of wild production. Most fishes undergo ontogenetic migrations from nursery areas to wider adult rearing habitats, where fish from many nursery areas or stocks may be mixed. The fishing effort attracted to (or permitted because of a high total abundance in these mixing areas because of the presence of hatchery fish) can cause widespread overfishing of the remaining wild stocks.

Enhancement has sometimes been proposed as an "interim measure" to sustain fishing opportunities and values (e.g., employment) until effective means of reducing fishing mortality rates can be discovered or implemented (e.g., by the retirement of current fishers). That is, the pretense is that fishers have to be supported now, but there is an intent to reduce the scale and impact of the fishery gradually or at some future time. What is really going on in these cases can be very nicely summed up by the old joke "the check is in the mail." The real intent is to avoid hard regulatory decisions now so as to place the burden of making those decisions on someone else, sometime. We have not seen a single instance of a management agency following through on such promises.

There are cases in which it appears possible to use permanent enhancement facilities to "supplement" wild production, by increasing the total recruitment while strictly regulating fishing mortalities so as to protect and rebuild wild production. In these cases, another trade-off arises: we have to accept the "waste" of some of the potential harvest of hatchery fish, assuming these fish can withstand higher fishing mortality rates. A management challenge is then to find ways to avoid the waste through selective fisheries for enhanced fish, e.g., by locating "terminal fisheries" near hatcheries and by selectively marking hatchery fish so that they can be recognized and retained by fishers that are required to avoid killing wild fish.

Do Careful Stock Assessments to Show That the Target Stock is Recruitment Overfished or Can No Longer Rear Successfully in the Wild

If wild stock protection or restoration is a priority (the usual case), it is critical to establish that the wild stock is not already recruiting at rates independent of spawning abundance, i.e., on the flat part of the stock-recruitment relationship, where recruitment is limited through density-related changes in egg-juvenile mortality rates. Demonstrating that the adult stock is, in fact, low enough to limit recruitment (recruitment overfishing has actually occurred) is not as simple as it might sound. Indeed, demonstrating recruitment overfishing has been one of the most difficult challenges in fisheries-population dynamics.

It is not enough just to show that the adult population is much smaller than it used to be or would be if not fished. We know that spawning abundance can often be reduced greatly without recruitment effects. Low adult stock size (relative to unfished) is, in fact, expected for most populations when they have been managed to levels that will produce a maximum sustainable yield, due to the "normal" erosion of age and size structure (reduction in survivorship to old ages) that inevitably accompanies fishing. Low catch per effort or catch per fisher is also not a reliable indication of recruitment overfishing, because these indices reflect the combined effect of stock-size reduction (exploitation) and competition with other fishers for the fish. The competition effects on fishing success rates can be particularly large in recreational fisheries, where only a small proportion of the fish are likely to be behaviorally reactive to fishing gear at any moment, so fishers are competing for only this small proportion of the total stock.

Likewise, it is not enough just to show that very many fewer eggs are being laid, or that there are very many fewer larvae and small juveniles, than there used to be or would be if the stock were not fished. Decreases in these early life–stage abundances are expected because of the effects of even sustainable fishing on spawning-stock size. For example, we have seen very substantial declines in the spawning runs of coho salmon in British Columbia, and coho fry densities in our streams are much lower now than we would have expected based on earlier spawning runs. Yet until very recently we saw no obvious declines in the numbers of smolts actually surviving the full freshwater-nursery stage and leaving for the ocean, indicating strong density-dependent improvements in fry-to-smolt survival as fry numbers declined. Stocking fry into these streams that were already producing "carrying capacity" numbers of smolts would have been an economic waste (no benefit from the costs of producing the fry) and would have depressed the number of wild smolts leaving the streams.

One possible expression of recruitment overfishing for stocks that use a variety of localized nursery areas (reefs, small streams, estuarine and salt marsh areas) is for some of the localized areas to produce nearly normal recruitments, while fish disappear entirely from many other areas. That is, overfishing can be expressed as an erosion in the spatial stock structure or spatial diversity. This can happen if there is either localized overfishing of adult fish or a variation in productivity among the rearing areas so that some of these "substocks" or "microstocks" are more sensitive to wide-scale fishing. Such effects should be evident in the estimated total recruitment rates from regional-scale stock assessments but imply a quite different enhancement strategy than would be used if the regional stock were, in fact, a single reproductive unit. In the single-unit case, enhancing recruitment anywhere would contribute to rebuilding spawning abundance everywhere. But in the more structured case, enhancement should be restricted to those substocks that have suffered the most severe losses. It can be hard to convince proponents and operators of enhancement facilities to target the most severely impacted substocks, because it may be hard to get enough eggs from these and there may be suspicions about the suitability of currently barren areas for restocking at all.

Show That Enhanced Fish Can Recruit Successfully in the Wild

This is the easiest step in enhancement-policy development. Scientifically, the most effective way to accomplish it is to do the same thing recommended in the next section, namely, to mark all of the enhanced fish before release. This ensures the largest possible sample size for field survival surveys and fishery contribution assessments and removes any ambiguity about which fish sampled at older ages have, in fact, been produced by enhancement. The cost of marking many fish is not a valid excuse for bypassing this assessment step. We have plenty of relatively inexpensive methods for mass-marking fish, ranging from thermal shock treatments in hatcheries (which cause distinctive marks on bones like otoliths) to sprayed dye-marking to coded-wire tagging. Where possible, the marking should obviously be done with visible marks (e.g., fin clips) to reduce the costs of later mark detection in recapture sampling. Where visible marking is impractical (e.g., very small fish) or unnecessary, there are some really neat new technologies that we can use to determine the probable origins and past rearing locations of fish. These range from genetic markers (or, ultimately, DNA fingerprinting of many individuals) to elemental ratio signatures in bones using microscopic laser ablation spectroscopy.

Marking all enhanced fish may also create an opportunity for a less painful policy response should it be discovered later in the program development that the uncontrolled fishing effort responses to the presence of enhanced fish are causing worse overfishing for wild fish. This is the opportunity for "selective mark fisheries," involving gear that can safely return wild fish to the water while retaining enhanced fish (see points in section 13.1).

Show That Total Abundance Is at Least Initially Increased by the Hatchery Fish Contribution

The same strategy of mass marking all hatchery fish as mentioned above is an important first step in demonstrating whether the overall abundance has been increased by hatchery production. To our knowledge, the first large-scale application of the "mark every hatchery fish" approach was in restoration programs for lake trout (*Salvelinus namaycush*) in the Laurentian Great Lakes. In that case, mark-rate sampling in research surveys and fishery catches allowed a completely unambiguous determination that wild lake trout were, in fact, recovering in one lake (Superior) but that there was no recovery at all in two others (Michigan, Huron) (SLIS 1980; STOCKS 1981). Had less than 100% of the hatchery fish been marked, you can bet that there would have been all sorts of arguments about how successful the stocking program was at initiating wild reproduction in Lakes Michigan and Huron.

It is crucial at this stage in enhancement program evaluation to avoid the mistake of treating the increasing proportional (percentage) or absolute contribution of hatchery fish to catches and survey abundances as evidence of program success. These increases mean only that hatchery fish can survive in the wild, not that they can do so without severe negative impacts on wild fish via juvenile competition. Survey methods and stock-reconstruction methods based on catch statistics also need to be good enough to demonstrate whether and how the wild population is responding to enhancement.

Absent some marking method that demonstrates clearly that enhanced juveniles do at least survive in the wild, it is impossible to determine whether the lack of an overall recruitment increase following enhancement is due to a carrying-capacity problem (hatchery fish surviving but replacing wild fish) or to hatchery fish not surviving in the first place. It is frustrating to see excellent long-term, spatial data like the red drum data presented by Scharf (2000), in which there are strong hints at competitive interaction effects but no way to be sure that hatchery fish have ever survived well or long enough to contribute significantly to competition-driven changes in the juvenile survival rates.

Show That Fishery Regulations Are Adequate to Prevent Continued
Overfishing of the Wild Population, Unless There Has Been an Explicit
Decision to "Write Off" the Wild Population

This is again a matter of using the best available stock-assessment techniques for the wild stock, to address the particular risk that wild fish may be subject to increased fishing mortality due to the effort attracted by the enhanced fish. As discussed in chapter 4, after much sad experience with existing assessment methods, we now feel that it is important to measure fishing mortality rates directly whenever that is physically and economically possible. This means using tagging programs and/or swept-area assessments based on spatial data on fishing-effort patterns and the associated sampling of fish distribution patterns. It is just too risky to calculate fishing mortality rates by dividing reliable catch statistics by highly unreliable estimates of total abundance based on an analysis of abundance trend and the composition information. This is especially true in enhancement situations, in which fishing efforts and the impacts on wild fish can change rapidly in response to the initial success at producing fish. It is hardly comforting to know that methods like VPA will eventually give "converged" estimates of what happened a decade ago as enhancement was developing, when a further decade of continued and undetected overfishing could have huge long-term costs. For those who do not believe that such risks and costs are a real concern, we advise they pay a visit sometime to a Newfoundland cod fisher.

A monitoring approach that has been very useful in Pacific salmonid enhancement programs is to conduct marking programs for juveniles from "indicator" wild stocks in parallel with the marking of hatchery releases. The absolute estimates of total survival and exploitation rates from such marking programs are dubious because of uncertainties about the impact of the marking process on wild juveniles and about the sampling rates for the marked fish (e.g., tag reporting rates from fisheries). But relative estimates, particularly trends in tag recovery rates over time and among stocks, have been very useful in helping to plan harvest-regulation programs (like joint U.S./Canada efforts to rebuild chinook and coho salmon stocks). Obviously, such marking programs can also provide much useful information on dispersal and migration patterns. The key to making them work is to have routine, annual marking for a fairly wide and representative range of wild indicator stocks/areas. Once-off marking of wild fish from a single indicator or "control" area is a waste of research and monitoring effort.

*Show That the Hatchery Production System Is Actually Sustainable over
the Long Run, When It Is to Be a Permanent Component of the
Production System*

Perhaps the single biggest mistake that can be made in enhancement program development is to assume at the outset that it is going to work, and
to invest accordingly in capital facilities intended for long-term use. When
Canada's Pacific salmonid enhancement program (SEP) was started in the
early 1970s, one unpublished government document noted, "There was a
commitment, at the outset of the program, to an adaptive management process." That is, some fisheries people saw the program as quite risky, and
we spoke out in favor of not only careful monitoring but also the initial use
mainly of small-scale, temporary facilities that could be easily abandoned
(e.g., using cheap plastic swimming pools for rearing, rather than expensive
concrete raceways). Despite these recommendations, we saw the actual
capital-development planning and project-selection process become dominated by large-scale, supposedly "cost-effective" hatchery proposals. There
was continued lip-service to caution and adaptive management, but this
quote from another planning document shows how grossly that concept was
misunderstood: "Continual monitoring as 'actual' results become available
will provide more and more concrete evidence of the success of SEP."
  We got plenty of concrete all right, but the evidence part has largely been
ignored. We now have concrete dinosaurs continuing to pump out large
numbers of fish despite considerable evidence of risk to wild stocks, and
even into areas where the fisheries have been largely shut down because of
conservation concerns about those wild stocks.

13.3   MONITORING AND EXPERIMENTAL REQUIREMENTS

So we have seen that there are four main ways that a marine enhancement
program can end up doing more harm than good. In order of immediacy or
likely timing of appearance during the development of an enhancement program, these are: (1) the replacement of wild with hatchery recruitment, with
no net increase in the total stock available for harvest (competition/predation
effects); (2) unregulated fishing-effort responses to the presence of hatchery
fish, that cause overfishing of the wild stock; (3) "overexploitation" of the
forage resource base for the stocked species, with attendant ecosystem-scale
impacts; and (4) genetic impacts on the long-term viability of the wild stock.
It is important to understand that these are not just hypothetical or occasional difficulties; at least (1) and (2) have plagued salmonid enhancement
programs in the Pacific Northwest, (3) has been a central concern in the
restoration of Great Lakes fishes, and (4) is becoming a topic of major research investment if not sad practical experience.

Enhancement programs are notoriously difficult to stop once they have become established in management institutions. They generally have strong public support despite any evidence of nonperformance and are an important employment opportunity within management agencies (plenty of people are eager to defend their jobs). This leads to a politically powerful coalition of interests wherever there is an established enhancement program. To prevent such coalitions from delaying or totally preventing needed corrective changes in enhancement programs, it is critical to monitor the impacts of enhancement very carefully as any program develops so as to have nearly incontrovertible evidence at hand when/if debate about the efficacy of the program does surface.

Here is a summary from the above sections of the key monitoring and experimental requirements for preventing harmful programs from becoming permanent.

- *Mark all (or at least a high and known proportion of) fish released from hatcheries.*

  An excellent example of the value of this tactic is lake trout stocking in the Great Lakes. Having all hatchery fish marked removes all debate about whether wild reproduction has contributed to any apparent stock recoveries that might accompany the enhancement program.

- *Mark as many wild juveniles as possible at the same sizes/locations as hatchery fish are being released.*

  Unfortunately, this tactic has not yet been followed in any major enhancement program. But where we have been "lucky" enough to have wild index stocks marked in areas where enhancement has occurred (e.g., wild coho index stocks in southern British Columbia), we have seen parallel trends in wild and hatchery survival that could indicate either competitive interaction effects or shared response to environmental factors. The point is that marking wild fish is critical to determine whether wild survival to recruitment is depressed as hatchery production increases, but it cannot demonstrate conclusively that the survival depression is, in fact, due to competition/predation effects (a parallel decline would be expected from shared environmental effects as well).

- *Experimentally vary hatchery releases over a wide range from year to year and from area to area, probably in on/off alternation (temporal blocking) so as to break up the confounding of competition/predation effects with shared environmental effects.*

  In the Pacific Northwest, there have been parallel declines in wild and hatchery coho and Chinook salmon survival rates, correlated with increased hatchery releases. But survival declines have continued even after hatchery releases stopped increasing, prompting the argument that the declines were due not to competition but to environmental factors in the first place. We have strongly recommended varying hatchery releases in order to resolve this debate (Walters 1994) and to determine whether there is a particularly damaging environment-competition interaction effect (competition with hatchery fish may exaggerate the effects of environmental change on wild stock survival).

- *Monitor changes in total recruitment to, production of, and fishing effort in impacted fisheries, not just the percentage contribution of hatchery fish to production.*

  This apparently obvious admonition is to help detect whether hatchery fish add to the total production or just replace wild production. Unfortunately, it has been an extremely difficult recommendation to implement, especially where the main harvest is by spatially diffuse recreational fisheries where catch and effort monitoring is expensive.

- *Monitor changes in the fishing mortality rates of both wild and hatchery fish directly, through carefully conducted tagging programs that measure short-term probabilities of capture.*

  This is a key need for improving fish stock assessment and harvest management in general. It is particularly important in conjunction with hatchery programs, to measure whether increased fishing attracted by hatchery fish is causing an increased fishing impact on wild stocks.

- *Monitor reproductive performance of hatchery-origin fish and hatchery-wild hybrid crosses in the wild.*

  The main genetic concern is whether a selection for hatchery-performance characteristics may somehow "pollute" the genetic structure of the wild stock. Operationally, that problem should be expressed most immediately through impacts on the viability of the offspring produced in the wild by fish that were not subject to natural selection for wild characteristics. It is not enough just to show that hatchery fish have some different genetic characteristics (e.g., lower heterozygosity) than wild fish, since these characteristics may or may not have significant ecological and evolutionary consequences over the long term.

Taken together, these monitoring requirements would grossly increase the cost of most enhancement programs compared with the usual production-cost accounting used to justify enhancement programs. But the risk of net deleterious effects of large enhancement programs is certainly great enough to justify considerable investment.

Options for Sustainable Ecosystem Management

THERE ARE SOME THINGS that most stakeholders in marine ecosystem management would agree are needed. No one wants to see ecosystems that are so grossly overfished as to produce much less economic and recreational value than is possible. No one wants to see nontarget, especially rare and endangered, species disappear through competition with fisheries and through direct bycatch/culling impacts. No one wants to see productive capacity degraded by fishing practices and coastal-zone developments that destroy the habitat basis for productivity, especially the benthic flora and fauna that are likely critical to recruitment processes for many species. So we will almost inevitably see some major policy changes in the direction of better ecosystem management, whether or not careful technical analysis and modeling drives that change. Fishing fleet sizes (overcapacity) will be reduced, some of the most destructive and nonselective practices will be banned, tactics and devices will be found to reduce direct by-catch impacts, some areas will be protected from fishing entirely, and there will be progressively more restrictive coastal zoning to prevent further inshore and estuarine habitat loss.

But what there is as yet no consensus about is how to manage the painful economic and social changes associated with reducing fishing fleets to sustainable sizes and applying more selective fishing practices, and how to trade-off between sometimes conflicting objectives related to the productivity of alternative fisheries and creatures that are not harvested. This book has reviewed a variety of modeling tools that offer promise for making useful predictions about, or at least a crude screening for, policy options ranging from simple exploitation-rate controls to complex spatial-effort management, while accounting for unpredictable and uncontrollable variations in future aquatic environmental conditions. There are now integrated software packages, especially Ecosim, that incorporate virtually all the relationships or "submodels" discussed in previous chapters into single simulation programs aimed at encouraging policy exploration and gaming. When such software systems started to be developed in the early 1970s as aids for the design of adaptive management policies, one of the first lessons that we learned (Holling 1978; Walters 1986) about using them is that the main product of analysis/simulation is typically not precise quantitative predictions but, rather, alternative "visions" of possible management strategies. Indeed, about the only adaptive management plans that have actually been implemented have been ones in which stakeholders involved in the modeling and policy analysis have been able to cut through the tactical details of modeling and day-to-day management, so as to reach a consensus (vision) about desired ecological states, about the most critical uncertainties, and about broad experimental management "treatments" for resolving the uncertainties (Walters 1997).

So far, we are not seeing the development of such alternative visions of restoration and sustainable management for aquatic ecosystems. Instead, we are seeing lists of ill-defined, broad objectives, much complaint about how complicated nature is, much quibbling about precisely how to model and regulate both single-species and multispecies harvest dynamics, and much hand-wringing and debate about how environmental variability creates short-term unpredictability and confounds our ability to detect the effects of human activities like fishing. There is no clear consensus about what constitutes "ecosystem management" beyond improving single-species management by reducing overcapacity and fishing effort (Sissenwine 2001). In some cases, we are seeing progressively more complex regulatory schemes, that critics of bureaucratic management often call "fire fighting" and that fishers call "the death by a thousand cuts." Partly in reaction to such schemes, at an opposite extreme we are seeing a tendency toward simplistic policy recommendations based on the same sort of protected area tactics that have dominated terrestrial conservation activities, justified by arguments of the form "we cannot regulate fisheries and so should have large protected areas where fishing is not allowed at all." Hidden in such arguments is not only the presumption that traditional harvest-regulation systems have failed but also the presumption that more natural is always better, i.e., that ecosystem management and ecosystem restoration should be basically the same thing. Further, there is often a barely hidden contempt for fishers and fishing communities and their rights and values, as though fishers were capable only of rapacious and greedy behavior.

This closing chapter offers our views about a few alternative visions of future management, based on experience with Ecosim policy exploration by a variety of people (Pitcher and Cochrane 2002). Based on these visions, we end the book with a wish list of critical needs for more effective regulation. These are mainly not needs for better modeling methods but, rather, better incentives and institutional arrangements for regulation and information gathering.

14.1   ALTERNATIVE VISIONS OF ECOSYSTEM STRUCTURE

When we run formal harvest-policy optimizations or gaming scenarios with models like Ecosim, it quickly becomes apparent that sustainable ecosystem management might take a variety of strategic directions, i.e., might involve some very different visions about how to achieve and maintain long-term productivity. The objectives and goals for ecosystem management are sometimes presented as though the only sustainable option were the restoration of natural diversity and abundances, as though any other ecosystem configuration would necessarily lead to unpredictable changes and unacceptable patterns of variation. There is, in fact, little empirical evidence that diversity and stability are strongly linked in aquatic systems, despite evidence that the presence of predators and competitors may be necessary to any expression of stabilizing density dependence in the recruitment rates (Carr et al. 2002). What the typically vague arguments about how restoration is "needed" or "necessary" for sustainability are really about is disguising the promotion of particular existence values (e.g., scientists who enjoy observing and studying various creatures) behind the pretense that such values have strong economic and social implications as well. This is not to say that stakeholders who promote restoration and diversity should be ignored, but that we should not try to justify such values by the scare tactic of pretending that no other options are sustainable. In the following subsections, we review some of the alternative "visions" of potentially sustainable ecosystems that arise when conflicting values and objectives are given different weights in searching for optimum ecosystem configurations.

The "Back to the Future" Option

Probably the clearest articulation of the vision that the overall restoration of historical abundance patterns may produce far more valuable marine ecosystems has been the Pitcher (2001) "back to the future" argument (see also Pitcher et al. 1999, 2002; Sumaila et al. 2001). He argues that a wide variety of scientific and traditional knowledge sources, along with models like Ecopath and Ecosim, can be used to reconstruct natural patterns of abundance and productivity for most large marine systems. These reconstructions can then be used to demonstrate just how much potential production (and biodiversity) has been lost through past mismanagement or neglect, and to predict potential sustainable production due to restored ecosystem structure. For a series of test ecosystems (North Sea, Newfoundland, Northern British Columbia), Pitcher and his colleagues have demonstrated that there is considerable potential for increase in both consumptive and existence values. In this approach, there is some danger of confusing abundance and production (high abundances may be sustainable only if most biological production is "used" to sustain the abundances rather than to produce yields), but the approach can at least be used to define clear ranges of possible future states as alternative targets for management.

   Discussions about the "back to the future" vision of historical abundances and biodiversity have so far stopped short of making specific policy recommendations about how to move from the present, depleted circumstances to more natural ones (T. Pitcher, UBC, pers. comm.). Most proponents of major restoration appear to have adopted the view that we should reduce fishing as quickly as possible, move toward more selective and less efficient fishing gears, then let "nature take its course." It does not take much policy gaming with population and ecosystem models to demonstrate that even modest fishing can grossly retard or prevent recovery of at least some (particularly long-lived) ecosystem components. So if some variant on the "back to the future" vision is adopted, the most bitter debates will likely not be about the vision itself but, rather, about how to achieve it in an economically and socially reasonable way.

The "Accelerated Restoration" Option

In the last few decades there has been a peculiar concatenation of circumstances such that there has been a commitment to protect and restore marine mammal and bird populations; at the same time there has been continued or more severe overfishing of the fish stocks that support these creatures. So in systems like the Newfoundland shelf and the Georgia Strait, we are seeing an apparently strong trophic "imbalance," with abundant and growing mammal populations combined with depleted fish stocks (Bundy 2001; see also Georgia Strait history in fig. 12.6). Further, we are seeing at least some evidence of "cultivation-depensation" effects, such that previously dominant fish stocks are showing poor recruitment performance after severe depletion that might be due to increases in competitor-predator populations (Swain and Sinclair 2000; Walters and Kitchell 2001).

In such situations, models like Ecosim make very ambiguous predictions about whether letting nature take its course will ever result in the restoration of natural abundance and biodiversity. Instead, the models predict that ecosystems may be "trapped" for relatively long periods in low-diversity, low-abundance "equilibria" even if fishing is stopped entirely. The best empirical example of this possibility is the Newfoundland cod stock, which has not started to recover after a decade of fishery closure. This stock may well be subject to both depensatory predation impacts from marine mammals (seals) and to cultivation-depensation effects.

A possible policy recommendation would then be to engage in more active management of trophic interactions and stock rebuilding, by initially reducing (culling) natural predator and competitor populations. This is very different from the idea that mammalian and bird predators should be culled on a regular and sustained basis for the sake of fishery production or to prevent direct competition with fishers (Yodzis 2001). The idea would be to accelerate declines in predator abundance that may be inevitable anyway on the road to ecosystem recovery, with the full intention to eventually see increased abundances of both the fishes and the predators. We have essentially no empirical, field experience with such transitional policies and would be basing any application of them on calculations/predictions of changes in mortality and population growth rates using various models. Models like Ecosim do, in fact, predict considerably faster ecosystem rebuilding for systems like the Georgia Strait under active culling policies, but these predictions are "guesstimates" at best.

However abhorrent such active intervention policies may be to many scientists and stakeholders, there is one thing that we can agree upon. If such

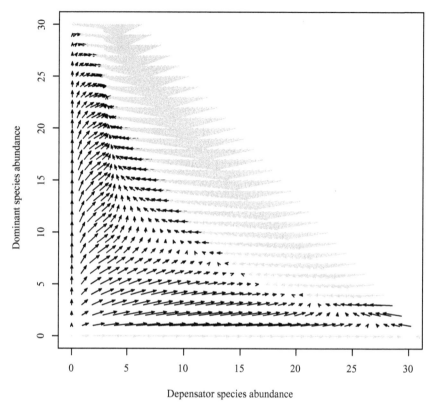

FIGURE 14.1: The dynamic pattern of state-space transitions expected for a system in which a "dominant" species can be impacted at low abundances by a "depensator" species that can, in turn, become abundant only when the dominant declines for some reason. The black arrows represent points in the state space where the dominant predator increases in abundance under conditions of no fishing, and the grey arrows represents points at which the depensator species negatively impacts the dominant species.

policies are not implemented, depensatory predation and competition impacts will certainly delay the ecosystem-recovery process, perhaps for time scales of several decades.

Simple optimization models support the idea that active intervention, in the form of at least occasional culls, might improve long-term fisheries value. For example, figure 14.1 shows the state-space "flow" pattern predicted for a simple model of cultivation-depensation effects, represented by two population models linked through mutual negative-recruitment effects. The "dominant" or large species does best when the "depensator" or small species is rare and is able to keep the depensator species down through predation on its juveniles. But when the dominant species has been reduced for some reason (e.g., overfishing), the depensator species increases enough to cause (e.g., by egg/larval predation) decreases in recruitment of the dominant species such that there can be an uneasy "equilibrium" where the depensator species remains abundant and prevents recovery of the dominant species even if fishing on the dominant is stopped. Figure 14.2 shows estimates obtained by dynamic programming of the optimum stationary harvest policy for the dominant species, as a function of the abundance of both the dominant and the depensator species, and the optimum annual "cull" of the depensator. The objective function for this example was just the discounted sum of catches over time for the dominant, minus a cost factor times the cull of the depensator for each time step. The feedback policy is quite simple to interpret: provided the dominant is abundant, harvest it with a simple feedback control that cuts back sharply on harvesting whenever its abundance starts to drop, and meanwhile, ignore the depensator species (do not cull it). But if the dominant population collapses for any reason, and if the depensator species becomes abundant, then cull it hard for a short time to move the system into a "recovery domain" for the dominant.

Should we ever become capable of computing and visualizing optimum-feedback policies for much more complex ecosystem models, we will very likely see qualitative patterns similar to those in figure 14.2. That is, we will see domains or regions in the state space where the best policy is simple feedback harvest management of valued species. We will see other domains where more extreme "remedial" interventions, including culling and perhaps other measures like stocking and fertilization, would be justified as a means to accelerate a recovery into the "healthy" regions of the state space. Lacking both optimization methods and precise enough predictive models for finding such policy domains with any numerical precision, a key challenge today is to see whether we can at least map out reasonable (and productive) domains for applying alternative management prescriptions.

The "Ecosystem Cultivation" or "Ecosystem Engineering" Option

A common exercise with Ecosim has been to use nonlinear search procedures to seek optimum combinations of fleet sizes for complex, multiple-gear fisheries that target various parts of ecosystems (e.g., see case studies in Pitcher and Cochrane 2002). A particularly interesting pattern typically results when the fishing gears are treated as though they were able to work co-operatively, i.e., to pool all incomes and costs so as to operate in a way that maximizes the total profit summed over all gear types (to operate as though the fishery had a sole owner), and when the total profit is taken to be the only measure of ecosystem-management performance. The fishing-effort policy that maximizes total profit typically "uses" some gears to fish much harder than would be economical for those gears considered in isolation, not only to obtain value but also to deliberately control less valued species so as to increase the economic value of particular species. The

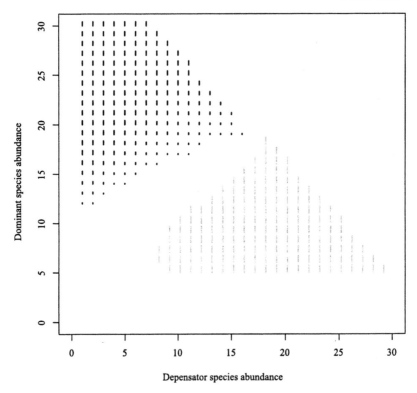

FIGURE 14.2: The stationary optimum policy for the joint harvesting of the dominant species in figure 14.1, along with culling (costly harvest without direct economic benefit) of the depensator species. For each combination of current or initial abundance of the two species, the policy specifies the best harvest to take of the dominant species (black bars) and the best cull to take of the depensator species (grey bars). Note that the state-space transition pattern in figure 14.1 implies that dominant species will recover fairly rapidly after each cull of the depensator, since each cull is assumed to move the depensator state to low abundance.

most dramatic example of this effect has been in models that represent the increase in shrimp production that has accompanied declines in cod stocks in the Atlantic (Bundy 2001; Vasconcellos et al. 2002). In some areas, shrimp fisheries are apparently now more valuable than the cod fisheries that they have replaced by "fishing down the food chain" (Pauly et al. 2000), and under the simple total profit objective, Ecosim optimization suggests that this replacement or deliberate "farming" for shrimp should be encouraged on a sustainable basis. Other analysts have also begun to question whether the usual single-species rebuilding objective for cod is really the best long-term option (Jakobsson and Stefansson 1998).

There is a long human tradition of converting complex terrestrial food webs into simpler, productive food chains. We call this "agriculture," and most people have at least some faith that human ingenuity can keep up with the various problems (e.g., unexpected pest outbreaks) that such simplification is bound to create. So it is not surprising that there is commonly support from fishing interests for doing the same thing in aquatic environments, through cultivation practices ranging from predator control to the deliberate alteration of seabeds to fertilization and seeding through stocking programs. When we use simple economic objective functions in ecosystem models, we cannot, in fact, generally demonstrate that it would be impossible or unwise to follow the terrestrial lead.

Despite what has been found with models to date, here are two main counterarguments to the idea that we can safely treat marine ecosystems like agricultural systems.

Risk spreading and variance reduction. Farmers can respond quickly to variations in the price and productivity of alternative crop species by planting different mixes, so as to spread risks and reduce variance in income. Indeed, one of the best ways to be a successful fisher or fishing company is to develop a diversified portfolio of licenses and fishing tactics. But once a natural aquatic ecosystem has been deliberately simplified in any one direction, we cannot just "plant" or quickly restore species that have been deliberately reduced in abundance; natural population recovery dynamics typically occur on time scales of decades. That is, mistakes are much more costly to correct than in systems with an annual production cycle and much opportunity for spatial experimentation and risk spreading via tactics like crop mixes and crop rotation.

Escalating control costs. It is quite possible that every ecosystem has natural "vampires in the basement," species that are normally so rare as to go unnoticed (and not modeled) due to trophic interactions with the normal dominant species, but are capable of an explosive and costly increase following ecosystem simplification. For example, the very simple "ecosystems" represented by fish and shrimp farming have been plagued by previously rare disease organisms. It is not that such "vampires" cannot be controlled as they are discovered (people are very inventive about responding to such threats) but, rather, that control costs may escalate so much as to make the cultivation policy uneconomical. Should such a point be reached, management is in big trouble, because of erosion in the capacity to reverse the simplification over a reasonable time.

Notice that neither of these arguments is a claim about the certitude of ecological "failure" for cultivation policies. In view of our lack of empirical experience with highly simplified marine ecosystems, any such claim would be foolish. Rather, the arguments are about prudent economic risk management, beyond the simple notion of managing risk just by being "precautionary." We remind the reader that there is as yet absolutely no standard of, or guidelines for, prudence in public decision-making and resource management in general, let alone in fisheries.

The "Balanced Values" Option

When we have used ecosystem models in multiplayer "gaming" situations, in which the players represent a diversity of consumptive and nonconsumptive interests, it is not common for the players or interest groups to demand any extreme restoration or cultivation policy. Rather, we typically see a sense of "fair play" from most actors, with some willingness to seek compromise policy options that maintain a diversity of values (and options for the future). The simulated ecosystem configurations that result from such options are often far from natural in terms of biodiversity and population structure and, in effect, represent new visions of future ecosystems. By "new" here we mean apparently sustainable configurations of abundance and production for which there is no historical experience or precedent; our historical experience for most ecosystems consists mainly of anecdotal information about natural states, along with much data about transient change under progressively more severe fishing impacts.

While these gaming exercises have involved purely scientific players to date, presumably it is possible in most real management jurisdictions to develop institutional processes for consensus-building or adaptive policy design that would lead to similar agreements among real stakeholders about the desirable "mix" of future ecosystem values, and about how to design monitoring and regulatory structures to meet various needs and objectives in trying to move toward such mixes. This is, in fact, where we seem to be heading in systems like the Bering Sea, where complicated public review and judicial processes are gradually leading to consensus about long-term objectives and experimental management plans for meeting many existence values and fishery objectives (NRC 2002).

As such consensus-building processes are developed, probably the most difficult things for most participants to understand will be the multiple time scales of ecological response (why desirable states cannot be achieved immediately), and the tactics for resolving critical uncertainties through large-scale management experiments. Modeling can help with the first of these difficulties and can help expose both the most critical uncertainties and the inevitability of learning about these the hard way whether or not there is careful planning of management "treatments" so as to learn which ones do not work as quickly as possible. But there will doubtless be much misunderstanding, both about how much to trust the models and about how much to fear them.

The "Bionomic Equilibrium" Option

Under this option, we would allow economic investment/disinvestment and fishing decision processes to play out under pure, myopic economic responses to changes in profitability. This option is not, in fact, likely to lead to the complete ecological destruction of all fishery values (because it would in general not be profitable to chase down the last few of all harvested species) but is, instead, likely to lead to "uneasy" bionomic equilibrium with at least some fisheries persisting through the economic feedback effects on harvest rates of declining profits and increasing costs as target species become rare. The equilibrium is quite likely to involve "fishing down the food chain," particularly if lower trophic–level species (like shrimp) are more productive and valuable than long-lived piscivores. Further, any such equilibrium would be uneasy in the sense that it would be sensitive to innovations in technology and market demand, with a general tendency toward progressively more severe "sustained overfishing" with improvements in technology and growth in demand for fish products.

Ecosim runs with simulated fishing fleet and effort dynamics, but no simulated regulatory response to changes in fishing mortality rates, warn us that the single biggest risk in counting on bionomic factors to prevent wholesale overfishing is probable density dependence in the catchability coefficients. If catchabilities (proportions of stocks taken per unit of fishing effort) increase greatly at low stock sizes, as typically occurs when efficient gears use nonrandom search tactics to target shrinking aggregations of fish, then the bionomic "signals" of stock declines are lost even for the dominant or "target" fish stocks.

There is much intuitive appeal in the idea of using the "invisible" hand of fisher decision-making as a substitute for complex and often ineffective regulatory and allocation processes. The resulting bionomic equilibria could be made more productive and stable by using some relatively simple economic controls to prevent strong density dependence in the catchability coefficients. Such controls include (1) removal of perverse incentives for investment and continued fishing at low stock levels that have been created in many major fisheries through public subsidy programs, (2) direct increase in the costs of fishing through license and landing taxes, (3) limitations on fishing technology to prevent fishers from maintaining high catch rates at low stock sizes (e.g., technologies for increasing the search efficiency), and (4) permanent closures of fishing grounds that are more costly to fish so as to prevent fishers from moving onto these grounds when/if costs are reduced through technological innovation. Indeed, the successful application of such controls would go a long way toward meeting the claim of some scientists and managers that all we really need for successful ecosystem management is better management of individual fisheries.

The fundamental problem with this approach is not that it would be impossible to implement controls leading to a productive and stable bionomic equilbrium but, rather, that the myopic decision-making of individual fishers and fleets would fail to account for the very real indirect costs that each fishery might cause for other fisheries, and for other public values like marine mammals and birds. In particular, "bulk" (low-value) fisheries targeted on abundant, lower trophic–level species would in most cases appropriate enough of the productivity of these species to severely reduce the capacity of the ecosystem to produce value from fisheries directed at higher trophic levels. An interesting and completely open question is whether it would be possible (and wise) to implement policies for "internalizing" such hidden costs by requiring fishers to directly compensate other fishers (and other stakeholders) for estimated losses in production and abundance.

14.2 MOVING TOWARD SUSTAINABLE ECOSYSTEM MANAGEMENT

We cannot claim to know what all the ingredients and steps should be for sustainable ecosystem management, nor is it even clear what vision of long-term ecosystem configuration should motivate and shape that management. But we can at least present a list of things that we believe should be priority items for management agencies that are pretending to at least move in the direction of ecosystem management, in some more substantive way than just claiming to protect fish habitats and fisheries values. There are two kinds of items in the following wish list: changes in how fisheries are regulated, and changes in the way information is gathered to detect and correct the inevitable mistakes that will be made in making regulatory changes.

Develop clear ecosystem-scale exploitation-rate goals and contingency plans for varying exploitation rates during periods of extreme ecosystem states (target and limit reference points).

There are very few marine ecosystems or management areas for which there has even been a simple compilation or list of target exploitation rates for the major fish stocks, let alone any assurance that such target rates have been estimated with models that give reasonable consideration to trophic interaction (food supply and predation impact) effects. There is no case in which scientists have gone beyond such a list to identify the domains of ecosystem states (like fig. 14.2) for which it might be prudent to alter harvest policies radically so as to encourage system recovery. That is, there are no ecosystem-scale feedback policies or decision rules. Without a commitment to such rules, there is no way to judge objectively whether management agencies are achieving regulatory objectives or to prioritize objectively the needs for regulatory changes. It does not really matter whether we can put the "right" numbers into such policy rules; in fact, we can be absolutely certain that whatever numbers we do develop will be wrong to at least some degree. But in the long term, adaptive processes for improving management will surely fail if we cannot base our learning on clear commitments to trial policies and clear expectations about what to expect if such trial policies are or are not working.

Develop systems of fishing rights that create incentives for fishers to co-operate in information-gathering and regulation.

To judge whether even single species, let alone ecosystem, management is working, we need much more detailed spatial, temporal, and multispecies data than is now being gathered through fishery independent survey programs and through information capture from fishing activities. Further, under most regulation systems there are strong disincentives for fishers to report their impacts, particularly captures of tagged fish and bycatches of nontarget and protected species. Where rights like transferable quotas have been implemented, so fishers see the value in protecting those rights (e.g., a quota as a source of retirement income), there has been a radical change in willingness to invest and cooperate in scientific information-gathering. For a wonderfully witty example of this effect, see Wickham's (2002) discussion of the British Columbia black cod (sablefish) fishery. Probably fishing rights should not be denominated in terms of catch quotas for most fisheries, due to the high stock-assessment costs and risks associated with output control of harvesting. But surely we can devise other bundles of fishing rights that offer security and the promise of future value, without the pretense of precise quantitative quotas for the future.

Encourage novel approaches to meeting exploitation-rate and ecosystem-recovery goals through practices other than annual fishing limits.

Particularly where marine habitats have been severely degraded by past fishing practices, we have little experience upon which to base any predictions and recommendations for system recovery. Further, it may be impractical or unnecessary to fully protect such areas, and we may, instead, learn much and eventually develop sustainable fisheries by abandoning simplistic views about fishing as an annual, extensive process. For example, a strategic alternative would be to use ideas like crop rotation and fallow periods from agriculture (Myers et al. 2000), which would provide for both habitat-recovery dynamics and useful experimental contrasts for scientific study. Such novel approaches to management should be encouraged and tested experimentally whenever practical.

Develop a concerted international attack through both governments and NGOs on the most destructive, overcapitalized, and subsidized fishing fleets.

Much of the destructive fishing worldwide is being done by a few fleets of large vessels operating with the blessing of and even subsidy from home nations, that were developed over the past few decades during a heyday of investment in "bigger is better" fishing technology. These fleets haunt the continental shelf margins and high seas off every nation's waters and have been notoriously difficult to monitor and regulate. They will not be stopped until there is a concerted international effort to do so, backed by the threats of market sanctions that NGOs have proven effective at providing.

Develop objective approaches to the evaluation of habitat-management options.

The practice of habitat "management" in fisheries has largely been driven by simplistic assumptions that "natural is always better," implying that management is equivalent to protection and restoration. But when we look objectively at fish abundance and production data from disturbed versus natural habitats, it is not uncommon to see at least some fish species thrive in disturbed areas (coho salmon in logged watersheds in the Pacific North-west, snook in disturbed coastal estuaries in Florida, trout in the Colorado River in the Grand Canyon). Further, there are cases in which we have even deliberately introduced fertilization programs to restore the productivity of systems that were previously "cleaned up" through the control of anthro-pogenic nutrient-loading sources (e.g., Kootenay Lake, British Columbia, Ashley et al. 1997). Too often, the presumption of need for fish protec-tion has been used as one reason or excuse for habitat protection in general, and this can in essence represent a dishonest argument for protection (fish should not be used as an excuse for protection when the economic benefits of protection would otherwise be unclear). If we are honest about evalu-ating both potential benefits and damages to fish, we will very likely find many win-win options for working with other resource interest groups (e.g., forest-harvesting interests) to mutual benefit.

Reverse information-gathering and regulatory priorities for new fisheries that are developing in the ashes of historical collapses, using ideas of proven production potential and titration experiments.

The traditional view of priorities for developing fisheries has been to concentrate first on gathering simple catch and effort data (for surplus-production assessments) while encouraging economic development, followed by the collection of more detailed biological information for size-age structured assessments and restriction of harvesting (Hilborn and Walters 1992; Perry et al. 1999). A very different view is that management agencies should "front-end load" intensive research on developing fisheries, especially on spatial-distribution patterns and biological characteristics that are good predictors of sustainable exploitation rates (growth, vulnerability, and nat-ural mortality patterns), while severely restricting harvest development to "proven production potentials" (Walters 1998, 1998a). In this alternative view, the spatial development pattern for each new fishery would be care-fully managed as a "titration experiment," with large areas initially closed to fishing so as to provide both a risk-management buffer and scientific refer-ence information about fishing impacts. Such areas would be progressively opened to fishing if/as sustainable-harvest policies (without unacceptable im-pacts on nontarget species) are demonstrated to be possible in the initial open areas.

Radically improve monitoring and information capture from fishing oper-ations, through Vessel Monitoring Systems (VMS) and on-board monitoring devices.

Modern fishing fleets cover staggering areas of the world's surface in search of fish. Future sustainable fisheries are very likely to involve both much more complex monitoring programs for harvest and nontarget species and also more complex space-time regulation schemes, e.g., marine protected areas. With modern information technology, there is an opportunity both to mon-itor such activities more closely (for enforcement purposes) and to capture vastly more information about both fishing activities and the ocean through which the fishers search. Imagine a VMS (Vessel Monitoring System) "black box" that records spatially referenced inputs ranging from acoustic scattering patterns (plankton and small pelagic fish data) to data from oceanographic sensors (temperature, salinity) to video images of fish net openings and fish passing through sorting equipment. Given such information from a fleet of fishing vessels, along with "hole-filling" scientific surveys for areas not fished, it should be possible to develop radically better spatial and temporal maps of marine ecosystem structure and change.

Create strong economic incentives for fishers to develop selective fishing practices.

Most fishing directly impacts multiple stocks with widely varying productivities. It does not take much frustrating modeling and analysis (e.g., Walters and Bonfil 1999) to show that such fishing is inherently wasteful and creates trade-offs that are nearly impossible to manage. Either less productive stocks are overfished or productive potential is "wasted" by not fishing to the levels that more productive stocks can withstand. There has been much scientific research on how to make fishing more selective and how to avoid bycatch. But in the end, it has been mainly inventive fishers faced with possible regulatory and market sanctions who have come up with effective means to reduce the bycatch (e.g., to avoid catching dolphins in tuna purse-seines or turtles in shrimp trawls). There should be major programs to further harness this inventiveness by making it progressively more costly to fish in wasteful and nonselective ways. The simplest incentive system would be to charge fishers for the value of the creatures that they capture but are not allowed to market. Given the high value of some of the creatures that now go to waste, like steelhead trout taken in the Skeena River gillnet fishery for sockeye salmon in British Columbia, it is a safe bet that being charged for the waste of these creatures would very quickly lead to the invention of ways to avoid such costs. An alternative, cruder strategy would be to close "dirty" fisheries and reopen the grounds to any fishing-license holder willing to demonstrate the use of a more selective gear or practice.

Make coordinated, international investments in the expensive research needed to develop improved tagging technologies for direct monitoring of exploitation rates, using in-situ tagging and tags that can be automatically detected on vessels and in ports.

For many ecosystems it will be uneconomical to invest in all the data-gathering needed to develop even good single-species assessments for all potentially important species. Further, for reasons discussed earlier in this book, the various methods for estimating the exploitation rates and fishing impacts from alternative data sources (abundance trend and composition sampling) are prone to failure, especially during stock declines. For both these reasons, there is a desperate need in fisheries for relatively inexpensive ways to directly monitor exploitation rates, presumably by monitoring the rates of recovery of tags from tagged cohorts of "known" (up to tag loss and tag-induced mortality) size. Tagging approaches have been plagued by difficulties in applying tags without causing mortality, hence the need for in-situ tagging methods. They have also been plagued, especially for recreational fisheries, by difficulties in recovering tags from fishers due to obvious incentives for nonreporting. To solve this second problem, we need tags that report themselves. Imagine, e.g., a tag the size of a grain of rice that has a small radio receiver and a burst transmitter, with a long enough battery life so that it could transmit its tag code just once (after years at sea) to a black box when its receiver is interrogated by a small transmitter in that box. Such tags could be applied to a wide variety of fish, and the black-box interrogation systems could be placed in locations ranging from fishing vessels to boat docks and ramps. They would quickly replace coded-wire and PIT tags, especially for monitoring the survival and harvest of smaller fish, due to the much lower cost of sampling-tag recaptures. Technologies like this could be developed, but the development costs would be prohibitive for any single management agency.

Make management agency staff personally accountable for all of these things, with extensive use of independent public review and auditing processes to detect nonperformance.

In sharp contrast to other professional disciplines, fisheries scientists and managers in public agencies are almost never held accountable for the consequences of their assessments and recommendations. Indeed, it is not far off the mark to say that one of the best ways to get promoted is to really foul up a fishery, i.e., to say that people who make the worst mistakes are the ones most likely to be promoted to positions of higher responsibility. This sad state of affairs is not going to change just by making people accountable or liable on paper for their mistakes. There is a need for both the statutory recognition of accountability and the development of more effective public review, "watchdog," and performance-auditing systems that bring independent and objective expertise to bear on the complicated tangles of data and paperwork that typically surround fisheries decision-making. Such review processes are needed not only to make public employees accountable for their mistakes but also to identify and correct poor policy choices and mistakes as quickly as possible.

Definitions for Mathematical Symbols

α	Intercept of Ford-Brody growth model used in delay-difference models; also used by Myers and colleagues as recruitment curve slope parameter
a	Rate of effective search in predation models (area or volume searched per unit predator abundance per time); also used as subscript index for age
$\acute{a}$	Power parameter in stock-recruitment power model
A	Total area over which interactions (predation, fishing) are distributed
b	Density-dependence parameter in stock-recruitment models; also used for biomass per unit area when biomass B is discussed as $B = Ab$, A = area
B	Biomass-state variable, with subscripts dependent on context (subscripts: t for time, s for biomass per school, i for species or biomass pool in ecosystem models)
β	Power parameter representing hyperstability/hyperdepletion in relative-abundance indices
c_t	Possibly time-varying relative carrying-capacity parameter in recruitment models
C_t	Catch, as number or biomass of individuals removed from a population over a defined time
C	Large-scale mean food concentration outside foraging arenas (in areas not accessible to feeding juveniles
δ_t	An unexplained deviation or "process error" from some model prediction, caused by unexplained system dynamics rather than observation/measurement error
Δ	A small increment (in numbers, biomass, or time, depending on context)
D	Proportion of assimilated intake lost to specific dynamic action in bioenergetics models
D_t	Total number of deaths over a defined interval
DC_{ij}	Diet composition or proportion of prey i in the diet of predator j in Ecopath models
E_t	Total fishing effort over some defined period, e.g., one year

ϵ_t	Total egg production in one spawning season, summed over all spawning females
f	Individual mean fecundity or egg production; also used in effort models to denote the instantaneous number of vessels fishing at any moment
F, F_t	Instantaneous fishing mortality rate, ratio of catch over some defined period (usually one year) to biomass at risk to catch, if biomass were held constant over the period
h	Handling time per prey item in type II functional response equations
H	Handling-time factor in predation-rate equations; $1/H$ is the proportion of time spent searching for prey
k	Used for various scaling constants in mathematical expressions; definition depends on context
K	Metabolic parameter in von Bertalanffy growth model $l_a = l_\infty(1 - e^{-Ka})$, l_a = length at age a
K	Carrying-capacity parameter (maximum equilibrium population size absent harvest) in logistic population growth/production models; also used to represent cumulative catch in simple population depletion models
κ	Recruitment compensation parameter: the ratio of the slope of the stock-recruitment curve at very low stock size to the slope of a line passing through the mean recruitment at unfished stock size; also the ratio of maximum juvenile survival rate at low densities to juvenile survival rate in an unfished population
l_a	Body length at age a (usually years)
lx_a	Survivorship: the probability of surviving from age 1 to age a ($lx_1 = 1$)
l_∞	Asymptotic body length in von Bertalanffy growth model
$logL$	Log likelihood function in statistical estimation
M	Instantaneous natural mortality rate or rate component, measured as the number or biomass of individuals lost to the rate divided by the mean number at risk to that loss
MSY	Maximum sustainable yield as an average annual value from a population managed so as to maximize annual surplus production

mw	Weighting factor used to set proportions of animals moving to different cells (and staying in same cell) in spatial-movement models
n_s	Number of schools in a population of schooling fish (effort dynamics models), number of observations (statistical likelihood functions)
$N, N_t, N_{a,t}$	Number of individuals in a defined population or population age-size class
NA	Number of prey attacked per unit time in predation functional response derivations
$\widehat{NA}$	Number of aggregations or schools of fish encountered
p	A proportion or probability, e.g., p_i is a proportion of type i individuals, $p(x)$ is a probability density function for x
P	Instantaneous predation mortality rate (risk) in foraging arena models, or price in profitability calculations
$P(i|j)$	State transition probability matrix; probability of moving from state j after harvest one year to state i before harvest the next year
PB	Production per biomass in Ecopath models
pr	Profitability of fishing (price × catch per effort - cost of a unit effort), either as a local value or the average over fishing opportunities that would be expected under Ideal Free Distribution (IFD) conditions
q	Catchability coefficient: the proportion of stock taken by one unit of fishing effort, or the proportion of stock observed by one unit of survey effort
Q	Total quantity harvested or eaten by predators per unit time, as a quota in fisheries yield calculations or predator consumption in trophic interaction calculations
qb	Per-capita or per-biomass food-consumption rate Q/B; called QB in Ecopath models
r	Relative rate of population growth; intrinsic or maximum rate in logistic growth models, or annual ratio next year to this year's population size in simple balance models
R_t	Total number of recruits entering a population over a defined period

s_0 — Slope of stock-recruitment curve; interpreted as the survival rate from egg to recruitment if stock is measured as total egg production, or as a product of survival times fecundity if stock is measured as numbers or biomass of adults

$s_{(j)}, s_{(a)}$ — Annual survival rates of juvenile (first year) and adult (one-year-old and older) individuals in simple population balance models

s — Survival rate (proportion of individuals surviving some defined period of mortality, sometimes designated S if no confusion would result about the common use of S as spawning-stock biomass; lowercase s also used to designate log of spawning stock in stock-recruitment power-model fitting, chapter 7

S — Spawning biomass in stock-recruitment equations

τ — Relative time spent foraging and at risk to predation in foraging arena predation models; scaled so that $\tau = 1$ when predation variables (rates, abundances) are at input base levels

u, U — Exploitation rate: proportion of individuals harvested over some defined period of harvest

v_t — Observation or measurement error in stock-assessment models

v_a — Relative-vulnerability parameter in discrete time models (e.g., v_a denotes the proportion of age-a animals vulnerable to harvest in some year)

$v, \acute{v}$ — Instantaneous vulnerability exchange rates for the movement of organisms between invulnerable and vulnerable states (predation interactions, vulnerability to fishing)

ϕ_x — Botsford incidence function representing the sum of survivorships to age times x at age; function used to sum egg productions, fecundities, biomass, etc.

w_i — Relative value weight placed on value component i in a utility or value function; as w_a, used to denote mean body weight of age-a individuals

V — A measure of the total value from a fishery (profit, weighted sum of utility components, etc.). Expected value (over uncertain possible values) denoted by EV

V	Density of prey vulnerable to predation or fishing at any moment in foraging arena theory and effort-response models
ξ	A standardized environmental index, e.g., ocean temperature; standardized by subtracting mean and dividing by standard deviation of the environmental data series
Y	Index of abundance that hopefully varies as qN, where q is catchability and N is absolute abundance
z	Log of ratio of index value Y to state X being indexed, i.e., $z = ln(Y/X)$ for lognormal observation models
Z	Instantaneous total mortality rate, summed over fishing (F) and natural mortality (M) components

Abrahams, M. 1989. Foraging guppies and the ideal free distribution: the influence of information on patch choice. *Ethology* 82:116–126.

Abrams, P. 1984. Foraging time optimization and interactions in food webs. *Am. Nat.* 124:80–96.

Abrams, P. 1993. Why predation rates should not be proportional to predator density. *Ecology* 74:726–733.

Akaike, M. 1973. Information theory and an extension of the maximum likelihood principle. In B. Petrov and F. Csaki, eds., *Proc. 2d International Symposium on Information Theory*, pp. 267–281. Budapest, Hungary: Akademia Kiado.

Allen, K. 1973. The influence of random fluctuations in the stock-recruitment relationships on the economic return from salmon fisheries. In B. Parrish, ed., *Fish stocks and recruitment* volume 164, pp. 350–359. Rapp. Et Proces-Verbaux Reun. Cons. Int. Explor Mer.

Allen, M., and W. I. Pine. 2000. Detecting fish population responses to a minimum length limit: effects of variable recruitment and duration of evaluation. *N. Am. J. Fish. Man.* 20:672–682.

Anderson, E., ed. 2002. *The Raymond J. H. Beverton lectures at Woods Hole, Massachusetts. Three lectures on fisheries science given May 2–3, 1994* volume 54. U.S. Dept. Commer. NOAA Tech. Memo NMFS-F/SPO.

Anderson, J., and G. Rose. 2001. Offshore spawning and year-class strength of the northern cod (2J3KL) during the fishing moratorium, 1994–1996. *Can. J. Fish. Aquat. Sci.* 58, 1386–1394.

Anderson, L. 2000. The effects of ITQ implementation: a dynamic approach. *Nat. Res. Modelling* 13, 435–470.

Anderson, P., and J. Piatt. 1999. Community reorganization in the Gulf of Alaska following ocean climate regime shift. *Mar. Ecol. Prog. Ser.* 189:117–123.

Anholt, B., and E. Werner. 1998. Predictable changes in predation mortality as a consequence of changes in food availability and predation risk. *Evolutionary Ecology* 12, 729–738.

Argue, A., R. Hilborn, R. Peterman, M. Staley, and C. Walters. 1983. The Strait of Georgia chinook and coho fisheries. *Can. Bull. Fish. Aq. Sci.* 211, 91 pp.

Ashley, K., L. Thompson, D. Lasenby, L. Mceachern, K. Smokorowski, and D. Sebastian. 1997. Restoration of an interior lake ecosystem: the Kootenay Lake fertilization experiment. *Water Qual. Res. J. Can.* 32:295–323.

Ault, J. S., J. Luo, S. G. Smith, J. E. Serafy, J. D. Wang, R. Humston, and G. A. Diaz. 1999. A spatial dynamic multistock production model. *Can. J. Fish. Aquat. Sci.* 56 (suppl. 1):4–25.

Ault, J., J. Bohnsack, and G. Meester. 1998. A retrospective (1979–1996) multispecies assessment of coral reef fish stocks in the Florida Keys. *Fish. Bull.* 96:395–414.

Ault, J., J. Luo, and J. Wang. 2003. A spatial ecosystem model to assess spot-ted seatrout population risks from exploitation and environmental change. In S. Bortone, ed., *Biology of the spotted seatrout*, 267–300. New York: CRC Press.

Ault, J., and D. Olson. 1996. A multicohort stock production model. *Trans. Am. Fish. Soc.* 125:343–363.

Baird, J., C. Bishop, W. Brodie, and E. Murphy (1992). An assessment of the cod stock in NAFO divisions 2J3KL. *CAFSAC Res. Doc.* 92/75, 76.

Baretta-Becker, J., and J. Baretta, eds. 1997. *Special issue: European Regional Seas Ecosystem Model II*, volume 38. *J. Sea Res.*

Barrowman, N. J., and R. Myers. 2000. Still more spawner-recruitment curves: the hockey stick and its generalizations. *Can. J. Fish. Aquat. Sci.* 57:665–676.

Bartell, S., W. Cale, R. O'Neill, and R. Gardner. 1988. Aggregation error: research objectives and relevant model structure. *Ecol. Modelling* 41:157–168.

Bartell, S., G. Lefebvre, G. Kaminski, M. Carreau, and K. Campbell. 1999. An ecosystem model for assessing ecological risks in Quebec rivers, lakes, and reservoirs. *Ecol. Modelling* 124:43–67.

Bax, N. 1998. The significance and prediction of predation in marine fisheries. *ICES J. Mar. Sci.* 55:997–1030.

Beauchamp, D., C. Baldwin, J. Vogel, and C. Gubala. 1999. Estimating diel, depth-specific foraging with a visual encounter rate model for pelagic piscivores. *Can. J. Fish. Aquat. Sci.* 56 (supplement 1):128–139.

Bellman, R. 1957. *Dynamic programming.* Princeton: Princeton Univ. Press.

Berenboim, B., A. Dolgov, V. Korzhev, and N. Yaragina. 2000. The impact of cod on the dynamics of Barents Sea shrimp (*Pandalus borealis*) as determined by multispecies models. *J. Northw. Atl. Fish. Sci.* 27:69–75.

Berryman, A. 1992. The origins and evolution of predator-prey theory. *Ecology* 73:1530–1535.

Bertignac, M., P. Lehodey, and J. Hampton. 1998. A spatial population dynamics simulation model of tropical tunas using a habitat index based on environ-mental parameters. *Fish. Oceanogr.* 7:326–334.

Beverton, R., and S. Holt. 1957. *On the dynamics of exploited fish populations.* London: Chapman and Hall.

Beverton, R., 1992. Patterns of reproductive strategy parameters in some marine teleost fishes. *J. Fish. Biol.* 41 (supplement B):137–160.

Beyer, J. 1989. Recruitment stability and survival-simple size-specific theory with examples from the early life dynamics of marine fish. *Dana* 7:45–157.

Blanchard, J., K. Frank, and J. Simon. 2003. Effects of condition on fecundity and total egg production of eastern Scotian Shelf haddock (*Melanogrammus aeglefinus*). *Can. J. Fish. Aquat. Sci.* 60:321–332.

Bodini, A. 1998. Representing ecosystem structure through signed digraphs. Model reconstruction, qualitative predictions and management: The case of a freshwater ecosystem. *Oikos* 83:93–106.

Bodini, A. 2000. Reconstructing trophic interactions as a tool for understanding and managing ecosystems: application to a shallow eutrophic lake. *Can. J. Fish. Aquat. Sci.* 57:1999–2009.

Borgstroem, R., J. Heggenes, and T. Northcote. 1993. Regular, cyclic oscillations in cohort strength in an allopatric population of brown trout, *Salmo trutta. L. Ecol. Freshwat. Fish.* 2:8–15.

Botsford, L. W., and R. Hobbs. 1995. Recent advances in the understanding of cyclic behavior of Dungeness crab (*Cancer magister*) populations. *ICES Mar. Sci. Symp.* 199:157–166.

Botsford, L. W., and D. E. Wickham. 1979. Population cycles casued by inter-age, density-dependent mortality in young fish and crustaceans. In E. Naybr and R. Hartnoll, eds., *Cyclic phenomena in marine plants and animals,* 73–82. Permagon, New York: Proceedings of 13th European Marine Biology Symposium.

Botsford, L., and W. Johnston. 1983. Effort dynamics of the northern California Dungeness crab (*Cancer magister*) fishery. *Can. J. Fish. Aquat. Sci.* 40:337–346.

Botsford, L. 1981a. Optimal fishery policy for size-specific density-dependent population models. *J. Math. Biol.* 12:265–293.

———. 1981b. The effects of increased individual growth rates on depressed population size. *Am. Nat.* 117:38–63.

Bradford, M., R. Myers, and J. Irvine. 2000. Reference points for coho salmon (*Oncorhynchus kisutch*) harvest rates and escapement goals based on freshwater production. *Can. J. Fish. Aquat. Sci.* 57:677–686.

Brett, M., and C. Goldman. 1996. A meta-analysis of the freshwater trophic cascade. *Proc. Nat. Acad. Sci. USA* 93:7723–7726.

Brown, C., and R. Day. 2002. The future of stock enhancements: lessons for hatchery practice from conservation biology. *Fish and Fisheries* 3:79–94.

Bundy, A.. 2001. Fishing on ecosystems: the interplay of fishing and predation in Newfoundland-Labrador. *Can. J. Fish. Aquat. Sci.* 58:1153–1167.

Burnham, K., and D. Anderson. 1998. *Model selection and inference: a practical information-theoretic approach.* New York: Springer-Verlag.

Butterworth, D. S., K. L. Cochrane, and J.A.A. DeOliviera. 1997. Management procedures: a better way to manage fisheries? The South African experience. *Am. Fish. Soc. Symp.* 20:83–90.

Caddy, J., and J. Gulland. 1983. Historical patterns of fish stocks. *Mar. Policy* 7:267–278.

Caddy, J., and J. Seijo. 1998. Application of a spatial model to explore rotating harvest strategies for sedentary species. *Can. Sp. Publ. Fish. Aquat. Sci.* 125:359–365.

Caddy, J., 1975. Spatial model for an exploited shellfish population, and its application to the Georges Bank scallop fishery. *J. Fish. Res. Bd. Canada* 32:1305–1328.

Campbell, R. A., A.D.M. Smith, and B. D. Mapstone. 2001. Evaluating large-scale experimental designs for management of coral trout on the Great Barrier Reef. *Ecol. Appl.* 11:1763–1777.

Caputi, N. 1988. Factors affecting the time series bias in stock-recruitment relationships and the interaction between time series and measurement error bias. *Can. J. Fish. Aquat. Sci.* 45:178–184.

Carpenter, S. R., and J. Kitchell. 1993. *The trophic cascade in lakes.* Cambridge: Cambridge University Press.

Carpenter, S. 2001. Alternate states of ecosystems: evidence and its implications. In M. Press, N. Huntly, and S. Levin, eds., *Ecology: Achievement and Challenge*, 357–383. London: Blackwell.

Carpenter, S. 2002. Ecological futures: building an ecology of the long now. *Ecology* (in press).

Carrick, N. 1988. The role of the industry and fleet manipulation studies in the Spencer Gulf prawn fishery. *SAFISH* 12:4–9.

Carr, M., T. Anderson, and M. Hixon. 2002. Biodiversity, population regulation, and the stability of coral-reef fish communities. *Proc. Nat. Acad. Sci. USA* 99:11241–11245.

Carscadden, J., K. Frank, and W. Leggett. 2001. Ecosystem changes and the effects on capelin (*Mallotus villosus*), a major forage species. *Can. J. Fish. Aquat. Sci.* 58:73–85.

Castleberry, D. T., J. Cech Jr., D. C. Erman, D. Hankin, M. Healey, G. M. Kondolf, M. Mangel, M. Mohr, P. B. Moyle, J. Nielsen, T. P. Speed, and J. C. Williams 1996. Uncertainty and instream flow standards. *Fisheries* 21(8):20–21.

Cave, J., and W. Gazey. 1994. A preseason simulation model for fisheries on Fraser River sockeye salmon (*Oncorhynchus nerka*). *Can. J. Fish. Aquat. Sci.* 51:1535–1549.

Charnov, E. 1996. *Life history invariants: some explorations of symmetry in evolutionary ecology.* Oxford: Oxford Univ. Press.

Christensen, V., and D. Pauly. 1992. ECOPATH II—a software for balancing steady state models and calculating network characteristics. *Ecol. Modelling* 61:169–185.

Clark, C. 1976. *Mathematical bioeconomics: the optimal management of renewable resources.* New York: John Wiley.

———. 1985. *Bioeconomic modeling and fisheries management.* New York: John Wiley.

Claytor, R., and J. Allard. 2001. Properties of abundance indices obtained from acoustic data collected by inshore herring gillnet boats. *Can. J. Fish. Aquat. Sci.* 58:2502–2512.

Claytor, R. 1996. In-season management of Atlantic salmon (*Salmo salar*): an example using southern Gulf of St. Lawrence rivers. *Can. J. Fish. Aquat. Sci.* 53:1535–1549.

Cochrane, K. 2000. Reconciling sustainability, economic efficiency and equity in fisheries: the one that got away? *Fish and Fisheries* 1:3–21.

Collie, J., and H. Gislason. 2001. Biological reference points for fish stocks in a multispecies context. *Can. J. Fish. Aquat. Sci.* 58:2167–2176.

Collie, J., R. Peterman, and C. J. Walters. 1990. Experimental harvest policies for a mixed-stock fishery: Fraser River sockeye salmon *Oncorhynchus nerka*. *Can. J. Fish. Aquat. Sci.* 47:145–155.

Collie, J., and P. Spencer. 1993. Management strategies for fish populations subject to long-term environmental variability and depensatory predation. In *Proc. Internat. Symp. on management strategies for exploited fish popuations, Oct. 21–24, 1992, Anchorage Alaska.*, 629–650. Alaska Sea Grant Program.

Conover, D., and S. Munch. 2002. Sustaining fisheries yields over evolutionary time scales. *Science* 297:94–96.

Coronado, C., and R. Hilborn. 1998. Spatial and temporal factors affecting survival in coho salmon (*Oncorhynchus kisutch*) in the Pacific Northwest. *Can. J. Fish. Aquat. Sci.* 55:2067–2077.

Cosner, C., D. DeAngelis, J. Ault, and D. Olson. 1999. Effects of spatial grouping on the functional response of predators. *Theor. Pop. Biology* 56:65–75.

Cox, S. P., T. E. Essington, J. F. Kitchell, S.J.D. Martell, C. J. Walters, C. Boggs, and I. Kaplan. 2002. Reconstructing ecosystem dynamics in the central Pacific Ocean, 1952–1998. II. A preliminary assessment of the trophic impacts of fishing and effects on tuna dynamics. *Can. J. Fish. Aquat. Sci.* 59:1736–1747.

Cox, S., T. Beard, and C. Walters. 2002. Harvest control in open-access sport fisheries: hot rod or asleep at the reel? In *Bull. Mar. Sci. 3rd Mote Symp. Issue* (in press).

Cox, S., and C. Walters. 2002. Modeling exploitation in recreational fisheries and implications for effort management on British Columbia rainbow trout lakes. *N. Am. J. Fish. Man.* 22:21–34.

Crowder, L., S. Lyman, W. Figueira, and J. Priddy. 2000. Source-sink population dynamics and the problem of siting marine reserves. *Bull. Mar. Sci.* 66:799–820.

Cury, P., A. Bakun, R. Crawford, A. Jarre, R. Quinones, L. Shannon, and H. Verheye. 2000. Small pelagics in upwelling systems: patterns of interaction and structural changes in "wasp-waist" ecosystems. *ICES J. Mar. Sci.* 57:603–618.

Cyr, H., and M. Pace. 1993. Allometric theory: extrapolation from individuals to communities. *Ecology* 74:1234–1245.

Davis, A. 1996. Barbed wire and bandwageons: a comment on ITQ fisheries management. *Rev. Fish Biol. Fish.* 6:97–107.

Dayton, P. 1998. Reversal of the burden of proof in fisheries management. *Science* 279:821–822.

Day, E., and G. Branch. 2002. Effects of sea urchins (*Parechinus angulosus*) on recruits and juveniles of abalone (*Haliotis midae*). *Ecol. Monogr.* 72:133–149.

Deriso, R. B. 1980. Harvesting strategies and parameter estimation for an age-structured model. *Can. J. Fish. Aquat. Sci.* 37:268–282.

Deriso, R., T. J. Quinn II, and P. Neal. 1985. Catch-age analysis with auxiliary information. *Can. J. Fish. Aquat. Sci.* 42:815–824.

Deriso, R. 1987. Optimal $F_{0.1}$ criteria and their relationship to maximum sustainable yield. *Can. J. Fish. Aquat. Sci.* 44 (suppl. 2):339–348.

De Roos, A., and L. Persson. 2001. Physiologically structured models—from versatile technique to ecological theory. *Oikos* 94:51–71.

de Valpine, P., and A. Hastings. 2002. Fitting population models incorporating process noise and observation error. *Ecol. Monogr.* 72:57–76.

DFO (1999). Coho salmon in the coastal waters of the Georgia basin. *DFO. Science Stock Status Report D6* (07).

Dill, L., M. Heithaus, and C. Walters. 2002. Behaviorally-mediated indirect species interactions in marine communities and their importance to conservation and management. In *Ecology* (in press).

Donovan, G., ed. 1989. *The comprehensive assessment of whale stocks: the early years* volume (special issue). Cambridge, UK: Rep. Int. Whaling Comm.

Doubleday, W. 1976. A least squares approach to analzing catch at age data. *Res. Bull. Int. Comm. Northwest Atlantic Fish. Comm.* 12:69–81.

Drinkwater, K. F., and R. A. Myers. 1987. Testing predictions of marine fish and shellfish landings from environmental variables. *Can. J. Fish. Aquat. Sci.* 44:1568–1573.

Eales, J., and J. Wilen. 1986. An examination of location choice in the pink shrimp fishery. *J. Mar. Res. Econ.* 2:331–351.

Edwards, A., and J. Brindley. 1999. Zooplankton mortality and the dynamical behavior of plankton population models. *Bull. Math. Biol.* 61:303–339.

Einum, S., and I. Fleming. 2000. Selection against late emergence and small offspring in Atlantic salmon (*Salmo salar*). *Evolution* 54:628–639.

Eythorsson, E. 2000. A decade of ITQ-management in Icelandic fisheries: consolidation without concensus. *Mar. Policy* 24:483–492.

FAO. 1995. Code of conduct for responsible fisheries. *Food and Agriculture Organization of the United Nations.* Rome, 41pp.

———. 2001. *Expert consultation on economic incentives and responsible fisheries*, volume 638. FAO Fish. Rep.

Fleming, I., and S. Einum. 1997. Experimental tests of genetic divergence of farmed from wild Atlantic salmon due to domestication. *ICES J. Mar. Sci.* 54:1051–1063.

Fleming, I., A. Ian, K. Hindar, I. Mjolnerod, B. Jonsson, T. Balstad, and A. Lamberg. 2000. Lifetime success and interactions of farm salmon invading a native population. *Proceedings of the Royal Society Biological Sciences* Series B 267(1452) 7:1517–1523.

Fleming, I., B. Jonsson, M. Gross, and A. Lamberg. 1996. An experimental study of the reproductive behaviour and success of farmed and wild Atlantic salmon (*Salmo salar*). *J. Applied Ecology* 33:893–905.

Fournier, D. A., J. Hampton, and J. R. Sibert. 1998. MULTIFAN-CL: a length-based, age-structured model for fisheries stock assessment, with application to South Pacific albacore, *Thunnus alalunga*. *Can. J. Fish. Aquat. Sci.* 55:2105–2116.

Fournier, D. A., J. R. Sibert, J. Majkowski, and J. Hampton. 1990. Multifan: a likelihood-based method for estimating growth parameters and age composition from multiple length frequency data sets illustrated using data for Southern bluefin tuna (*Thunnus maccoyii*). *Can. J. Fish. Aquat. Sci.* 47:301–317.

Fournier, D., and C. Archibald. 1982. A general theory for analyzing catch at age data. *Can. J. Fish. Aquat. Sci.* 39:1195–1207.

Fournier, D., J. Sibert, and M. Terceiro. 1991. Analysis of length frequency samples with relative abundance data for the Gulf of Maine northern shrimp (*Pandalus borealis*) by MULTIFAN method. *Can. J. Fish. Aquat. Sci.* 48:591–598.

Fretwell, S. D., and J. Lucas, H. L. 1970. On territorial behavior and other factors influencing habitat distribution in birds. *Acta Biotheoretica* 19:16–36.

Fryxell, J., and P. Lundberg. 1998. Individual behavior and community dynamics. Chapman and Hall: London.

Fuiman, L., and R. Werner, eds. 2002. *Fishery science, the unique contributions of early life stages.* Oxford: Blackwell Pub.

Fulton, E. 2001. The effects of model structure and complexity on the behaviour and performance of marine ecosystem models. Ph.D. dissertation, Univ. of Tasmania.

———. 2002a. Effect of complexity on ecosystem models. *Mar. Ecol. Prog. Ser.* 253:1–6.

———. 2002b. The effect of physiological detail on ecosystem models I: the generic behaviour of a biogeochemical ecosystem model. *Ecol. Modelling* (in press).

Fu, C., and T. J. Quinn II. 2000. Estimability of natural mortality and other population parameters in a length-based model: *Pandalus borealis* in Kachemak Bay, Alaska. *Can. J. Fish. Aquat. Sci.* 57:2420–2432.

Gaard, E., B. Hansen, B. Olsen, and J. Reinert. 2002. Ecological features and recent trends in the physical environment, plankton, fish stocks, and seabirds in the Faroe shelf ecosystem. In K. Sherman and H. Skjoldal, eds., *Large Marine Ecosystems of the North Atlantic*, 245–265. Elsevier.

Gavaris, S. 1988. An adaptive framework for the estimation of population size. *Can. Atl. Fish. Scientific Adv. Comm. (CAFSAC) Res. Doc.* 88(29), 12.

Gelb, A., ed. 1974. *Applied optimal estimation.* Cambridge, MA: MIT Press.

Gilliam, J., and D. Fraser. 1987. Habitat selection under predation hazard: test of a model with foraging minnows. *Ecology* 68:1856–1862.

Goodyear, C. P. 1977. Assessing the impact of power plant mortality on the compensatory reserve of fish populations. In W. Van Winkle, ed., *Proceedings of the conference on assessing the effects of power plant induced mortality on fish populations.* New York: Permagon Press.

Goodyear, C. P. 1980. Compensation in fish populations. In C. Hocutt and C. J. Stauffer, eds., *Biological monitoring of fish.* Lexington, MA.: Lexington Books, D.C. Heath Co.

Goodyear, C. P. 1989. Spawning stock biomass per recruit: the biological basis for a fisheries management tool. *ICCAT Working Document SCRS 89/82*, 10.

Grand, T., and L. Dill. 1997. The energetic equivalence of cover to juvenile coho salmon (*Oncorhynchus kisutch*): ideal free distribution theory applied. *Behavioral Ecology* 8:437–447.

Grand, T. 1997. Foraging site selection in juvenile coho salmon (*Oncorhynchus kisutch*): ideal free distributions of unequal competitors. *Animal Behaviour* 53:185–196.

Guenette, S., T. Pitcher, and C. J. Walters. 2000. The potential of marine reserves for the management of northern cod in Newfoundland. *Bull. Mar. Sci.* 66:729–743.

Gulland, J. 1961. Fishing and the stocks of fish at Iceland. *U.K. Ministry of Agriculture and Fisheries, Food, Fisheries Investment (Series 2) 23*(4).

Gunderson, L., and C. Holling, eds. 2001. *Panarchy: understanding transformations in human and natural systems*. Washington, D.C.: Volume Island Press.

Gunderson, L., and L. J. Pritchard, eds. 2002. *Resilience and the behavior of large-scale systems*. London: Island Press.

Hall, S. 1999. *The effects of fishing on marine ecosystems and communities*. Fish Biology and Aquatic Resources Series I. London: Blackwell Science.

Hampton, J., and J. Gunn. 1998. Exploitation and movements of yellowfin tuna (*Thunnus albacares*) and bigeye tuna (*T. obesus*) tagged in northwestern Coral Sea. *Mar. Freshw. Res.* 49:475–489.

Harley, S., R. Myers, and A. Dunn. 2001. Is catch-per-effort proportional to abundance? *Can. J. Fish. Aquat. Sci.* 58:1760–1772.

Hartman, K., and S. Brandt. 1995. Comparative energetics and the development of bioenergetics models for sympatric estuarine piscivores. In *Can. J. Fish. Aquat. Sci.* 52:1647–1666.

Hedrick, P. 2001. Invasion of transgenes from salmon or other genetically modified organisms into natural populations. *Can. J. Fish. Aquat. Sci.* 58:841–844.

Helminen, H., and J. Sarvala. 1994. Population regulation of vendace (*Coregonus albula*) in Lake Pyhaejaervi, southwest Finland. *J. Fish. Biol.* 45:387–400.

———. 1997. Responses of Lake Pyhaejaervi (southwest Finland) to variable recruitment of the major planktivorous fish, vendace (*Coregonus albula*). *Can. J. Fish. Aquat. Sci.* 54:32–40.

Helu, S., D. Sampson, and Y. Yin. 2000. Application of statistical model selection criteria to the Stock Synthesis assessment program. *Can. J. Fish. Aquat. Sci.* 57:1784–1793.

Henderson, P., and M. Corps. 1997. The role of temperature and cannibalism in interannual recruitment variation of bass in British waters. *J. Fish. Biol.* 50:280–295.

Higgins, K., A. Hastings, and L. W. Botsford. 1997. Stochastic dynamics and deterministic skeletons: population behavior of Dungeness crab. *Science* 267:1431–1435.

Hightower, J., J. Jackson, and K. Pollock. 2001. Use of telemetry methods to estimate natural and fishing mortality of striped bass in Lake Gaston, North Carolina. *Trans. Am. Fish. Soc.* 130:557–567.

Hilborn, R., and D. Eggers. 2000. A review of the hatchery programs for pink salmon in Prince William Sound and Kodiak Island, Alaska. *Trans. Am. Fish. Soc.* 129:333–350.

Hilborn, R., and W. Luedke. 1987. Rationalizing the irrational: a case study in user group participation in pacific salmon management. *Can. J. Fish. Aquat. Sci.* 44:1796–1805.

Hilborn, R., and M. Mangel. 1997. *The ecological detective: confronting models with data.* Princeton: Princeton Univ. Press.

Hilborn, R., and J. R. Sibert. 1998. Adaptive management of developing fisheries. *Mar. Policy* 12:112–121.

Hilborn, R., and C. J. Walters. 1987. A general model for simulation of stock and fleet dynamics in spatially heterogeneous fisheries. *Can. J. Fish. Aquat. Sci.* 44:1366–1369.

———. 1992. *Quantitative fisheries stock assessment: choice, dynamics, & uncertainty.* Chapman & Hall: London.

Hilborn, R., and J. Winton. 1993. Learning to enhance salmon production: lessons from the Salmonid Enhancement Program. *Can. J. Fish. Aquat. Sci.* 50:2043–2056.

Hilborn, R. 1979. Comparison of fisheries control systems that utilize catch and effort data. *J. Fish. Res. Bd. Canada* 36:1477–1489.

———. 1985. Fleet dynamics and individual variation: why some people catch more fish than others. *Can. J. Fish. Aquat. Sci.* 42:2–13.

———. 2001. Risk analysis for salmon spawning reference levels. In E. Prevost and G. Chaput, eds., *Stock, recruitment and reference points assessment and management of Atlantic salmon*, 177–193. Paris: INRA.

———. 2002. Population management in stock enhancement and sea ranching. In *Proc. Second International Symposium in stock enhancement and sea ranching, Kobe Japan.* London: Blackwell Science (in press).

Hixon, M., and M. Carr. 1997. Synergistic predation, density dependence, and population regulation in marine fish. *Science* 277:946–949.

Hixon, M., and B. Menge. 1991. Species diversity: prey refuges modify the interactive effects of predation and competition. *Theor. Pop. Biology* 39:178–200.

Holland, D., and J. Sutinen. 1999. An empirical model of fleet dynamics in New England trawl fisheries. *Can. J. Fish. Aquat. Sci.* 56:253–264.

Holland, D. 2000. A bioeconomic model of marine sanctuaries on Georges Bank. *Can. J. Fish. Aquat. Sci.* 57:1307–1309.

Holling, C. 1959. The components of predation as revealed by a study of small mammal predation of the European pine sawfly. *Can. Ent.* 91:293–320.

———. 1973. Resilience and stability of ecological systems. *Ann. Rev. of Ecol. and Syst.* 4:2–23.

———. ed. 1978. *Adaptive environmental assessment and management.* New York: John Wiley and Sons.

Hollowed, A., N. Bax, R. Beamish, J. Collie, M. Fogarty, P. Livingston, J. Pope, and J. Rice. 2000. Are multispecies models an improvement on single-species models for measuring fishing impacts on marine ecosystems? *ICES J. Mar. Sci.* 57:707–719.

Holtby, L. 1988. Effects of logging on stream temperatures in Carnation Creek, British Columbia, and associated impacts on the coho salmon (*Oncorhynchus kisutch*). *Can. J. Fish. Aquat. Sci.* 45:502–515.

Houston, A., and J. McNamara. 1987. Switching between resources and the ideal free distribution. *Animal Behaviour* 35:301–302.

Humston, R., J. Ault, M. Lutcavage, and D. Olson. 1999. Schooling and migration of large pelagic fishes relative to environmental cues. *Fisheries Oceanography* 9:136–146.

Hutchings, J., and M. Ferguson. 2000. Temporal changes in harvesting dynamics of Canadian inshore fisheries for northern cod, *Gadus morhua. Can. J. Fish. Aquat. Sci.* 57:805–814.

Hutchings, J., R. A. Myers, and G. Lilly. 1993. Geographic variation in the spawning of Atlantic cod, *Gadus morhua*, in the Northwest Atlantic. *Can. J. Fish. Aquat. Sci.* 50:2457–2467.

Hutchings, J., and R. Myers. 1994. What can be learned from the collapse of a renewable resource? Atlantic cod, *Gadus morhua*, of Newfoundland and Labrador. *Can. J. Fish. Aquat. Sci.* 51:2126–2146.

Hutchings, J. 2000. Collapse and recovery of marine fishes. *Nature* 406, 882–885.

Isawa, Y., V. Andreasen, and S. Levin. 1987. Aggregation in model ecosystems I: perfect aggregation. *Ecol. Modelling* 37:287–302.

Jackson, J., M. Kirby, W. Berger, K. Bjorndal, L. Botsford, B. Bourque, R. Bradbury, R. Cooke, J. Erlandson, J. Estes, T. Hughes, S. Kidwell, C. Lange, H. Lenihan, J. Pandolfi, C. Peterson, R. Steneck, M. Tegner, and R. Warner. 2001. Historical overfishing and the recent collapse of coastal ecosystems. *Science* 293:629–637.

Jakobsson, J., and G. Stefansson. 1998. Rational harvesting of the cod-capelin-shrimp complex in the Icelandic marine ecosystem. *Fisheries Research* 37:7–21.

James, M., P. Armsworth, L. Mason, and L. Bode. 2002. The structure of reef fish metapopulations: modeling larval dispersal and retention patterns. *Proc. R. Soc. Lond., Ser. B* 269:2079–2086.

Janssen, M., and S. Carpenter. 1999. Managing the resilience of lakes: a multi-agent modeling approach. *Conservation Ecology* 3 (on line, www.consecol.org).

Jones, D., and C. Walters. 1976. Catastrophe theory and fisheries regulation. *J. Fish. Res. Bd. Canada* 33:2829–2833.

Jorgensen, S. 1995. *Handbook of environmental and ecological modeling.* New York: CRC Press.

————. 2000. *Handbook of ecosystem theories and management.* New York: CRC Press.

Kacelnik, A., J. Krebs, and C. Bernstein. 1992. The ideal free distribution and predator-prey populations. *Trends in Ecology & Evolution* 7:50–55.

Keeney, R., and H. Raiffa. 1976. *Decisions with multiple objectives.* New York: John Wiley.

Kevorkian, J., and J. Cole. 1996. *Multiple scale and singular perturbation methods.* Berlin: Springer-Verlag.

Korman, J., R. Peterman, and C. J. Walters. 1985. Empirical and theoretical analysis of correction of time series bias in stock-recruitment relationships of sockeye salmon (*Oncorhynchus nerka*). *Can. J. Fish. Aquat. Sci.* 52:2174–2189.

Krebs, J., and N. Davies. 1981. *An Introduction to Behavioral Ecology.* Sunderland, Massachusetts: Sinauer Associates, Inc.

Kruse, G., and A. Tyler. 1989. Exploratory simulation of English sole recruitment mechanisms. *Trans. Am. Fish. Soc.* 118:101–118.

Laevastu, T., and F. Favorite. 1988. *Fishing and stock fluctuation.* Farnham, United Kingdom: Fishing Books Ltd.

Lane, D., and R. Stephenson. 1998. Fisheries co-management: organization, process, and decision support. *J. Northwest Atl. Fish. Sci.* 23:251–265.

Lane, P., and R. Levins. 1977. The dynamics of aquatic systems. 2. The effects of nutrient enrichment on model plankton communities. *Limnol. Oceanogr.* 22:454–471.

Lane, P. 1986. Symmetry, change, perturbation, and observing mode in natural communities. *Ecology* 67:223–239.

Larkin, P. 1996. Concepts and issues in marine ecosystem management. *Rev. Fish Biol. Fish.* 6, 139–164.

Larson, R. and J. Casti 1978. *Principles of dynamic programming. Part I: Basic analytic and computational methods.* New York: Dekker Pub. Co.

Laslett, G., J. Eveson, and T. Polachek. 2002. A flexible maximum likelihood approach for fitting growth curves to tag-recapture data. *Can. J. Fish. Aquat. Sci.* 59:976–986.

Law, R., and D. Grey. 1989. Evolution of yields from populations with age-specific cropping. *Evolutionary Ecology* 3:343–359.

Law, R. 2000. Fishing, selection, and phenotypic evolution. *ICES J. Mar. Sci.* 57:659–668.

Leber, K. 2002. Advances in marine enhancement: shifting emphasis to theory and accountability. In R. Stickney and J. McVey, eds., *Responsible Marine Aquaculture*, 79–90. Oxon, United Kingdom: CAB International.

Leslie, P., and D. Davis. 1939. An attempt to determine the absolute number of rats on a given area. *J. Anim. Ecol.* 8:94–113.

Levins, R. 1974. The qualitative analysis of partially specified systems. *Ann. N.Y. Acad. Sci.* 231:123–138.

Levin, P., R. Zable, and J. Williams. 2001. The road to extinction is paved with good intentions: negative association of fish hatcheries with threatened salmon. *Proc. R. Soc. Lond., Ser. B* 268:1153–1158.

Ley, J., I. Halliday, A. Tobin, R. Garrett, and N. Gribble. 2002. Ecosystem effects of fishing closures in mangrove estuaries of tropical Australia. *Mar. Ecol. Prog. Ser.* 245:223–238.

Liermann, M., and R. Hilborn. 2001. Depensation: evidence, models, and implications. *Fish and Fisheries* 2:33–58.

Lilly, G., D. Parsons, and D. Kulka. 2000. Was the increase in shrimp biomass on the northeast Newfoundland shelf a consequence of a release in predation pressure from cod? *J. Northw. Atl. Fish. Sci.* 27:45–61.

Link, J., and R. Peterman. 1998. Estimating the value of in-season estimates of abundance of sockeye salmon (*Onchorhynchus nerka*). *Can. J. Fish. Aquat. Sci.* 55:1408–1418.

Lipcius, R., and W. van Engel. 1990. Blue crab population dynamics in Cheasapeake Bay. Variation in abundance (York River, 1972–1988) and stock-recruit functions. *Bull. Mar. Sci.* 46:180–194.

Li, H., and P. B. Moyle 1981. Ecological analysis of species introductions into aquatic systems. *Trans. Am. Fish. Soc.* 110:772–782.

Ludwig, D., M. Mangel, and B. Haddad. 2001. Ecology, conservation, and public policy. *Ann. Rev. Ecol. Syst.* 32:481–517.

Ludwig, D., and C. J. Walters. 1981. Measurement errors and uncertainty in parameter estimates for stock and recruitment. *Can. J. Fish. Aquat. Sci.* 38:711–720.

———. 1985. Are age structured models appropriate for catch-effort data? *Can. J. Fish. Aquat. Sci.* 42:1066–1072.

MacCall, A. 1990. *Dynamic geography of marine fish populations.* Seattle: Univ. of Washington Press.

Mace, P., and M. Sissenwine. 1993. How much spawning per recruit is necessary? In S. Smith, J. Hunt, and D. Rivard, eds., *Risk evaluation and biological reference points for fisheries management,* volume 120, 101–118. Can. Sp. Publ. Fish. Aquat. Sci.

Mace, P. 1994. Relationship between common biological reference points used as thresholds and targets for fisheries management strategies. *Can. J. Fish. Aquat. Sci.* 42:1066–1072.

Mackinson, S. 2001. Integrating local and scientific knowledge: an example in fisheries science. *Environmental Management* 27:533–545.

Magnusson, K. 1995. An overview of the multispecies vpa-theory and applications. *Rev. Fish. Biol. Fisheries* 5:195–212.

Mangel, M. 1985. *Decision and control in uncertain resource systems.* New York: Academic Press.

———. 2000. Trade-offs between fish habitat and fishing mortality and the role of reserves. *Bull. Mar. Sci.* 66:663–674.

Mangel, M., and C. Clark. 1988. *Dynamic modeling in behavioral ecology.* Princeton: Princeton Univ. Press.

Martell, S.J.D., C. J. Walters, and S. S. Wallace. 2000. The use of marine protected areas for conservation of lingcod (Ophiodon elongatus). *Bull. Mar. Sci.* 66(3):729–743.

Martell, S.J.D., and C. J. Walters. 2002. Implementing harvest rate objectives by directly monitoring exploitation rates and estimating changes in catchability. *Bull. Mar. Sci.* 70(2):695–713.

Martell, S.J.D. 2002. Variation in pink shrimp populations off the west coast of Vancouver Island: oceanographic and trophic interactions. Ph.D. dissertation, University of British Columbia.

Marten, G., and J. Polovina, eds. 1982. *A comparative study of fish yields from various tropical ecosystems,* volume 105 of *Theory and management of tropical fisheries.* Manila: ICLARM Contrib.

Mason, D., and S. Brandt. 1999. Space, time, and scale: new perspectives in fish ecology and management. *Can. J. Fish. Aquat. Sci.* 56 (suppl. 1):1–3.

Mathews, J. 1992. *Numerical methods for Mathematics, Science, and Engineering* (2d ed.). Englewood Cliffs, NJ: Prentice-Hall, Inc.

McAllister, M., and J. Ianelli. 1997. Bayesian stock assessment using catch-at-age data and the sampling importance resampling algorithm. *Can. J. Fish. Aquat. Sci.* 54:284–300.

McAllister, M., and G. Kirkwood. 1998. Using Bayesian decision analysis to help achieve a precautionary approach for managing developing fisheries. *Can. J. Fish. Aquat. Sci.* 55:2642–2661.

McCann, K., and B. Schuter. 1997. Bioenergetics of life history strategies and the comparative allometry of reproduction. *Can. J. Fish. Aquat. Sci.* 54:1289–1298.

McCarthy, M., L. Ginzburg, and H. Akcakaya. 1995. Predator interference across trophic chains. *Ecology* 76:1310–1319.

McNamara, J., and A. Houston. 1994. The effect of a change in foraging options on intake rate and predation rate. *Am. Nat.* 144:978–1000.

Meester, G., J. Ault, S. Smith, and A. Mehrotra. 2001. An integrated simulation modeling and operations research approach to spatial management decision making. *Sarsia* 86:543–558.

Meester, G. 2000. A mathematical programming and simulation-based approach to determining critical factors in the design of effective marine reserve plans for coral reef fishes. Ph.D. dissertation, University of Miami.

Methot, R. 1990. Synthesis model: an adaptable framework for analysis of diverse stock assessment data. *Int. North Pac. Fish. Comm. Bull.* 50:259–277.

Meyer, R., and R. Millar. 2000. *Baysian dynamic modeling of stock-recruitment relationships.* Tech. Rep. STAT 0004. Dept. Statistics, Univ. Auckland, Auckland, NZ.

Millar, R., and R. Meyer. 2000. Bayesian state-space modeling of age-structured data: fitting a model is just the beginning. *Can. J. Fish. Aquat. Sci.* 57:43–50.

Millisher, L., D. Gaschuel, and A. Biseau. 1999. Estimation of the overall fishing power: a study of the dynamics and fishing strategies of Brittany's industrial fleets. *Aquatic Living Resources* 12:89–103.

Moloney, C., and J. Field. 1989. General allometric equations for nutrient uptake, ingestion and respiration in plankton organisms. *Limnol. Oceanogr.* 34:1290–1299.

Morishima, G., and K. Henry. 2000. The history and status of Pacific Northwest Chinook and coho salmon ocean fisheries and prospects for sustainability. In E. Knudson, C. Steward, D. MacDonald, J. Williams, and D. Reiser, eds., *Sustainable fisheries management: Pacific salmon*, 219–236. Boca Raton: CRC Press.

Moyle, P. 1996. Biological invasions of fresh water: empirical rules and assembly theory. *Biol. Conservation* 78:149–161.

Muir, W., and R. Howard. 1999. Possible ecological risks of transgenic organism release when transgenes affect mating success: sexual selection and the Trojan gene hypothesis. *Proc. Nat. Acad. Sci. USA* 96:13853–13856.

Munk, K. M. 2001. Maximum ages of groundfishes in waters off Alaska and British Columbia and considerations of age determination. *Alaska Fisheries Research Bulletin* 8(1):12–21.

Munro, J., and R. Thompson. 1983. The Jamaican fishing industry. In J. Munro, ed., *Caribbean coral reef fish resources*. Manila: ICLARM.

Murray, A., and J. Parslow. 1999a. Modeling of nutrient impacts in Port Phillip Bay-a semi-enclosed marine Australian ecosystem. *Mar. Freshw. Res.* 50:597–611.

———. 1999b. The analysis of alternative formulations in a simple model of a coastal ecosystem. *Ecol. Modeling* 119, 149–166.

Myers, R. A., and N. J. Barrowman. 1995. Time series bias in the estimation of density-dependent mortality in stock-recruitment models. *Can. J. Fish. Aquat. Sci.* 52(1):223–232.

———. 1996. Is fish recruitment related to spawner abundance? *Fish. Bull.* 94:707–724.

Myers, R. A., K. G. Bowen, and N. J. Barrowman. 1999. Maximum reproductive rate of fish at low populations sizes. *Can. J. Fish. Aquat. Sci.* 56:2404–2419.

Myers, R. A., and N. Cadigan. 1993. Density-dependent juvenile mortality in marine demersal fish. *Can. J. Fish. Aquat. Sci.* 50:1576–1590.

Myers, R. A., S. Fuller, and D. Kehler. 2000. A fisheries management strategy robust to ignorance: rotational harvest in the presence of indirect fishing mortality. *Can. J. Fish. Aquat. Sci.* 57:2357–2362.

Nielsen, A. 1894. Operations at Dildo Island marine hatchery. In *Annual report of the Newfoundland Department of Fisheries, for the year 1894*, 20–35. St. John's, Nfld.

Nihoul, J. 1998. Optimum complexity in ecohydrodynamic modeling: an ecosystem dynamics standpoint. *J. Mar. Systems* 16:3–5.

Noakes, D., R. Beamish, R. Sweeting, and J. King. 2000. Changing the balance: interactions between hatchery and wild Pacific coho salmon in the presence of regime shifts. *North Pacific Anadromous Fish Comm. Bull.* 2:155–163.

Nordwall, F., I. Naslund, and E. Degerman. 2001. Intercohort competition effects on survival, movement, and growth of brown trout (*Salmo trutta*) in Swedish streams. *Can. J. Fish. Aquat. Sci.* 58:2298–2308.

NRC. 1996. *The Bering Sea ecosystem.* Washington, D.C.: National Academy Press.

———. 1998. *Improving fish stock assessments.* Washington, D.C.: National Academy Press.

———. 1999. *Sustaining marine fisheries.* Washington, D.C.: National Academy Press.

———. 2001. *Marine protected areas: tools for sustaining ocean ecosystem.* Washington, D.C.: National Academy Press.

———. 2002. *The decline of the steller sea lion in Alaskan waters: untangling food webs and fishing nets.* Washington, D.C.: National Academy Press.

Nystrom, M., C. Folke, and F. Moberg. 2000. Coral reef disturbance and resilience in a human-dominated environment. *TREE* 15:413–417.

Nystrom, M., and C. Folke. 2001. Spatial resilience of coral reefs. *Ecosystems* 4:406–417.

O'Malley, R. 1974. *Introduction to singular perturbations.* New York: Academic Press.

O'Neill, R., D. DeAngelis, J. Waide, and T. Allen. 1986. *A hierarchical concept of ecosystems.* Princeton: Princeton Univ. Press.

O'Neill, R., and B. Rust. 1979. Aggregation error in ecological models. *Ecol. Modelling* 7:91–105.

Parma, A. M., and R. B. Deriso 1990a. Dynamics of age and size composition in a population subject to size-selective mortality: effects of phenotypic variability in growth. *Can. J. Fish. Aquat. Sci.* 47:274–289.

———. 1990b. Experimental harvesting of cyclic stocks in the face of alternative recruitment hypotheses. *Can. J. Fish. Aquat. Sci.* 47(3):595–610.

Pascual, M. A., and O. O. Iribarne. 1993. How good are empirical predictions of natural mortality. *Fisheries Research* 16:17–24.

Patten, B., ed. 1971. *Systems analysis and simulation in ecology*, volume 1. New York: Academic Press.

Patterson, K., R. Cook, C. Darby, S. Gavaris, L. Kell, P. Lewy, B. Mesnil, A. Punt, V. Restrepo, D. Skagen, and G. Stefansson. 2001. Estimating uncertainty in fish stock assessment and forecasting. *Fish and Fisheries* 2:125–157.

Patterson, K. 1992. Fisheries for small pelagic species: an empirical approach to management targets. *Rev. Fish Biol. Fisheries* 2:321–338.

Pauly, D., V. Christensen, J. Dalsgaard, R. Froese, and F. T. Jr. 1998. Fishing down marine food webs. *Science* 279:860–863.

Pauly, D., V. Christensen, R. Froese, and M. Palomares. 2000. Fishing down aquatic food webs. *Am. Sci.* 88:46–51.

Pauly, D., P. Muck, J. Mendo, and I. Tsukayama, eds. 1989. *The Peruvian Upwelling Ecosystem: Dynamics and Interactions*. Instituto del Mar del Peru (IMARPE).

Pauly, D. 1980. On the interrelationships between natural mortality, growth parameters, and mean environmental temperature in 175 fish stocks. *J. Cons. Int. Explor. Mer.* 39:175–192.

———. 1995. Anecdotes and the shifting baseline syndrome of fisheries. *TREE* 10(10):430.

Pearse, P., and C. Walters. 1992. Harvest regulation under quota management systems for ocean fisheries: Decision making in the face of natural variability, weak information, risk and conflicting incentives. *Marine Policy* 16:167–182.

Pella, J. 1993. Utility of structural time series models and the Kalman filter for predicting the consequences of fishery actions. In G. Kruse, ed., *Proceedings of the International Symposium on Management Strategies for Exploited Fish Populations*, 571–593. Fairbanks: Alaska Sea Grant College Program Report No. 93-02.

Perry, I., C. Walters, and J. Boutillier. 1999. A framework for providing scientific advice for the management of new and developing fisheries. *Rev. Fish Biol. Fisheries* 9:1–26.

Peterman, R. M., B. J. Pyper, M. F. Lapointe, M. D. Adkison, and C. J. Walters. 1998. Patterns of covariation in survival rates of British Columbian and Alaska sockeye salmon (*Oncorhynchus nerka*) stocks. *Can. J. Fish. Aquat. Sci.* 55:2503–2517.

Peterman, R. M. 1991. Density-dependent marine processes in north Pacific salmonids: lessons for experimental design of large scale manipulations of fish stocks. *ICES J. Mar. Sci. Symp.* 192:69–77.

Peterman, R., B. Pyper, and J. Grout. 2000. Comparison of parameter estimation methods for detecting climate-induced changes in productivity of Pacific salmon (*Oncorhynchus* spp.). *Can. J. Fish. Aquat. Sci.* 57:181–191.

Peterman, R. 1981. Form of random variation in salmon smolt-to-adult relations and its influence on production estimates. *Can. J. Fish. Aquat. Sci.* 38:1113–1119.

Pimm, S., and J. B. Hyman. 1987. Ecological stability in the context of multispecies fisheries. *Can. J. Fish. Aquat. Sci.* 44:84–94.

Pitcher, T., and K. Cochrane. 2002. Use of ecosystem models to investigate multispecies management strategies for capture fisheries. *Fisheries Centre Research Reports* 10(2):156.

Pitcher, T., N. Haggan, D. Preikshot, and D. Pauly. 1999. "Back to the Future": a method employing ecosystem modelling to maximise the sustainable benefits from fisheries. In *Ecosystem approaches for fisheries management*, volume AK-SG-99-01, 447–466. Fairbanks: University of Alaska Sea Grant.

Pitcher, T., R. Watson, N. Haggan, S. Guenette, R. Kennish, U. Sumaila, D. Cook, K. Wilson, and A. Leung. 2000. Marine reserves and the restoration of fisheries and marine ecosystems in the South China Sea. *Bull. Mar. Sci.* 66:543–566.

Pitcher, T. 2001. Fisheries managed to rebuild ecosystems: reconstructing the past to salvage the future. *Ecol. Appl.* 11(2):601–617.

Porch, C., C. Wilson, and D. Nieland. 2002. A new growth model for red drum (*Sciaenops ocellatus*) that accommodates seasonal and ontogenic changes in growth rates. *Fish. Bull.* 100:149–152.

Possingham, H., and J. Roughgarden. 1990. Spatial population dynamics of a marine organism with a complex life cycle. *Ecology* 71:973–985.

Post, J. R., E. A. Parkinson, and N. T. Johnston. 1999. Density dependent processes in structured fish populations: interaction strengths in whole-lake experiments. *Eco. Mono.* 69:155–175.

Post, J., M. Sullivan, S. Cox, N. Lester, C. Walters, E. Parkinson, A. Paul, L. Jackson, and B. Shuter. 2002. Canda's recreational fisheries: the invisible collapse? *Fisheries* 27(1):6–17.

Press, W., S. Teukolsky, W. Vetterling, and B. Flannery. 1996. *Numerical Recipies in Fortran 77, The Art of Scientific Computing* (2nd ed.). Cambridge: Cambridge University Press.

Prince, J. 1989. *The fisheries biology of the Tasmanian stocks of Haliotis rubra.* Ph.D. dissertation, University of Tasmania.

Puccia, C., and R. Levins. 1985. *Qualitative modeling of complex systems: an introduction to loop analysis and time averaging.* Cambridge, MA: Harvard Univ. Press.

Punt, A. E., and R. Hilborn. 1997. Fisheries stock assessment and decision analysis: the Bayesian approach. *Reviews in Fish Biology and Fisheries* 7:35–63.

Punt, A. E., and A.D.M. Smith. 1999a. Harvest strategy evaluation for the eastern stock of gemfish (*Rexea solandri*). *ICES J. Mar. Sci.* 56:860–875.

———. 1999b. Management of long-lived marine resources: a comparison of feedback-control management procedures. *Am. Fish. Soc. Symp.* 23:243–265.

Punt, A., A.D.M. Smith, A. Davidson, B. D. Mapstone, and C. Davies. 2001. Evaluating the scientific benefits of spatially explicit experimental manipulations of common coral trout (*Plectropomus leopardus*) populations on the Great Barrier Reef, Australia. In G. H. Kruse, N. Bez, A. Booth, M. Dorn, S. Hills, R. Lipcius, D. Pelletier, C. Roy, S. Smith, and D. Witherells, eds., *Spatial Processes and Management of Marine Populations*, volume 17, 67–103. Fairbanks: Univ. of Alaska Sea Grant, Lowell Wakefield Fish. Symp.

Punt, A. 1993. The comparative performance of production-model and ad hoc tuned VPA based feedback-control management procedures for the stock of Cape hake off the west coast of South Africa. *Can. Sp. Publ. Fish. Aquat. Sci.* 120:283–299.

Quinn, T. J., and R. Deriso. 1999. *Quantitative Fish Dynamics.* Oxford: Oxford University Press.

Quinn, T., C. Turnbull, and C. Fu. 1998. A length-vased population model for hard-to-age invertebrate populations. In F. Funk, T. J. Quinn, J. Heifetz, J. N. Ianelli, J. E. Powers, J. F. Schweigert, P. J. Sullivan, and C. I. Zhang, eds., *Fishery Stock Assessment Models*, 531–556. Fairbanks: Alaska Sea Grant College Program Report No. AK-SG-98-01, University of Alaska.

Raiffa, H. 1982. *The Art and Science of Negotiations*. Cambridge, MA: Harvard University Press.

Rasteller, E., A. King, B. Cosby, G. Hornberger, R. O'Neill, and J. Hobbie. 1992. Aggregating fine-scale ecological knowledge to model coarser-scale attributes of ecosystems. *Ecol. Appl.* 2:55–70.

Reed, W. J., and C. M. Simons. 1996. Analyzing catch-effort data by means of the Kalman filter. *Can. J. Fish. Aquat. Sci.* 53:2157–2166.

Reed, W. J. 1979. Optimal escapement levels in stochastic and deterministic harvesting models. *J. Env. Econ. Mgmt.* 6:350–363.

Research, O. 1994. *An introduction to AD Model Builder for use in nonlinear modeling and statistics*. Nanaimo, B.C.: Otter Research Ltd.

Restrepo, V., G. Thompson, P. Mace, W. Gabriel, L. Low, A. McCall, R. Methot, J. Powers, B. Taylor, P. Wade, and J.F. Witzig. 1998. Technical guidelines on the use of the precautionary approach to implementing National Standard of the Magnuson-Stevens Fishery Conservation and Management Act. *NOAA Technical Memorandum NMFS-F/SPO* 31:54.

Ricker, W. 1958. Maximum sustained yields from fluctuating environments and mixed stocks. *J. Fish. Res. Bd. Canada* 15:991–1006.

———. 1963. Big effects from small causes: two examples from fish population dynamics. *J. Fish. Res. Bd. Canada* 20:257–264.

———. 1973. Two mechanisms that make it impossible to maintain peak period yields from Pacific salmon and other fishes. *J. Fish. Res. Bd. Canada* 30:1275–1286.

———. 1975. *Computation and interpretation of biological statistics of fish populations*, volume 191. Can. Dept. Environment, Fish. and Marine Service, Bull.

Rivot, E., E. Prevost, and E. Parent. 2001. How robust are Bayesian posterior inferences based on a Ricker model with regards to measurement errors and prior assumptions about parameters? *Can. J. Fish. Aquat. Sci.* 58:2284–2297.

Robins, C., Y. Wang, and D. Die. 1998. The impact of global positioning systems and plotters on fishing power in the northern prawn fishery, Australia. *Can. J. Fish. Aquat. Sci.* 55:1645–1651.

Rosenzweig, M., and R. MacArthur. 1963. Graphical representation and stability conditions of predator-prey interactions. *Am. Nat.* 97:209–223.

Rosenzweig, M. 1971. Paradox of enrichment: destabilization of exploitation ecosystems in ecological time. *Science* 171:385–387.

Rose, K., J. J. Cowan, K. Winemiller, R. A. Myers, and R. Hilborn. 2001. Compensatory density dependence in fish populations: importance, controversy, understanding and prognosis. *Fish and Fisheries* 2:293–327.

Rudstam, I., G. Aneer, and M. Hilden. 1994. Top-down control in the pelagic Baltic ecosystem. *Dana* 10:105–129.

Russ, G. 2002. Marine reserves as reef fisheries management tools: yet another review. In P. Sale, ed., *Coral reef fishes: new insights into their ecology*. New York: Academic Press.

Sainsbury, K. J., A. E. Punt, and A.D.M. Smith. 2000. Design of operational management strategies for achieving fishery ecosystem objectives. *ICES J. Mar. Sci.* 57:731–741.

Sainsbury, K. J. 1998. Living marine resource assessment for the 21st century: what will be needed and how will it be provided? In F. Funk, T. Quinn II, J. Heifetz, J. Ianelli, J. Powers, J. Schweigert, M. Sullivan, and C.-I. Zhang, eds., *Fishery Stock Assessment Models*, 1–40. Fairbanks: Alaska Sea Grant Program Report No. AK-SG-98-01.

Samb, B., and D. Pauly. 2000. On "variability" as a sampling artefact: the case of *Sardinella* in north-western Africa. *Fish and Fisheries* 1:206–210.

Sampson, D. 1992. Fishing technology and fleet dynamics: predictions from a bioeconomic model. *Mar. Resource Econ.* 7:37–58.

Scandol, J. 1999. CotSim—an interactive *Acanthaster planci* metapopulation model for the central Great Barrier Reef. *Marine Models Online, 29 Jan 1999 (URL: http://www.elsevier.com/gej-ng/10/31/42/show/).*

Scharf, F. 2000. Patterns in abundance, growth, and mortality of juvenile red drum across estuaries on the Texas coast and implications for recruitment and stock enhancement. *Trans. Am. Fish. Soc.* 129:1207–1222.

Scheffer, M., C. Carpenter, J. Foley, C. Folke, and B. Walker. 2001. Catastrophic shifts in ecosystems. *Nature* 413:591–596.

Scheffer, M., and R. De Boer. 1995. Implications of spatial heterogeneity for the paradox of enrichment. *Ecology* 76:2270–2277.

Scheffer, M. 1990. Multiplicity of stable states in freshwater systems. *Hydrobiologia* 200/201:475–486.

Schindler, D., and L. Eby. 1997. Stochiometry of fishes and their prey: implications for nutrient cycling. *Ecology* 78:1816–1831.

Schnute, J. T., and A. R. Kronlund. 1996. A management oriented approach to stock recruitment analysis. *Can. J. Fish. Aquat. Sci.* 53:1281–1293.

———. 2002. Estimating salmon stock-recruitment relationships from catch and escapement data. *Can. J. Fish. Aquat. Sci.* 59:433–449.

Schnute, J. T. 1994. A general framework for developing sequential fisheries models. *Can. J. Fish. Aquat. Sci.* 51:1676–1688.

———. 1985. A general theory for analysis of catch and effort data. *Can. J. Fish. Aquat. Sci.* 42:414–429.

———. 1987. A general fishery model for a size-structured fish population. *Can. J. Fish. Aquat. Sci.* 44:924–940.

Schwarz, G. 1978. Estimating the dimension of a model. *Ann. Stat* 6:461–464.

Scott, A. 1979. Development of economic theory on fisheries regulation. *J. Fish. Res. Bd. Canada* 36:725–741.

Seber, G. 1982. *The estimation of animal abundance and related parameters.* London: Griffin.

Shampine, L. F. 1994. *Numerical Solution of Ordinary Differential Equations.* New York: Chapman & Hall.

Shelton, P., and B. Healey. 1999. Should depensation be dismissed as a possible explanation for the lack of recovery of the northern cod (*Gadus morhua*) stock? *Can. J. Fish. Aquat. Sci.* 56:1521–1524.

Shepherd, J. 1982. A versatile new stock-recruitment relationship for fisheries, and the construction of sustainable yield curves. *J. Cons. CIEM* 40:67–75.

Shima, M., A. Hollowed, and G. VanBlaricom. 2000. Response of pinniped populations to directed harvest, climate variability, and commercial fishery activity: a comparative analysis. *Reviews in Fisheries Science* 8:89–124.

Sibert, J., J. Hampton, D. Fournier, and P. Bills. 1999. An advection-diffusion-reaction model for the estimation of fish movement parameters from tagging data, with application to skipjack tuna (*Katsuwonus pelamis*). *Can. J. Fish. Aquat. Sci.* 56:925–938.

Sinclair, A., D. Swain, and J. Hanson. 2002a. Disentangling the effects of size-selective mortality, density, and temperature on length-at-age. *Can. J. Fish. Aquat. Sci.* 59:372–382.

———. 2002b. Measuring changes in the direction and magnitude of size-selective mortality in a commercial fish population. *Can. J. Fish. Aquat. Sci.* 59:361–371.

Sissenwine, M. 2001. The concept of fisheries ecosystem management: current approaches and future research needs. *ICLARM, Conf. Proc.* 56:21–22.

Skud, B. 1975. Revised estimates of halibut abundance and the Thompson-Burkenroad debate. *Int. Pac. Halibut Comm. Scientific Report No. 56.*

SLIS. 1980. Proceedings of the Sea Lamprey International Symposium. *Can. J. Fish. Aquat. Sci.* 37(11):1585–2214.

Smith, M. D. 2002. Two econometric approaches for predicting the spatial behavior of renewable resource harvesters. *Land Economics.* 78:522–538.

Smith, P. 1999. Genetic resources and fisheries policy aspects. In R. Pullin, D. Bartley, and J. Kooiman, eds., *Towards policies for conservation and sustainable use of aquatic genetic resources*, volume 59:43–62. ICLARM Conf. Proc.

Somers, I., and Y.-G. Wang. 1997. A simulation model for evaluating seasonal closures in Australia's multispecies northern prawn fishery. *N. Am. J. Fish. Man.* 17:114–130.

Somers, I. 1994. Species composition and distribution of commercial penaeid prawn catches in the Gulf of Carpentaria, Australia, in relation to depth and bottom type. *Aust. J. Mar. Freshwat. Res.* 45:317–335.

Spencer, P., and J. Collie. 1997. Patterns of population variability in marine fish stocks. *Fish. Oceanogr.* 6:188–204.

Spencer, P. 1997. Optimal harvesting of fish populations with nonlinear rates of predation and autocorrelated environmental variability. *Can. J. Fish. Aquat. Sci.* 54:59–74.

Sprout, P., and R. Kadawaki. 1987. Managing the Skeena River sockeye salmon (*Oncorhynchus nerka*)—the process and the problems. *Can. Sp. Publ. Fish. Aquat. Sci.* 96:385–395.

Steele, J., and E. Henderson. 1992. The role of predation in plankton models. *J. Plankton Res.* 14:157–172.

Stickney, R., and J. McVey, eds. 2002. *Responsible marine aquaculture.* New York: CABI Publishing.

STOCKS. 1981. Proceedings of the 1980 Stock Concept International Symposium. *Can. J. Fish. Aquat. Sci.* 38:1457–1921.

Sullivan, P. J., H. Lai, and V. Gallucci. 1990. A catch-at-length analysis that incorporates a stochastic model of growth. *Can. J. Fish. Aquat. Sci.* 47:184–198.

Sullivan, P. J. 1992. A Kalman filter approach to catch at age analysis. *Biometrics* 48:237–257.

Sumaila, R., T. Pitcher, N. Haggan, and R. Jones. 2001. Evaluating the benefits from restored ecosystems: a back to the future approach. In A. Shriver and R. Johnston, eds., *10th International Conference of the International Institute of Fisheries Economics and Trade*, Corvallis, Oregon, USA, 1–7.

Sutherland, W. 1996. *From individual behaviour to population ecology.* Oxford: Blackwell.

Su, Z., and M. Adkinson. 2002. Optimal inseason management of pink salmon given uncertain run sizes and seasonal changes in economic value. *Can. J. Fish. Aquat. Sci.* (in press).

Swain, D., and A. Sinclair. 2000. Pelagic fishes and the cod recruitment dilemma in the Northwest Atlantic. *Can. J. Fish. Aquat. Sci.* 57:1321–1325.

Tegner, M., and P. Dayton. 1999. Ecosystem effects of fishing. *TREE* 14:261–262.

Thedinga, J., M. Murphy, J. Heifetz, K. Koski, and S. Johnson. 1989. Effects of logging on size and age composition of juvenile coho salmon (*Oncorhynchus kisutch*) and density of premolts in Southeast Alaska streams. *Can. J. Fish. Aquat. Sci.* 46:1383–1391.

Thompson, L. 1999. Abundance and production of zooplankton and kokanee salmon (*Oncorhynchus nerka*) in Kootenay Lake, British Columbia during artificial fertilization. Ph.D. dissertation, University of British Columbia.

Thom, R. 1975. *Structural stability and morphogenesis: an outline of a general theory of models.* Reading, MA: Benjamin.

Tregenza, T. 1995. Building on the ideal free distribution. *Adv. Ecol. Res.* 26:253–307.

Tyler, A., W. Gabriel, and W. Overholtz. 1982. Adaptive management based on structure of fish assemblages of northern continental shelves. *Can. Sp. Publ. Fish. Aquat. Sci.* 59:149–156.

Tyler, J., and S. Brandt. 2001. Do spatial models of growth rate potential reflect fish growth in a heterogeneous environment? A comparison of model results. *Ecol. Freshwat. Fish* 10:43–56.

UN. 1996. Agreement for the Implementation of the Provisions of the United Nations Convention on the Law of the Sea of 10 December 1982 Relating to the Conservation and Management of Straddling Fish Stocks and Highly Migratory Fish Stocks. In *United Nations General Assembly*, New York, A/RES151/35.

Ursin, E. 1982. Stability and variability in the marine ecosystem. *Dana* 2:51–65.

Van Winkle, W., C. C. Coutant, H. I. Jager, J. Mattice, D. Orth, R. G. Otto, S. F. Railsback, and M. J. Sale. 1997. Uncertainty and instream flow standards; prespectives based on hydropower research and assessment. *Fisheries* 22(7): 21–22.

Vasconcellos, M., S. Heymans, and A. Bundy. 2002. The use of Ecosim to investigate multispecies harvesting strategies for capture fisheries of the Newfoundland-Labrador shelf. *Fisheries Centre Research Reports* 10(2):69–72.

Walters, C. J., and J. S. Collie. 1988. Is research on environmental factors useful to fisheries management? *Can. J. Fish. Aquat. Sci.* 45:1848–1854.

———. 1989. An experimental management strategy for groundfish management in the face of large uncertainty about stock size and production. *Can. J. Fish. Aquat. Sci.* 108:13–25.

Walters, C. J., N. Hall, R. Brown, and C. Chubb. 1993. A spatial model for the dynamics of the Western Australian rock lobster fishery. *Can. J. Fish. Aquat. Sci.* 50:1650–1662.

Walters, C. J., C. Hannah, and K. Thomson. 1991. A microcomputer program for simulating effects of physical transport processes on fish larvae. *Fisheries Oceanography* 1(1):11–19.

Walters, C. J., R. Hilborn, R. M. Peterman, and M. Staley. 1978. Model for examining early ocean limitation of Pacific salmon production. *J. Fish. Res. Bd. Canada* 35(10):1303–1315.

Walters, C. J., and R. Hilborn. 1976. Adaptive control of fishing systems. *J. Fish. Res. Bd. Canada* 33:145–159.

Walters, C. J., and F. Juanes. 1993. Recruitment limitations as a consequence of natural selection for use of restricted feeding habitats and predation risk taking by juvenile fishes. *Can. J. Fish. Aquat. Sci.* 50:2058–2070.

Walters, C. J., and J. Korman. 1999. Linking recruitment to trophic factors: revisiting the Beverton-Holt recruitment model from a life history and multispecies perspective. *Rev. Fish Biol. Fish.* 9:187–202.

Walters, C. J., and D. Ludwig. 1981. Effects of measurement errors on the assessment of stock and recruitment relationships. *Can. J. Fish. Aquat. Sci.* 38: 704–710.

Walters, C. J., and J. Maguire. 1996. Lessons for stock assessment from the northern cod collapse. *Rev. Fish Biol. Fisheries* 6:125–137.

Walters, C. J., and S.J.D. Martell. 2002. Stock assessment needs for sustainable fisheries management. *Bull. Mar. Sci.* 70(2):629–638.

Walters, C. J., J. H. Prescott, R. McGarvey, and J. Prince. 1998. Management options for the South Australian rock lobster (*Jasus edwardsii*) fishery: a case study of co-operative assessment and policy design by fishers and biologist. In G. S. Jamieson and A. Campbell, ed. *Proceedings of the North Pacific Symposium on Invertebrate Stock Assessment and Management. Can. Spec. Publ. Fish. Aquat. Sci.* 125:377–383.

Walters, C. J., and K. J. Sainsbury. 1990. *Design of a large-scale experiment to measure effects of fishing on the Great Barrier Reef.* Townsville: Great Barrier Reef Marine Park Authority.

Walters, C. J., M. Stocker, A. Tyler, and S. Westerheim. 1986. Interaction between Pacific cod (*Gadus macrocephalus*) and herring (*Clupea harengus pallasi*) in Hecate Strait, British Columbia. *Can. J. Fish. Aquat. Sci.* 43:830–837.

Walters, C., V. Christensen, and D. Pauly. 1997. Structuring dynamic models of exploited ecosystems from trophic mass-balance assessments. *Reviews Fish Biology and Fisheries* 7:1–34.

———. 1999. ECOSPACE: prediction of mesoscale spatial patterns in trophic relationships of exploited ecosystems, with emphasis on the impacts of marine protected areas. *Ecosystems* 2:539–554.

Walters, C., and S. Cox. 1999. Maintaining quality in recreational fisheries: how success breeds failure in management of open access sport fisheries. *Fisheries Centre, University of British Columbia, Vancouver, Research Reports* 7:22–29.

Walters, C., R. Goruk, and D. Radford. 1993. Rivers Inlet Sockeye salmon; an experiment in adaptive management. *North Am. J. Fish. Mgmt.* 13:253–262.

Walters, C., and J. Kitchell. 2001. Cultivation/depensation effects on juvenile survival and recruitment: implications for the theory of fishing. *Can. J. Fish. Aquat. Sci.* 58:39–50.

Walters, C., and D. Ludwig. 1994. Calculation of Bayes posterior probability distributions for key population parameters. *Can. J. Fish. Aquat. Sci.* 51:713–722.

Walters, C., and A. M. Parma. 1996. Fixed exploitation rate strategies for coping with effects of climate change. *Can. J. Fish. Aquat. Sci.* 53:148–158.

Walters, C., D. Pauly, V. Christensen, and J. Kitchell. 2000. Representing density dependent consequences of life history strategies in aquatic ecosytems: EcoSim II. *Ecosystems* 3:70–83.

Walters, C., and P. Pearse. 1996. Stock information requirements for quota management systems in commercial fisheries. *Rev. Fish. Biol. Fisheries* 6:21–42.

Walters, C., and J. Post. 1993. Density dependent growth and competitive asymmetry in size-structured fish populations: a theoretical model and recommended field experiments. *Trans. Am. Fish. Soc.* 122:34–45.

Walters, C., and B. Ward. 1998. Is solar radiation responsible for declines in marine survival rates of anadromous salmonids that rear in small streams? *Can. J. Fish. Aquat. Sci.* 55(12):2533–2538.

Walters, C. J. 1969. A generalized computer simulation model for fish population studies. *Trans. Am. Fish. Soc.* 98:505–512.

———. 1975. Optimal harvest strategies for salmon in relation to environmental variability and uncertain production parameters. *J. Fish. Res. Bd. Canada* 32:1777–1784.

———. 1978. Some dynamic programming applications in fisheries management. In M. Puterman, ed., *Dynamic programming and its applications*, 233–248. New York: Academic Press.

———. 1985. Bias in the estimation of functional relationships from time series data. *Can. J. Fish. Aquat. Sci.* 42:147–159.

———. 1986. *Adaptive management of renewable resources.* New York: Macmillan Publishing Co.

———. 1987. Nonstationarity of production relationships in exploited populations. *Can. J. Fish. Aquat. Sci.* 44 (supplement II):156–165.

———. 1989. Value of short term recruitment forecasts for harvest management. *Can. J. Fish. Aquat. Sci.* 51:2705–2714.

———. 1993. Dynamic models and large scale field experiments in environmental impact assessment and management. *Aust. J. Ecology* 18:53–61.

Walters, C. 1994. Use of gaming procedures in evaluation of management experiments. *Can. J. Fish. Aquat. Sci.* 51:2705–2714.

———. 1995. Use of gaming procedures in adaptive policy design. *Can. J. Fish. Aquat. Sci.* 51:2705–2714.

———. 1997. Challenges in the management of riparian and coastal ecosystems. *Conservation Ecology* 1(2): *http://www.consecol.org/vol1/iss2/art1.*

———. 1998. Designing fish management systems that do not depend on accurate stock assessment. In T. Pitcher, P. Hart, and D. Pauly, eds., *Reinventing Fisheries Management*, 279–288. London: Chapman & Hall.

———. 1998a. Evaluation of quota management policies for developing fisheries. *Can. J. Fish. Aquat. Sci.* 55(12):2691–2705.

———. 2000. Impacts of dispersal, ecological interactions, and fishing effort dynamics on efficacy of marine protected areas: how large should protected areas be? *Bull. Mar. Sci.* 66(3):745–757.

Walters, C., and R. Bonfil. 1999. Multispecies spatial assessment models for the B.C. groundfish trawl fishery. *Can. J. Fish. Aquat. Sci.* 56:601–628.

Wang, Y.-G., M. R. Thomas, and I. Somers. 1995. A maximum likelihood approach for estimating growth from tag-recapture data. *Can. J. Fish. Aquat. Sci.* 52:252–259.

Wang, Y.-G. 1998. An improved Fabens method for estimation of growth parameters in the von Bertalanffy model with individual asymptotes. *Can. J. Fish. Aquat. Sci.* 55:397–400.

Watson, R., D. Die, and V. Restrepo. 1993. Closed seasons and tropical penaeid fisheries: a simulation including fleet dynamics and uncertainty. *North Am. J. Fish. Mgmt.* 13:326–336.

Werner, E., and J. Gilliam. 1984. The ontogenetic niche and species interactions in size-structured populations. *Ann. Rev. Ecol. Syst.* 15:393–425.

Werner, E., and D. Hall. 1988. Ontogenetic habitat shifts in bluegill: the foraging rate-predation risk trade-off. *Ecology* 69:1352–1366.

Whipple, S., J. Link, L. Garrison, and M. Fogarty. 2000. Models of predation and fishing mortality in aquatic ecosystems. *Fish and Fisheries* 1:22–40.

Wickham, E. 2002. *Dead fish and fat cats.* Vancouver, B.C.: Fishermen's Wharf Market Publishing.

Wilen, J. 1979. Fisherman behavior and the design of efficient fisheries regulation programs. *J. Fish. Res. Bd. Canada* 37:855–858.

Wilen, J., M. Smith, D. Lockwood, and L. Botsford. 2002. Avoiding surprises: incorporating fisherman behavior into management models. In *Bull. Mar. Sci. (3rd Mote Symposium Issue).*

Williams, I., M. Bradford, K. Shortreed, J. Hurne, S. Macdonald, L. B. Holtby, G. Ennis, G. Logcan, and H. Stalberg. 1996. Productive capacity freshwater freshwater/anadramous Pacific Region, pages 88–94 *in* C. D. Levings, K. Minns, and F. Aitkens, eds. *Research priorities to improve methods for assessing productive capacity for fish habitat management and impact assessment.* Can. Tech. Rep. Fish. Aquat. Sci.

Wilson, J., J. Acheson, M. Metcalfe, and P. Kleban. 1994. Chaos, complexity and community management of fisheries. *Mar. Policy* 18:291–305.

Wilson, J., J. French, P. Kleban, S. McKay, and R. Townsend. 1991. Chaotic dynamics in a multiple species fishery: a model of community predation. *Ecol. Model.* 58:303–322.

Wilson, J., P. Kleban, S. McKay, and R. . Townsend. 1991. Management of multispecies fisheries with chaotic population dynamics. *ICES Mar. Sci. Symp. (Multispecies models relevant to management of living resources)* 193:287–300.

Wing, S., L. Botsford, and Q. J.F. 1998. The impact of coastal circulation on the spatial distribution of invertebrate recruitment, with implications for management. *Can. Spec. Publ. Fish. Aquat. Sci.* 125:285–294.

Worm, B., and R. A. Myers. 2003. Meta-analysis of cod-shrimp interactions reveals top-down control in oceanic food webs. *Ecology* 84:162–173.

Wright, S. 1981. Contemporary Pacific salmon fisheries management. *North Am. J. Fish. Mgmt.* 1:29–40.

———. 1993. Fishery management of wild Pacific salmon stocks to prevent extinctions. *Fisheries* 18(5):3–4.

Yodzis, P. 1984. The structure of assembled communities. *J. Theor. Biol.* 107:115–126.

———. 2000. Diffuse effects in food webs. *Ecology* 81:261–266.

———. 2001. Must top predators be culled for the sake of fisheries? *Trends in Ecology and Evolution* 16(2):78–84.

Zheng, J., M. C. Murphy, and G. H. Kruse. 1995. A length-based population model and stock-recruitment relationships for red king crab, *Paralithodes camtschaticus*, in Bristol Bay, Alaska. *Can. J. Fish. Aquat. Sci.* 52:1229–1246.

References to boxes, tables, and figures are indicated respectively by the suffixes *b, t,* and *f.*

trophic linkages, 81. *See also* fishing
mortality; trophic interactions
fishing mortality: and arena-scale
dynamics, 246; in catch and release
fishing, 78, 324; cumulative, 56*b*–57*b*,
186–87; defined, 44*f*; depensatory
effects and, 192–93, 200; in developing
fisheries, 16; and fecundity depression,
51, 156; as forcing input for model
fitting, 300; and growth-curve
estimation, 119; historical data for, 43,
44*f*; in input control, 69; for mixed
hatchery and wild fish, 316*b*, 325, 327,
330, 333; optimization of, 81; and
Pacific salmon, 14*b*–15*b*; recreational
vs. commercial fisheries, 209, 210*f*; in
single-species stock assessment, 90; and
spatial distribution, 225, 227*f*; and
spawning aggregations, 182; and
tagging, 40; and time-series bias,
168–70, 168*t*, 169*t*, 171*t*, 172*t*, 173*t*;
and "trophic triangles," 249–50. *See
also* fishing effort
fishing rights, 5, 29*b*, 335, 344–45
FISHMOD, 190–92
fixed-escapement rules, 46, 52*f*, 54–55.
See also output control
fixed-exploitation rules, 46, 48, 54–55,
101, 157–58. *See also* input control
fixed-quota rules, 46
fleet dynamics: and bionomic
development model, 203; capacity and
efficiency of, 38, 201–4; and the
destructive minority, 345; effort
responses and, 207*f*; expansion and
reduction in, 37–39, 204*f*, 334; and
individual-based models, 205–6; and
multi-fleet, multispecies management,
58. *See also* bionomic equilibrium;
fishing effort
food concentration dynamics, 136–37,
139–41, 147
food consumption rates, 235, 246
food webs. *See* ecosystem models;
foraging arena theory; trophic
interactions
foraging arena theory: and arena
development, 233*f*, 234*f*, 235*f*, 245;
and arena-scale dynamics, 127,
236–40; and behavioral adaptations,
124–28, 234–35; and Beverton-Holt
model, 127, 128–32, 135*f*, 136–47,
187–88; cohort effects in, 142–44;
diurnal movements vs. dispersal, 142;

and Ecosim model, 246–50, 252; and
ecosystem structure, 232–36; and
emergent stock-recruitment
relationships, 250; and enhancement
programs, 147–48, 308, 319–20; and
enrichment experiments, 260; and food
concentrations, 136–37; foraging-time
adjustments in, 240–44; from
individual-based spatial model, 284;
limitations of, 271; and mesoscale
models, 277; and predation rates, 265*f*;
recruitment research and predictions
in, 147–50; on trophic flows, 236–40,
292; on trophic interactions and
ontogeny, 244–46, 248–50, 252
Ford-Brody growth model, 107
freshwater shrimp (*Mysis relicta*),
274–75, 276*f*

gaming. *See* policy gaming
Gauss-Seidel iteration, 280*b*
gear technology: economics of, 33,
339–40; efficiency and catchability of,
69, 75, 207, 343; and gear selectivity,
21, 33, 109, 223, 347; in ideal
free-distribution models, 218–19; in
searching and handling, 216, 324; for
selective mark fisheries, 329. *See also*
selective fishing practices
genetic structure, 39, 83–85, 333
geographic information systems (GIS),
39, 82
Georgia Strait (British Columbia), 45, 66,
67*f*, 320–21
"goodness of fit" analysis, 116, 303–5
gradients, space-time, 223–24
gravity models, 194, 210–11
groundfish, 30*b*, 213–15, 227*f*
groupers, 61
growth curves, 50, 81, 84*f*, 116–19,
120*b*–21*b*, 290. *See also* growth rates;
von Bertalanffy growth models
growth overfishing, 53–54, 83–85
growth rates: in Beverton-Holt model,
107; for coho salmon, 36*f*;
fishery-induced changes in, 119; Ford-
Brody equation, 107; and growth
efficiency, 291*f*; of juvenile fish, 137,
139–40; in managing exploitation
rates, 81; and recruitment
relationships, 17–18; and stocking
density, 137, 139*f*; in trophic ontogeny
models, 250; variation in, 84, 84*f*, 119,

Milton Keynes UK
Ingram Content Group UK Ltd.
UKHW022130190524
442884UK00008B/491